Information Superhighways Revisited

The Economics of Multimedia

For a complete listing of the *Artech House Telecommunications Library,*
turn to the back of this book.

Information Superhighways Revisited

The Economics of Multimedia

Bruce L. Egan

Artech House
Boston • London

Library of Congress Cataloging-in-Publication Data
Egan, Bruce L.
   Information superhighways revisited : the economics of multimedia
/ Bruce L. Egan.
     p.   cm.
   Includes bibliographical references and index.
   ISBN 0-89006-903-4 (alk. paper)
   1. Telecommunication—United States.   2. Broadband communication
systems—United States.   3. Information superhighway—United States.
4. Multimedia systems industry—United States—Forecasting.
I. Title.
TK5102.3.U6E33   1996
384'.042'0973—dc20   96-35957
                                                  CIP

British Library Cataloguing in Publication Data
Egan, Bruce L.
   Information superhighways revisited : the economics of multimedia
   1. Information superhighway—Economic aspects 2. Information
   superhighway—Forecasting
   I. Title
   384.3'1

   ISBN 0-89006-903-4

Cover design by Darrell Judd

© 1996 ARTECH HOUSE, INC.
685 Canton Street
Norwood, MA 02062

International Standard Book Number: 0-89006-903-4
Library of Congress Catalog Card Number: 96-35957

10 9 8 7 6 5 4 3 2 1

For Kelly

Contents

Preface

With technology changing so fast and regulation being so unpredictable, it is risky to make prognostications about the future of telecommunications. But I have been in the business of forecasting for 20 years now, and it hasn't stopped me yet. I have concluded, however, that predicting technology is exceedingly difficult, and predicting demand is impossible. Still, the forecasting process goes on, and—fortunately—corporate forecasting budgets continue to grow. So here goes...

It has been over five years since I wrote my first book about the future of telecommunications. I will not bother to provide a detailed account on whether my predictions of future trends in that book ever came to pass. Suffice it to say that several did, and almost all of the others were well on their way to becoming realized when some (not totally) unanticipated events (including the regulatory push for the breakup of the local telephone monopolies) altered the initial conditions and undermined some key underlying assumptions.

This book, like my earlier effort, focuses on the future of integrated broadband network infrastructures. So much has changed in the last five years that revisiting the subject is certainly warranted. For example, in their complete overhaul of telecommunications law in America, federal lawmakers have, in one fell swoop, replaced the old monopoly model of regulation with one based on market competition. That single event significantly altered the institutional landscape and is certain to have a tremendous, potentially defining, impact on the direction of technology adoption in telecommunications.

In adition, at least two market megatrends have emerged that will have a lasting effect: the absolutely explosive growth of the Internet and the advance of digital wireless technology. Other important developments that should be taken into account are recent advances in digital signal compression technology, which have effectively redefined the term *broadband* telecommunications, and the growing popularity of multimedia *compact disc* (CD) technology. For many applications, the truly portable multimedia technology of CDs has the potential to displace more traditional telecommunication technologies, which rely on fixed (wired and wireless) transmission channels. After all, CDs have already become the market leader in *personal computer* (PC)–based multimedia presentations. Innumerable future applications that employ interactive and programmable CDs will become part and parcel of the multimedia revolution. It is a sure bet that CD players will become as complementary to televisions as *videocassette recorders* (VCRs) are today.

What is most interesting is that many of these multimedia phenomena are closely related, and any one of them may be used to bring broadband telecommunications into the living room of any household. The Internet, a fixed-channel medium, CDs, a portable medium, and digital signal compression, which enhances both fixed and portable media, are at the same time complements and substitutes for one another. That makes it difficult indeed to predict winners and losers in the race to become the multimedia technology of choice for the mass market.

In my earlier book, it was difficult enough to predict winners among competing networks and technological alternatives for transporting multimedia services *to* the home. Now there is also the daunting task of predicting which technological alternatives will be used for services *in* the home.

Usually, when we attempt to predict the near future, the devil is in the detail. In the case of predicting the future of multimedia telecommunications, things are happening so fast that the greater risk is in losing the forest for the trees; it is the big picture that is unknown. For example, while the Internet has been a remarkable success, there have been several remarkable failures, among them nearly every single trial to date to bring multimedia and broadband services to the home. Remember the hype surrounding such exciting new services as two-way cable (1970s), videotex (1980s), and interactive television (1990s)? It is easy to blame early market failures as being ahead of their time and ahead of the technology to make them work, but those excuses will not do. Either there is a market for services or there is not. It is not enough to say that if we just sit back and wait long enough technology will progress so that the average household will be an avid daily consumer of multimedia telecommunications.

In addition to market considerations, this book examines key regulatory issues. It is my firm belief that, as much as we would like for it to go away, regulation will be with us for some time to come. It is also my firm belief that regulation will be largely responsible for the structure of the industry, both in terms of the number of players and the way in which they operate. It would be nice to believe that technological progress in a free market setting is such a powerful force that it would inevitably overwhelm private interests, politics, and bureaucratic processes, but this will not be the case.

For that reason, I have dedicated a considerable portion of this book to an analysis of the critical regulatory issues that could materially affect both the path of technology adoption and the rate of infrastructure investment. Along the way, I unabashedly recommend the adoption of regulatory policies that I believe would promote the cost-effective deployment of a widely available and affordable public multimedia network infrastructure. I pay particular attention to two key societal goals that have become the focus of public policy in many countries and that are the cornerstones of the new telecommunications law: affordable universal service and network interconnection. Those two topics are more closely related than one might think. After all, even if affordable local telephone service is available to every household, from a societal perspective it is crucial that all local networks are effectively interconnected. In other words, universally available affordable local access lines should be coupled with universal and affordable network interconnection for ubiquitous call completions. What good is a digital phone line with fancy service capabilities if you cannot reasonably expect to be able to connect to the phone lines of others? That brings us to the world of perfect connectivity characterized by the four As: we would like to be able to call anyone, anytime, anywhere, anyhow. Like the economist's concept of perfect competition, perfect connectivity will exist only in theory, but the abstraction is useful as a social objective.

This book discusses and analyzes a broad spectrum of topics and presents fairly detailed discussions of the more important and complex issues. For that reason, the chapters have been written as self-contained modules, which may be read out of order to suit the reader's particular interests.

Chapter 1 provides the overall context for the discussion of broadband and multimedia network infrastructures and explains key concepts and terminology. Chapter 2 discusses trends in technology, and Chapter 3 discusses network deployment strategies and costs. Chapter 4 reports on the status and prospects for the deployment of new digital wireless network infrastructures, and Chapter 5 presents a comparison of wireless and wireline alternatives. Chapter 6 discusses economic issues of industry structure and regulation, and Chapter 7 focuses on the critical universal service issue. Chapter 8 discusses the unique situation facing rural areas. Finally, Chapter 9 discusses the political economy of telecommunication policy, synthesizes economic and institutional considerations, and makes recommendations for policy reform.

Acknowledgments

Several persons had a hand in helping me prepare this book, and some of them deserve special thanks. My friends Dr. Richard Emmerson and Prof. Dennis Weisman provided insights and ideas for major portions of Chapter 6, "Economics of Broadband Networks." Dr. Emmerson was also responsible for some key ideas on universal service, discussed in Chapter 7. Dr. Paul Shumate of Bellcore was gracious in providing some important network engineering information and references that I was able to incorporate into Chapters 3 and 4, which explore wireline and wireless technologies. J. T. Edelstein prepared the graphics with his usual efficiency and accuracy. Most of all, thanks to Kelly, my wife, who provided a substantial effort in the preparation and editing of the manuscript and who supported me from the very beginning of this project to its sweet completion.

Introduction

1

1.1 TECHNOLOGICAL CONVERGENCE

Convergence, the compelling buzzword of the information age, conjures up images of combining many different communications media using the common communications medium of digital signaling. Convergence potentially allows for effortless interactive communication among persons and machines, where voice, text, and images are integrated and carried on an integrated digital "information superhighway."

Thanks to rapid advances in *digital signal processing* (DSP) techniques, the technologies of computing and communications have effectively converged. Now it is up to entrepreneurs to discover mass market applications and develop user-friendly ways to make those applications work. Such systems will enable small businesses and individuals to take advantage of what convergence has to offer. That is what this book is about: exploring the types of multimedia delivery systems on the horizon, determining which of them is likely to actually be deployed, and explaining why. Because the government plays a role, this book also analyzes what types of regulations will promote investment in a multimedia network infrastructure that is both affordable and widely accessible to households nationwide.

The astonishing growth of the Internet (the "Net"), a global PC-based communications network, is by far the most illuminating example of convergence in multimedia telecommunications (see Figure 1.1). Through the use of a single common digital bitstream to transmit audio, video, and text simultaneously, the Internet's *World Wide Web* (WWW, or just the "Web") is truly the archetypal multimedia telecommunications network of the future. In just the few years since the development of inexpensive (or free) user-friendly point-and-click software interfaces (called Web *browsers*), the number of users of the Net in North America alone has grown to about 20 million. Contrary to the views of cynics, those 20 million users are not just playing computer games or engaging in other frivolous entertainment activities such as electronic pornography (much of which, pending judicial review, is banned by new federal laws). They are also seeking and finding a wide variety of information for their business and personal pursuits—many are even starting to make money at it.

Besides the multimedia Web, many other telecommunications market segments have been growing substantially, and all of them owe it to the advance of technology. Even though most non-Internet telecommunication networks still are not digital and therefore have not yet converged into a multimedia environ-

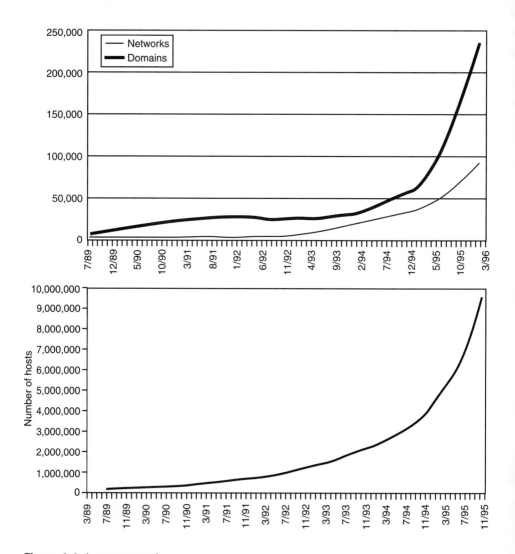

Figure 1.1 Internet growth.

ment, advances in digital technology and consumer demand will surely move them in that direction.

The United Nations *International Telecommunications Union* (ITU), based in Geneva, tracks telecommunications development for 205 countries with populations greater than 40,000 people (actually estimated to be 90% of the world market due to data availability). Figure 1.2 shows worldwide historical growth for a range of telecommunication services from basic telephone lines and broadcast television to digital satellite and Internet users.

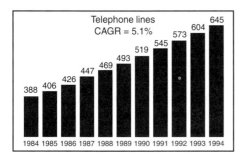

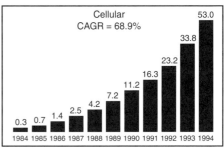

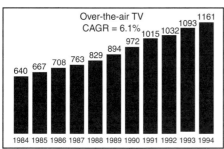

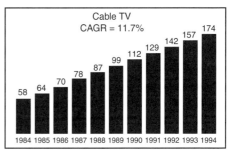

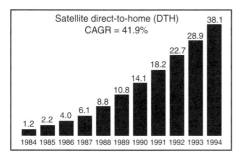

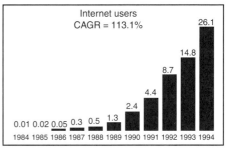

CAGR = Compound Annual Growth Rate.
All numbers are shown in millions.

Figure 1.2 ITU tracking of worldwide historical growth for various telecommunication services.

While basic analog telephone services have been growing at only 3–5% per year in the United States and most other developed countries, digital data telecommunication services have, for a number of years, been growing annually by double digits and, at least in the United States, now represent more than half of all telephone network usage. For example, U.S. digital fax services have grown at an average annual rate of 42% from 1988 to 1994. Worldwide, the growth for the period was 28%. Similarly, traditional broadcast television, in the United States has been flat, but new cable television services continue to grow, and the growth of new digital satellite broadcast services has been phenomenal.

Most of the world's cellular mobile systems developed from a relatively expensive and capacity-limited analog technology. DSP technology has brought the price of mobile subscriber handsets down by an order of magnitude in the decade from 1985 to 1995 (e.g., from $2,000 to $200). Over the same period, digital technology has allowed a three- to fourfold reduction in the incremental network investment costs. As cellular mobile telecommunications becomes digitized and system capacity expands, personalized portable service will become available and affordable for the mass market. Cellular mobile service worldwide has been expanding at over 40% per year from 1991 to 1995. As cellular service is increasingly used for fixed telephone services these growth rates can reasonably be expected to continue through the year 2000.

The societal implications from such growth rates are clear. Increasingly, wealth and power in modern society are derived from information and knowledge. Speedy access to superior information and knowledge is often the defining factor for determining winners and losers in the marketplace. All firms, especially those engaged in professional and business services, can gain an edge in productivity by investing in *information technology* (IT). Throughout the world, growth and productivity gains in the service sector far outstrip those of manufacturing, agriculture, and other less information intensive industries. IT generally, and telecommunications and computing technology specifically, have been largely responsible for the growth and productivity trends in the service sector.

For many years now, adoption of IT was the province of businesses that could afford it; technology investments have substantially increased the productivity of IT workers and increased their wages as well. Naturally, as computer technology is applied to many traditionally labor-intensive business activities, some workers will no longer be necessary. While in the short term some non-IT workers inevitably are going to be displaced, over the long term the net increase in total jobs and income throughout the entire economy will be enormous. IT-related service sector growth in employment and productivity has significant spillover effects that cause positive ripple effects in nearly all other sectors of the economy, including manufacturing and agriculture. Businesses have long benefited from the adoption of IT technology, but the benefits will extend far beyond the office door because mass market applications are imminent.

The vast amount of electronic information that is becoming available has the potential to dramatically improve the productivity not only of businesses but of households as well. As the use of IT continues to rise, prices have fallen dramatically, placing IT within the financial reach of many households. Consequently, most American households will own a computer before the turn of the century, and many of those computers will be equipped for telecommunications via modems. But IT is not just about computers in the home, at least not in the traditional sense of PCs. IT in the home is about the use of computing technology to enhance the communications capability of other everyday consumer devices including "smart" television sets, phones, and cars.

Because of the rapid transformation of traditional slow-speed analog telephone and cable television networks into high-speed public digital public network infrastructures, individual households will soon be able to reap the benefits of the information age. Witness the substantial growth in the number of people working at home, called "telecommuting" or "teleworking." It is estimated that some 36 million (about 40%) American households have an office in the home. Working at home is not the only way for households to gain from IT. In fact, IT is probably most useful for the nonpaid work activities in which household members engage. For the average household, quality of life is determined as much by the personal enjoyment of nonwork activities and the amount of leisure time as it is related to household income. A household's adoption of technology is one of the few ways in which its members can improve their lot in life without putting in more time in paid work activities to generate income. IT can save money, time, or both by increasing the productivity of household members in a wide range of everyday activities.

Over the years, individuals have been spending more and more time engaged in IT-related activities such as communications. IT technology increases the productivity of the time spent gathering and communicating information and knowledge. While most early adoption of IT technology in the home is in the form of electronic devices for entertainment, that surely will change as new devices and information databases are developed. Accessing information on the Internet from home via a modem-equipped PC is just the beginning.

Numerous electronic communication devices continue to penetrate the mass market. Among the newer ones, wireless and infrared remote control devices, PCs equipped with modems, CD players, cellular phones, and digital satellite broadcast systems are especially noteworthy as harbingers of the future of multimedia telecommunications.

No one knows what the full effect of technological convergence and multimedia will be on society, but the potential for a powerful new telecommunications infrastructure to create added value and wealth for businesses and households is truly staggering. In recognition of that potential, the federal government has passed sweeping legislation reforming the 60-year-old communications law, which was based on a monopoly model of telecommunications.[1]

1.2 WHAT IS BROADBAND?

In the context of telecommunications, the term *bandwidth* refers to the speed or frequency of a transmission or communications channel. The term *broadband* is used to describe a high-speed (or high-frequency) transmission signal or channel. It is the functional opposite of *narrowband*, which connotes relatively low speed. While large-scale telephone network trunk lines always operate at broadband speeds, local phone lines connecting households and small businesses to the trunk line network are limited to narrowband speeds. Cable television lines are already broadband connections, but they require costly specialized electronic equipment (e.g., modems) to allow them to transmit and receive two-way point-to-point signals the way telephone lines do.

The bandwidth, or speed, of a subscriber's local telecommunications connection is very important in the multimedia business because it is the primary limiting factor for the type and quality of telecommunication services that can be transmitted and received. The relevance for multimedia telecommunications is that while broadband communication channels can accommodate high-quality interactive multimedia services, narrowband channels may not (depending on the specific services desired).

The transmission speed of a broadband telecommunication channel is usually measured in *megabits per second* (Mbps), and it may be used for almost any type of service, potentially including video telephony. A low-speed narrowband channel, like today's basic phone line, is usually measured in *kilobits per second* (Kbps), and it may be used for traditional telephone service, low-speed data and image transmissions, or slow-scan video telephony. Compared to narrowband telecommunication channels, broadband channels require sophisticated and expensive electronics and DSP equipment because high-speed signals increase the risk of transmission errors.

The term *broadband* is often used loosely and has never been clearly defined (outside of technical standards–setting bodies). In casual discourse, *broadband* is used to describe a channel that is capable of delivering a video signal and is therefore synonymous with one-way television or two-way video telephony. Not too many years ago, it was popularly believed that multimedia

---

1. The Telecommunications Act of 1996, signed into law February 8, 1996, revises the Communications Act of 1934 (47 U.S.C. 151 et seq.) and defines itself as "an Act to promote competition and reduce regulation in order to secure lower prices and higher quality services for American telecommunications consumers and encourage the rapid deployment of new telecommunications technologies."

telecommunications, including the real-time transmission of moving images, required a broadband communications channel. Not any more. Advances in DSP called digital signal compression, continue to cram more and more information and images into less and less channel bandwidth. Thus, rather than referring to broadband, it is perhaps more meaningful to use the term *multimedia* (or, perhaps, *multifunctional*) to refer to high-speed general-purpose digital networks capable of video transmissions.

It has long been the dream of telecommunications engineers to cram a video signal down an ordinary telephone line. The huge popularity of the Web has added urgency to the task. In traditional uncompressed form, a television picture requires that 30 images, or frames, per second be delivered to the television set. Transmitting one television channel requires a bandwidth of 6 *megahertz* (Mhz) analog or about 45 Mbps digital bandwidth. Most modem-equipped PCs hooked up to a phone line can deliver a paltry 1 or 2 *frames per second* (fps), resulting in some very substandard video viewing. Some new digital video transmission techniques and software employ sophisticated DSP to deliver about 15 fps over a phone line, and the technology is still advancing. In addition, sophisticated cable modems will soon be available to exploit the broadband capacity of cable television lines to provide new interactive multimedia telecommunications.

About five years ago, national and international standards-setting bodies agreed that broadband telecommunications in public network infrastructures would be based on a building-block rate of about 50 Mbps and that basic service would consist of just over 150 Mbps. (No such national or international standards exist for the transmission of two-way broadband signals on cable television lines.) But that was when the technology for digital video transmission was in its infancy, a time when transmitting a single television channel required about 45 Mbps of bandwidth—way too much for a normal phone line to carry.

That was then. Now, with some electronic equipment added (similar to a computer modem) a normal phone line can deliver a VCR-quality digital television channel and still be used *at the same time* to make regular phone calls. Soon, advances in DSP technology and higher speed modems will enable delivery of 5 to 10 times that amount of information.

Thus, from the consumer's perspective, it is becoming less relevant to specify a single threshold level of bandwidth as broadband because the nominal bandwidth capacity (i.e., top signal speed) of the channel itself does not necessarily translate into a clear distinction as to which services can and cannot be obtained. However, from the service provider's standpoint, preselecting the bandwidth capacity of the transmission channel is important, because higher bandwidths (speed) cost more money to provide. As the next several chapters reveal, this creates tricky technology decisions for network operators: Considering all the different technologies from which they can choose, how much channel bandwidth capacity is enough to satisfy the (totally unknown) demands of most households for multimedia telecommunications?

Even though further advances in DSP make it difficult to specify a definitive channel bandwidth requirement for future broadband services, an integrated multimedia delivery system featuring multichannel real-time interactive video services and video telephony would require substantial bandwidth in both directions—somewhere between the current capacity requirements of video conferencing service (about 384 Kbps) up to the current capacity of modern cable television systems (550–750 MHz). Even then, an integrated network system would have to support a wide variety of potential services for voice, data, and video services, each with unique transmission requirements and inherent signal latency. Table 1.1 provides information on these characteristics for many types of digital services, including distributed computer applications, imaging, video telephony, and audio/video services.

Table 1.1
Different Applications—Different Needs

Criterion	Distributed Computing	High-Resolution Image	Conference Video	Audio/Video Database
Data type	Variable-length packets	Large data blocks	Continuous bitstream, 0.1–100 Mbps	Continuous bitstream, 0.1–100 Mbps
Delay requirement	Low latency (µs to ms)	Moderate latency (sec)	Low latency (<100 ms)	Moderate to high latency
Connectivity	Point to point, multipoint	Point to point	Point to point, multipoint	Point to point, multipoint
Symmetry	Symmetrical	Asymmetrical	Symmetrical	Asymmetrical

Source: [1].

1.3 CHARACTERISTICS OF MULTIMEDIA NETWORKS

In reality, the network transmission parameters in Table 1.1 are woefully inadequate to answer some of the most critical questions facing would-be multimedia infrastructure network operators, including: How much channel bandwidth is enough for most households? In an actual multimedia (multiservice) environment, any given household may want to use any number of different types of services at any one time, not to mention the difficulty facing a network service provider when many or most households want to use certain services at the same time. Indeed, this can be a network engineer's worst nightmare.

Network engineering considerations aside, successful multimedia systems must accommodate the nature of household demand. In many households, many different electronic communication devices, including stereos, televisions, PCs, phones, video games, and the like, often are used at the same time. That poses no serious problems as long as each device is connected to a separate network, as they are today. Furthermore, because most of today's nonvoice telecommunication is one way or "downstream" from the service provider to the subscriber, traditional networks are engineered specifically to handle asymmetrical traffic flows. Interactive multimedia networks, however, are a whole new ball game.

In the multimedia environment of the information superhighway, any number of services may be demanded and supplied simultaneously. But that is not as simple as broadcasting a downstream signal to as many households as may want to receive it (i.e., point-to-multipoint traffic pattern). Interactivity implies an upstream signal from the subscriber to an information service provider or to another subscriber (i.e., point-to-point traffic pattern). As more and more uses of the network for two-way telecommunications occur, the network traffic flows become more symmetrical and less asymmetrical. The type of network engineering required to handle symmetrical and point-to-point traffic flows is totally different and costs considerably more to supply than one-way traffic flows.

A familiar example can be instructive to reveal the magnitude of the engineering problems brought on by the advent of interactive multimedia. Current cable television systems feature nearly 100% asymmetrical traffic flows (the figure is nearly 100% because there is some minuscule upstream signaling for channel selection, including pay-per-view video where it is available).

Besides normal voice telephony, a good example of a symmetrical environment was the early Internet, which was used primarily for collaborative research and electronic mail (e-mail). Now the Net is used more intensively (in terms of the amount of data transmitted and received) for such asymmetrical applications as accessing multimedia information on the Web and other information services, bulletin boards, and databases than it is for more symmetrical applications such as e-mail, collaborative communications, and "chat" forums. Overall, the Internet exhibits about 100- or 200-to-1 asymmetrical (downstream) traffic flows. As more and more multimedia databases with fancy images are posted on the Web, the asymmetry of traffic flows is expected to rise even more.

It is interesting to note that in terms of total traffic flow on the Internet, the symmetrical two-way applications, e-mail and chat, while used much more frequently than the asymmetrical applications, require less bandwidth and therefore constitute much less in terms of total information being transmitted. However, at some point in the future, as more PCs and telephone and cable television lines are equipped with high-speed modems, two-way interactive traffic, including some upstream video transmissions, will decrease the ratio of downstream to upstream traffic, perhaps to only 20 to 1. Someday, as video telephony becomes commonplace, the ratio (at least for some subscribers) could equalize the way it tends to do on traditional telephone networks. Thus, the nature of the traffic flows has enormous implications for future multimedia networks. Network operators will have to design their networks flexibly to handle the shifting traffic flows.

Beyond issues of being able to provide the right amount of upstream and downstream bandwidth to meet subscriber demand, multimedia network operators should also consider the nature of the traffic in terms of the delay, or *latency*, of the applications and services. In the context of a telecommunications service, latency refers to the inherent delay in the actual transmissions for a given service or application. Latency is an important factor for multimedia networks for two reasons. First, the infrastructure multimedia network provider may save on costs by engineering the underlying infrastructure network to take advantage of the fact that some services and applications do not require continuous transmissions, thus providing idle capacity that is available for use by other services. Second, network-induced delays, as opposed to latency inherent to services themselves, becomes a critical quality variable for both users and network operators. Some users may accept longer delays to save money, while others may be willing to pay a premium for real-time service or service with otherwise shorter waiting times.

For example, a fast-action video program like a sports contest usually requires continuous signal transmissions because there is, quite literally, no break in the action. A typical phone call, on the other hand, exhibits periods of silence between transmissions. Data services are a mixed bag, with bursts of data and information intermingled with silent periods. Depending on the specific application, some computer data services involve continuous transmissions, while others may transmit data only in bursts. To cost effectively accommodate all such transmissions on a multimedia communications network or on a single subscriber's connection requires sophisticated engineering. In particular, if certain data applications, especially those where machines are communicating directly with other machines, do not require real-time two-way telecommunications (machines do not get annoyed with delays), then the network operator can take advantage of that. But when human beings are involved, multimedia network operators must try to accommodate all types of services without unduly annoying users. In a competitive market for multimedia services, network operators whose systems have annoying delays will not survive.

As a standardized packet data network, the Internet utilizes statistical sharing techniques to allow everything from voice to video transmissions to simultaneously share network transmission facilities on their way to their ultimate destinations. Each little packet has unique routing information within which it keeps track of where it came from, where it is going, and in what order it was sent. Because the network is statistically shared, each packet may hop on and hop off the network in no particular order. No matter how each packet ultimately gets to its destination, the user will receive the information error free and in the correct sequence. From the point of view of network engineers, this type of network can be extremely efficient because it cost effectively maximizes the sharing of physical network facilities across many users and applications. But what at first might seem to be a network engineer's dream could also be a network service provider's nightmare.

A statistically shared network like the Internet is a first-come first-served network. As soon as there is a space in time, a subscriber's message, in the form of a data packet, hops on. It then hops off when it gets to where it is going. When the Internet was used mostly for two-way transmissions of e-mail and text, statistical sharing was great, and, on average, everyone on the network waited about the same amount of time to get information. That dynamic changes when some users want to watch a football game and others are casually sending anonymous love letters. Waiting a few seconds for the former is probably more annoying than waiting a few seconds for the latter. It remains to be seen if a statistically shared network like the Internet will ever evolve into a reliable real-time interactive multimedia infrastructure, if it has to compete with niche network suppliers that guarantee satisfactory service because their networks are dedicated to serving only certain types of interactive user applications with similar latency characteristics and acceptable network-induced delays (e.g., interactive video games).

Unlike computer networks (e.g., the Internet) or popular *local area networks* (LANs) (e.g., Ethernets), cable television and telephone networks are not statistically shared, so they do not have the problem (yet) of determining how to provide for multiple services in real time. The telephone network is *circuit switched*, as opposed to *packet switched*. That means that every time a telephone transmission from point A to point B is requested and connected, the network dedicates that entire end-to-end circuit connection for the entire duration of the message or conversation. Obviously, there are times when the dedicated telephone connection is idle, which is much more costly for the network operator than circuit sharing. However, dedicated connections do not have the problem of service degradation and delay, which can result from statistical sharing.

The current competition among local Internet service providers is instructive in this regard. To a user connected to the Web, speed is everything, and long delay times are annoying. All good Web browser programs suppress the transmission of fancy images so users can get to the information they want in the form of text as quickly as possible. It is also why one of the key benchmarks in multimedia network performance is the delay time inherent to the network itself or to individual subscriber connections. Besides the average transmission delay times for the Internet's high-speed backbone network, the delay on any one subscriber's connection is usually much higher due to local network congestion.

No matter how much we know in the abstract about the latency characteristics of various telecommunications services and applications, in practice it is nearly impossible to do anything about it. "Bits is bits," and any given digital transmission at any given moment may be strictly unidentifiable to the multimedia network operator. Therefore, the network operator cannot know whether network delays are acceptable to the end user. That problem could be mitigated if there was a good way to allow a user to encode a unique identifier into each transmission to alert the network provider as to whether a delay is acceptable to that user.

A multimedia vendor would, obviously, want to accommodate as many services and users as possible to maximize revenue potential. On the other hand, people are already used to many separate electronic communications devices and networks operating at the same time and likely would not tolerate any hassles involved with time share among various activities. (Can you imagine having to tell fellow household members that they cannot watch a certain movie or play a certain video game at a certain time because you want to use the multimedia connection to make a video phone call or download some information from an online database?) The tension is already there as family members tie up the phone line by "surfing the Net" or tie up the cable television by playing video games. This situation will get worse when two-way cable modems increase the contention among household members for control of the television set(s). Thus, network bandwidths will potentially have to be engineered to simultaneously handle all of the telecommunications activities listed in Table 1.1 as well as other, as yet unidentified, applications. Ideally, a multimedia network will be capable of providing a range of services simultaneously, but, as will be demonstrated in later chapters, that is a truly expensive proposition for network operators.

Because network providers cannot accurately predict future demand for multimedia services, it may be a good strategy to deploy enough baseline capacity in the network system to make it capable of handling mass market demand for almost any type of service. The down side to that strategy is that it relies heavily on consumer purchases of applications-oriented devices and terminals to make it work. Still, it minimizes the types of services a network operator must presuppose and the cost of meeting unknown demand. In other words, the multimedia network would be more of a commodity and the value added services would rely on terminals and other devices that consumers would purchase in order to use them. In effect, the consumer terminals and devices would define the individual services themselves. That is the opposite of today's telecommunications delivery systems, which tend to have a unique distribution network for each category or type of service in a one-to-one correspondence with a specialized terminal device (e.g., video game player, VCR, television, stereo).

Nevertheless, it is risky for infrastructure network operators to assume that a single multimedia network can be all things to all people. Such a network deployment strategy could easily backfire if it becomes a bother for subscribers to use the integrated medium for many types of services. In the face of inconvenience, consumers will be tempted to stick with the status quo of separate networks and devices for each service they want. That is what makes the investment decision so difficult for would-be integrated and multimedia network operators. If it turns out that a given multimedia network technology or design can accommodate only a single multimedia application at a time, consumers might not be willing to buy it if other vendors' systems offer multiple services. Conversely, if consumers can obtain individual services cost effectively on limited networks from niche market suppliers, they might not be interested in paying a premium for the luxury of integrating service on a single multimedia platform. The key question then becomes: What is the "right" level of multimedia service for the average household?

That begs a fundamental question about the market viability of multimedia infrastructure networks. Is the market attraction of multimedia networks attributable to its occasional use for (often unanticipated) purposes, such as an individual's use of the Internet for e-mail or information on demand–type services? If so, then households will continue to use many nonintegrated or single media systems to obtain most services for which there is a more or less constant demand (e.g., entertainment television). Or is the main attraction of the multimedia network infrastructure its multifunctional general-purpose nature, which supports simultaneous use of multiple services by multiple users on demand, including traditional services like cable television and stereo audio? If so, then consumers would be willing to pay a premium for that service, and suppliers would select a sophisticated integrated multimedia infrastructure as the best investment. Of course, there is no answer to these questions—yet. Unfortunately, the network infrastructure investment decision necessarily involves committing huge amounts of funds to long-lived investments. Couple that with the risks and uncertainty of a competitive marketplace and it becomes absolutely critical that infrastructure network operators consider carefully the tradeoffs involved in construction cost and the level of network integration required to provide for multiple services and users.

Assuming that multimedia is going to become a reality, what different multimedia network infrastructures will be available to the mass market of residential users? The answer to that question is beginning to emerge in the form of new digital delivery systems. The most obvious candidates are digital telephone and cable television lines connected to television sets, PCs, and other multimedia devices in a household. The second most obvious possibilities are new digital wireless telecommunication systems. Many different flavors of digital radio systems are beginning to emerge, including intrapremises wireless and cordless telephony, cellular and noncellular terrestrial radio, and satellite systems. Then there are the truly portable media, such as CDs and videotapes.

It is probably safe to assume that all those solutions will evolve to become multimedia delivery system infrastructures, some more successful in penetrating the market than others. It is equally safe to assume that portable and fixed multimedia delivery systems will coexist, because it is not reasonable to assume that one is always going to be a perfect substitute for the other in the eyes of consumers. If that were so, it would be possible for one technology to totally displace another. But this case surely is not analogous to that of calculators supplanting slide rules. All of these alternative multimedia delivery systems are capable of delivering real-time, high-quality digital communication services. While CDs are increasingly displacing analog audio and videotapes, certain applications for new digital video tapes are likely to emerge as digital video production progresses and consumer market items such as digital videotape camcorders are introduced.

Digital satellite systems, introduced only two years ago, represent an interesting phenomenon that may dramatically affect the race for preeminence among competing multimedia delivery systems. Satellite delivery systems, while perhaps inferior in many respects to fixed terrestrial network systems, also exhibit some significant advantages. First, they are far and away the least expensive way to broadcast high-bandwidth transmissions to the maximum number of households. The early growth in terms of household penetration of the first digital *direct broadcast satellite* (DBS) systems for satellite television has outstripped that of VCRs and CD players before it. That growth is exceptional when we consider that over 90% of American households have had the option of purchasing wired cable television service for years. Had U.S. cable television operators not had such a huge head start, DBS technology would have taken off even more than it has. Most countries, especially those in Asia (the world's largest potential market), do not—and may not ever—have such well developed wired cable television infrastructures. Thus, new digital satellite systems may well dominate the scene for video broadcasts in Asia. In the United States, it is not expected that satellite delivery systems will ever become a dominant multimedia infrastructure technology, but it certainly seems likely that such systems will dominate some niche markets for mass market services, including some broadcasting services, maritime services, mobile paging, and other wide area locator services and for service to very remote locations or in areas with extremely rugged terrain.

Traditional land-based cable television and telephone network systems are a sure bet to become general-purpose multimedia delivery systems for the mass market. While both network types have relative technical advantages and disadvantages in the marketplace, they will coexist and compete over the long term. They are also likely to converge as the fundamental underlying digital technology used by each continues to converge. The primary unresolved issue for the multimedia infrastructure of the future is whether the wired connections to individual households will gradually migrate to digital wireless connections. That issue is investigated at length in later chapters.

1.4 DEMAND FOR MULTIMEDIA NETWORK INFRASTRUCTURES

Nobody knows what the future holds on the demand side of the multimedia equation. The amount of research literature on the subject is paltry and reeks of pure speculation. Little systematic formal research has been performed, mostly because, generally until the technology is made available, it is almost impossible to assess what consumers will do with it. One thing that is certain is that early prognostications were way off. Since the demise of early applications of interactive electronic media, there has been an almost ethereal atmosphere in the multimedia investment and market research community. Just like the hype for early biotechnology applications, backers of interactive electronic telecommunication systems have fallen on their faces.

To assess the demand for future multimedia delivery systems, some insight might be gained from past attempts to market interactive electronic systems. The earliest applications for interactive electronic media were for online news and information (i.e., electronic publishing), home banking, and other videotext services that were transaction oriented. The only successful "market" trial aimed at the mass market was in France, where the government gave away consumer video display terminals to be used in place of government-provided paper telephone directories. After a bout of healthy growth, the novelty wore off, and even that subsidized system is ailing. In the United States, media giants Hearst and Knight Ridder and others gave videotext a try and lost a lot of money. Banking giant Chemical and others got similar disappointing results. The highly touted IBM and Sears partnership resulted in the videotext service Prodigy, which has been a veritable money pit. Some other early online news, information, and electronic transaction services were moderately successful at penetrating the mass market, at least enough so as not to lose money. On the other hand, many niche electronic database and information systems targeted at business customers prospered, such as Mead Data Central's popular LEXIS/NEXIS online information service and West's Westlaw.

In the late 1980s and early 1990s, some in-home interactive multimedia systems were introduced. These expensive entertainment and information systems relied totally on inhome wiring and devices to make them work. For example, one such system by FROX offered interactive digital television programs based on an early videodisk format, along with CD-ROMs for stereo audio and electronic information services (e.g., electronic encyclopedias). These systems could easily cost $20,000, and the prices never fell enough to attract attention.

Major telephone and cable television companies have introduced many trials of broadband services to the home, with substantial doses of media hype and fanfare. To date, all of those service have failed to stimulate any considerable interest from paying consumers. The trials are either ended or perhaps extended, but no mass market rollout has resulted.

The most obvious reason that historical attempts to market interactive electronic media services have failed is that the costs of supply were too high compared to the value consumers perceive for them. But when it comes to public telecommunications network systems, individual users' perceived values of a system depend on the perceived values of others in the mass market, most of whom they do not even know. This is the curse and, at the same time, the windfall of network infrastructure businesses. That is, no matter how good your network system may be in an absolute sense, until it is perceived that way by many users, it is not going to sell; however, when it is perceived as ubiquitous, it literally sells itself.

Interactive systems doomed to failure usually have a low value for several reasons: (1) subscription to the system costs too much; (2) it is to hard to figure out how to use; (3) there is not enough good information content or programming; and (4) not enough people use it so that a critical threshold of demand is met. The fourth factor is absolutely critical to success in the mass market. The Internet phenomenon occurred and continues to grow because all those factors were met. By far, the best explanation for the explosive growth of the Internet is that its e-mail feature not only is versatile, easy, and cheap to use but also had reached a market penetration threshold that made it worth getting simply to communicate with those who already had it. Before any interactive electronic media service can be successful in the mass market, it must reach a critical mass of demand.

It is important to note that those same factors are responsible for the success and the failure of interactive electronic media that technically do not even rely on networks per se. For example, the relative market success of interactive electronic games for televisions and PCs is dictated largely by the perceptions of users generally, not by the few who would purchase them even if no one else did. Such fad effects can influence the purchases of many types of goods, but they are especially strong and long lived in the case of network systems where the value of the system to any one individual is directly affected by the use of the system by others.

Numerous nonleisure activities and early applications of multimedia technology will help drive the deployment of a multimedia infrastructure in such areas as telework, teleschool, telemedicine, university-based research and education networks, and a wide range of other business applications. However, even those uses will not be enough to drive the demand for a mass market public multimedia network infrastructure. The reason is simply that, except perhaps for telework and teleducation from the home, those applications require network connections to only a handful of locations, such as regional medical centers and hospitals, libraries, and schools. No "killer" application lurks among those specialized activities that will justify the investment in a nationwide public network infrastructure. Until an application becomes popular enough that more than half of all households will use it, it does not qualify as a mass market infrastructure.

Even the new federal legislation defines advanced universal services as those that "have through market forces been subscribed to by a majority of residential customers." In other words, until half of America's households choose to subscribe to a service it should not be considered as a public infrastructure offering that warrants government regulations to ensure that it continues to be made available to all households at affordable broadly averaged prices. After all, rates are only averaged so that "all consumers regardless of wealth and in rural and high-cost areas should have access to telecommunications and information services that are reasonably comparable to those provided in urban areas at reasonably comparable rates."

In summary, no matter how technologically advanced a given type of multimedia network is, potential network infrastructure operators should always keep in mind two golden rules that define success in any network business, including multimedia: (1) a network is not viable for the mass market until it reaches a critical mass of demand; and (2) everybody wants to be on the biggest network. The second rule means that once a critical mass has been reached, the financial success and sustainability of the multimedia enterprise will depend on absolute size. In any given market, other things being equal, it is always better for a subscriber to be connected to the biggest network. If other things were not equal, presumably everyone would want to join the best network anyway, and the result would be the same. That begs an important strategic question for investing in multimedia networks: Should the network be based on a proprietary standard or an open standard?

Recent evidence certainly favors network infrastructures based on open standards. Network openness is really a matter of degree. The ultimate in network openness is a standardized *plug-and-play* network. That type of network infrastructure has standardized interfaces such that competing network equipment suppliers and individual subscribers can simply plug in compatible network components and devices to obtain whatever services and functions they desire. From the point of view of users, the basic telephone network is a good example of a completely open system. Any individual can purchase telephone equipment or premises wiring from a multitude of vendors and plug it in to a standard telephone network interface. Cable television systems are open stand-ardized systems, but they are somewhat limited in that only approved sub-scriber equipment (e.g., set-top converter boxes) may be used, and usually such equipment must be obtained from the cable system operator.

Apple computers and the corresponding operating system are proprietary systems because, even though Apple users can purchase basic components that are compatible, they usually must be supplied by Apple itself. IBM-compatible PCs (PCs) are more open than Apple in that anyone may make or buy PC com-puters and basic components (e.g., keyboards, printers), and they will (usually) work together. Apple's proprietary system and early head start in the market for user-friendly computers yielded a high market share and high profit margins. Today, however, Apple's share is declining, and its margins are lower than other competing nonproprietary PC makers.

The motivation for proprietary systems is that there is money to be made from being the only supplier of a unique or superior system. However, once it can be imitated, an open system is more likely to be successful. For example, the early automatic teller machine systems, like the one offered by Citibank to its customers, were based on proprietary systems, but as soon as competing vendors like Cirrus and Plus offered interconnected, standardized, and more ubiquitous nonproprietary systems, the proprietary systems soon had to go along or risk losing customers. There are numerous examples of open systems ultimately displacing proprietary systems, suggesting that this will be a key fac-tor in determining the success of multimedia network providers.

1.5 THE [BROADBAND] TELECOMMUNICATIONS ACT OF 1996

Deregulation and competition are the broad objectives of the new telecommuni-cations law. However, the law also includes many statements that promote regulation to achieve universally available and affordable digital services, even in rural areas. For example, the new law states that:

> Access to *advanced* telecommunications and information services should be provided in *all* regions at just, reasonable, and affordable rates [emphasis added].

The key question is, what does the term "advanced telecommunications" mean? The answer is provided in Section 706 of the law:

> (c) DEFINITION—(1) ADVANCED TELECOMMUNICATIONS
> The term advanced telecommunications capability is defined, with-out regard to any transmission media or technology, as high-speed, switched, broadband telecommunications capability that enables us-

ers to originate and receive high-quality voice, data, graphics, and video telecommunications using any technology.

Taken at face value that definition could be either good news or bad news. The good news is that the letter of the law (nominally) mandates switched broadband multimedia telephony for all American households. The bad news is that such a proposition is extremely expensive (some would say prohibitively so), and someone has to pay for it. But the law also states that rates should be just, reasonable, and *affordable.* The inconsistency is obvious. What is not so obvious is how this advanced infrastructure can be paid for at "affordable" rates.

The current U.S. Congress also has decreed that there shall be no "unfunded mandates" handed down by the federal government. Given that, it was necessary to include in the new law a statement that the Federal Communications Commission (FCC) and state regulators must find a way to fund the advanced universally available infrastructure:

SPECIFIC AND PREDICTABLE SUPPORT MECHANISMS—There should be specific, predictable, and sufficient Federal and State mechanisms to preserve and advance universal service...

and

TELECOMMUNICATIONS CARRIER CONTRIBUTION—Every telecommunications carrier that provides interstate telecommunications services shall contribute, on an equitable and nondiscriminatory basis, to the specific, predictable, and sufficient mechanisms established by the Commission to preserve and advance universal service. The Commission may exempt a carrier or class of carriers from this requirement if the carrier's telecommunications activities are limited to such an extent that the level of such carrier's contribution to the preservation and advancement of universal service would be de minimis. Any other provider of interstate telecommunications may be required to contribute to the preservation and advancement of universal service if the public interest so requires.

Suffice it to say, those provisions are easier said than done; it was easy for Congress to mandate funding but the devil will undoubtedly be in the details of implementation.

The fundamental premises on which the new law is based are that the best way to achieve public policy objectives is to promote private sector investment to fund the construction of the information superhighway, called the *National*

Information Infrastructure (NII), and that a competitive market model is the way to attract the required investment. In the future, deregulation and competition will be the engine driving private sector investment in an advanced telecommunications network infrastructure. In that regard, perhaps the most significant aspect of the new law is what it does not do. In their zeal to satisfy the major concerns of numerous interest groups, lawmakers seem to have lost sight of the forest for the trees. While the law proposes a host of rules governing market entry and mergers and acquisitions, it does not address the indisputable and most fundamental market force of all: the profit motive. The opportunity to pursue profit is what provides private businesses with the incentive to invest in new technology. By retaining the old-fashioned profit regulation typically applied to monopoly public utilities rather than replacing it with a regulatory framework that would provide positive incentives for infrastructure network operators to invest in expensive new equipment, the law falls far short of achieving its primary objective.

Even though the new law's overall objective is deregulation and competition, it certainly calls for plenty of new regulations to replace the old—far too many. The ramifications of the new federal law will be discussed later. Suffice it to say that since the new law grants both federal and state regulators the power to exercise considerable control over the exact rules governing competitive market entry, pricing practices, and business operations, it remains to be seen just how far the implementation of the new law will actually go toward total deregulation. Nevertheless, whatever criticisms one might have regarding the level of continued regulation and the amount of political pork embedded in the new legislation, considering how long it took the government to finally change its pitifully outdated rules, the Telecommunications Act of 1996 is a giant step in the right direction.

But it will not be enough. The legislative effort must continue as it becomes clear that the current version of the law actually raises more policy questions than it answers. Part of the reason the new law was needed in the first place was that a lack of a clear national telecommunications policy under the old law had effectively put the authority for critical policy decisions in the hands of the nation's jurists. Whether or not one believes that the judicial branch of government has done a good job of implementing competition policy (starting in 1982 with the famous decision to break up AT&T), the fact of the matter is that the responsibility for formulating national competition policy was forced on them by a lack of federal statutes that clearly enunciated a unified national policy. Unfortunately, this is still the case with the new law, which is clear on the point that more competition is preferable to more regulation in future regulatory decisions but which provides few good suggestions as to how to realize that ideal. The problem, of course, is that no one really knows when competition truly exists in a market and when it does not. Since "perfect competition" does not exist, competition will always be in the eyes of the beholder, and court battles are sure to ensue when one beholder wants to enter another's market. The risk of not further clarifying various provisions in the new legislation as it stands now is that the courts will remain in the driver's seat, dictating the rules of the road for the new multimedia information super-highway.

1.6 RELATIONSHIP OF ECONOMICS, TECHNOLOGY, AND POLICY

The future of multimedia network infrastructures in the United States will be determined by the complex and sometimes competing forces of technological progress, economics, and public policy. Each of these forces may pull together or apart, depending on many factors, some of which can be controlled and some of which cannot.

Telecommunications technology is dynamic and quite advanced scientifically and tends not to be well understood by users of telecommunications services. Network technology is generally transparent to users, who might well understand the various applications for using certain telecommunication services but not the underlying technology. That creates something of a dichotomy between demand and supply for multimedia, which is largely the reason why known or perceived telecommunications demand is not very closely related to the ultimate decision to select one network technology over another. Basically, there are many different technological alternatives for providing any given telecommunications service or application. Historically, this was not the case. There tended to be only one clearly preferred technology for providing any given service or application. In other words, depending on the use intended, the technological choice was clear based on cost and performance criteria.

That is no longer true. Today's technological mix allows telecommunications technologies to be shared cost effectively across many different and distinct service applications. The ultimate technology decision is based on the relative long-run and short-run economics of alternative technologies. Or is it?

Unfortunately, in a world where there is no clear choice of technology for any given set of communications services, government policy plays a major role in both the speed of technology adoption in general and the choice of technology in particular. Government policy throws a monkey wrench into the marketplace, and it is the marketplace that would be the most efficient vehicle for sorting out such complex technological issues.

Thus, economics, technology, and public policy are all important ingredients in the process of communications infrastructure development. We cannot safely ignore any of them in trying to predict the future of telecommunications in the United States

The direction and progress of pure technical research in engineering, mathematics, physics, and chemistry establish the range of possibilities for telecommunications technology adoption. The net economic benefits, in terms of supply (cost) and demand, will determine private and public cash flows associated with each technological alternative, which, in a market setting, would ultimately determine the winners and the losers in the telecommunications technology race. Without economics, we have little guidance to help us select among technological alternatives. Public policy may significantly alter the private and public costs and benefits associated with any given technological alternative and therefore becomes a critical factor in the equation. Politics simply cannot be ignored. There are those who believe that politics may only have a transitory impact and that, in the long-run, technology and economics will prevail in selecting alternatives. The problem is that in the long-run we are all dead; there is no reason to believe that long-lived investments in telecommunications networks will not persist, if for no other reason than they represent sunk costs that have already been paid for. Other new technological alternatives, while perhaps cost effective at the margin, cannot compete with the embedded base. Political policies that favor any given technology may forever alter the dynamic of technology adoption from the one that a pure market setting may have produced.

The best public policy is one that does not favor any particular service provider, whether incumbent or new entrant, or one technological alternative over another. Most lawmakers and regulators proclaim that they do not want to be in the position of handicapping technological outcomes and dictating winners and losers in the technological race for multimedia networks. If that is true, then policymakers should develop technology and competition policies that are truly competitively neutral, thereby allowing the free market to sort it all out. Unfortunately, as later chapters on regulatory policy reveal, public policy has a long way to go beyond the new telecommunications law if it is to achieve any semblance of free-market outcomes.

Reference

[1]   Berthold, J., "Networking Fundamentals," OFC'96, San Jose, CA, Feb. 1996.

Multimedia Technology Trends

2.1 THE FUTURE OF MULTIMEDIA

Uncertainty reigns when it comes to predicting how the information super-highway is ultimately going to materialize. There are plenty of grand announcements, projections, and predictions about the future of telecommunications, cable television, and computers. Many experts see these three major sources of communication converging to form one supermode of instant, interactive, high-speed, affordable communication. They go so far as to predict that it will soon be in consumers' living rooms. Other experts see a high-stakes battle among telephone providers, cable companies, computer makers, software providers, and satellite companies, each fighting for its very life. What will actually happen is probably somewhere in between.

At stake is a potentially huge and lucrative global marketplace. A recent report by the ITU estimated that in 1994 alone the "info-communications" industry, which covers telecommunications (46%), computing (33%), and audio visual services (21%), generated revenues of about $1.43 trillion worldwide, or about 6% of the world's total economic output [1]. The ITU also estimated potential multimedia service revenue of over $40 per month per household.[1] Assuming that that figure is accurate, if only 50% of U.S. households subscribed, it would translate into annual revenues of over $20 billion. According to the ITU, *video on demand* (VOD) and online transaction services (e.g., shopping) represent the largest portion of the multimedia pie, each with 28% of the total $40 per month.

To gauge which countries are poised to see multimedia demand take off, the ITU has created indices of "teledensity" for 39 major world economies. Multimedia teledensity refers to the penetration rates for the primary methods of bringing multimedia to the mass market: telephones, TVs, and PCs. Table 2.1

---

1. Based on U.S. and Hong Kong data.

provides the results of the ITU teledensity study and shows that the United States is the uncontested leader of the pack. Table 2.2 shows the household penetration (1995) for a variety of consumer electronics gear, according to the *Electronic Industries Association* (EIA).[2] In 1994, 43% of the world's cable TV subscribers were in the United States, as were about 45% of the world's PCs and about two-thirds of estimated Internet users. In 1995, the United States had an estimated 37 million PC households, a third of which had modems. Interestingly, of the 250 million VCRs in the world, most are found in Japan and Hong Kong (each have an 87% penetration rate). The United States ranks fifth, with 79%, but still accounts for about 50% of all video rentals. Video games account for about $7.3 billion worldwide, and about 39% of U.S. households are equipped to play them.

Table 2.1
Results of ITU Teledensity Study

Rank	Country	Telephone Density	TV Density	PC Density
1	United States	59.5	79	29.7
2	Denmark	60.4	55	19.3
3	Canada	57.5	65	17.5
4	Sweden	68.3	48	17.2
5	Australia	49.6	48	21.7
5	France	54.7	58	14
5	Switzerland	59.7	41	28.8
8	Netherlands	50.9	48	15.6
9	Germany	48.3	55	14.4
10	Japan	47.8	64	12
11	United Kingdom	48.9	45	15.1
12	Austria	46.5	48	10.7
12	Belgium	44.9	47	12.9
12	Singapore	47.3	38	15.3
15	Hong Kong	54	36	11.3
16	Italy	42.9	45	7.2
17	Spain	37.1	42	7
18	Korea (Rep.)	39.7	32	11.2

2. Based on a survey by the EIA for the Consumer Electronics Manufacturers Association. In monthly surveys conducted randomly throughout 1995, the EIA interviewed a cumulative total of 25,000 adults by telephone in all 50 states. Because each survey focused on different areas, the margin of error is ±3 percentage points.

Rank	Country	Telephone Density	TV Density	PC Density
19	Taiwan-China	40	32	8.1
20	Hungary	17	42	3.4
21	Czech Republic	20.9	39	3.6
21	Israel	39.4	30	9.4
23	Greece	47.8	22	2.9
24	Portugal	35	25	5
25	Argentina	14.1	38	1.7
26	Poland	13.1	30	2.2
26	Russia	16.2	38	1
28	Malaysia	14.7	23	3.3
29	Chile	11	23	3.1
29	Turkey	20.1	27	1.1
31	Mexico	9.2	20	2.2
32	Brazil	7.4	29	0.9
33	Venezuela	10.9	18	1.3
34	South Africa	9.5	10	2.2
35	Thailand	4.7	19	1.2
36	China	2.3	23	0.2
37	Philippines	1.7	12	0.6
38	Indonesia	1.3	9	0.3
39	India	1.1	5	0.1
Average developed		52.3	63	18.7
Average developing		5.2	18	0.7
Overall average		14.5	27	4.3

Table 2.2
EIA Survey of U.S. Household Penetration of Consumer Electronics Gear, 1995

Consumer Electronics Product	Percentage of U.S. Households
Television	98%
Home radio	98%
Telephone (corded)	96%
VCR	88%
Set-top box (TV)*	70%
Answering machine	60%
Telephone (cordless)	59%

Table 2.2 (continued)

Consumer Electronics Product	Parecentage of U.S. Households
Home CD player	48%
Personal computer	40%
Video game controller*	39%
Computer printer	34%
Automobile alarm	25%
Camcorder	23%
Cellular telephone	21%
Computer with CD-ROM	19%
Modem	16%
Caller ID device	10%
Home fax machine	8%
Analog satellite dish*	5%
Digital satellite dish*	2%
Laserdisc player	2%

* Not part of the EIA survey.

Many experts in the field believe that the multimedia environment of the future will integrate many of the devices listed in Table 2.2. Some companies are betting their financial futures that their services and network delivery systems will ultimately become the multimedia standard. Given the wide range of service applications and alternative network delivery systems, that is a risky proposition. Some firms are diving headlong into the latest technology in an effort to be the first to deliver services. Others are taking a more cautious approach, waiting to see which technology will be the most affordable and user friendly. While the future remains murky, one thing that seems certain is that any major changes in telecommunications, cable TV, and computer operations are going to take time. While many observers say the great information future is just around the corner, it is becoming increasingly apparent that those predictions are more hype than reality. There are many problems inherent in creating a multimedia information superhighway, ranging from difficulty in getting the technology to work, to higher-than-expected costs, to lack of demand. Consumers may not experience the impact of the information revolution for another 10 to 20 years, rather than the 3 to 5 years that many pundits are touting.

It is anybody's guess as to which companies or industries will be the big winners. The biggest battle pits cable operators against telephone providers, with computer access providers playing the role of third-party spoilers. (Satellite and other "wireless" solutions cannot be counted out.) This section will analyze the relative strengths and weaknesses of the competing technologies in terms of their relative abilities to exploit the possibilities presented by the Internet and *interactive TV* (ITV), the two leading multimedia gateway options. Each technology has certain technological or geographical advantages over the other, yet, given the diversity of applications to which the technology is expected to be put, none is overwhelmingly superior.

Generally, telephone operators have an implicit advantage in that their service provides two-way communication, covers a wide customer base, and has greater market penetration. Cable operators, on the other hand, can deliver more information in the form of moving pictures and at much faster speeds than telephone providers, but, for the most part, they offer only one-way communication.

By comparison, computer access providers utilizing the phone network seemingly have much of what it takes to dominate the information superhighway. The service is ubiquitous, fast, affordable, and interactive. However, market penetration remains problematic, because computer costs, while dropping, are still too high to be affordable for many Americans. Even though about one-third of American households have computers, gaining ground from now on will be more difficult because so many affluent households already have PCs, and the less affluent ones will resist paying $2,000 or more for a computer. Based on a survey of 10,000 households, Dataquest estimated that two-thirds of all the PCs sold to households are bought by those with incomes of $40,000 or more. Yet census data show that only about one in three households have incomes of $40,000 or more. Unless prices drop drastically, some experts predict that it is unlikely that the PC will to find its way into more than 40 or 45 percent of American households.

The American public does not seem very interested in the high-technology hype. According to a recent survey of 1,000 home computer users, reported in *PC World* magazine, 41% of the respondents found computer manuals "so confusing, they may as well be written in a foreign language." And 31% of the persons surveyed said they spent more time trying to figure out how to use their computer than using it. While the growth of online computer service subscribers has been impressive, many subscribers are just experimenting with the new medium and then rarely using it. An estimated 60% of America Online's subscribers are on the service less than 5 hours a month. By contrast, the average American household has its TV on 5 hours a day.

In a 1993 poll, the *Institute of Electrical and Electronics Engineers* (IEEE) found that consumers seem relatively cool to the promise of new technologies. The poll found that only a small percentage felt that advanced electronics and computer developments would greatly affect their lives in the next decade. In addition, the poll found that: (1) only 15% believed that ITV, VOD, and *virtual reality* (VR) would affect their personal and business lives; and (2) only 7% believed that improvements in technology would influence their leisure time

[2]. Another poll, conducted by Dell Computer Corp., suggests that many Americans might not be up to the challenge of multimedia. That poll found that 25% of adults have never programmed a VCR to record a TV show. The survey also found that 25% lamented the passing of the typewriter. Supporting those findings was a Gallup survey commissioned by MCI Telecommunications that found that 32% of white-collar workers were afraid of cyberspace, 59% saying they would try new technology only after it was proved effective and 58% having never heard of the Internet.

Those statistics and others like them have not stunted the growing industry enthusiasm for multimedia. What has happened is that the focus, for now at least, has shifted from the never-quite-worked-out interactive TV to the here-and-readily-available World Wide Web—a sizable change. Only three years ago the picture painted by technology experts was one of futuristic consumers sitting in their living rooms, watching TV, and using the remote control to call up videos, shop in virtual malls, and even vote. The reality is that many of those activities are taking place in the here and now on the Web.

If the explosion of Web magazines, Web shopping, Web games, Web sports, and Web chat rooms is any indication, it seems that the Web has stepped in to fill the gap left by the much publicized but never materialized ITV. The charge to the Web has been led in part by the huge success of user-friendly Web browser software, like Netscape. Growth in the number of U.S. consumers using Web browsers has gone from about 5 million in 1994 to 15 million in 1996. According to Lycos, Inc., one of the most popular "search engines" for finding information on the Web, the number of consumers with Internet access is expected to top 30 million in 1998. Likewise, the number of pages of information available on the Web has grown from 5 million in 1994 to more than 25 million in 1996.

Anticipated Web growth can also be attributed, to some degree, to technological advancements that could end up making it look and act more like TV. Many cable companies are pushing hard to produce ultraspeedy modems that make navigating the Web quick and easy.[3] Those new modems, some of which are already on the market, can download information at 10 Mbps, which is more than 350 times faster than the 28.8-Kbps modems available for PCs today. However, such speedy cable modems can cost a whopping $500–$700 apiece. Still, telephone company modems that operate at a much lower 1.5–6.0 Mbps can cost from $600–$1,800, making the cable version look like a bargain.

---

3. That makers of cable television set-top converter boxes are rushing to transform their technology to make it work on the Web is a marked shift from their promises of a few years ago that digital set-top boxes would be the interactivity solution. It seems that cable operators underestimated the difficulty of the undertaking as well as the associated costs and financial returns required to pay for it, while grossly overestimating consumer demand.

Faster modems alone will not convert the Web into an ITV alternative. Infrastructure issues still must be worked out for both cable and telephone network applications. The problem is that the new modems are faster than the connecting networks and servers. For example, some video servers are unable to send information at speeds greater than 1.5 Mbps—far less than the cable modem can handle and on the low end of what the high-speed telephone modem can handle. Even putting the nominal speed of the public network delivery system aside, if many subscribers try to access the same programming and transmission links at the same time, the network capacity can quickly exhaust. Anyone who has tried to access a busy Web site during peak hours on the telephone network can vouch for that. Even with today's slow modem speeds of 14.4–28.8 Kbps, the actual transmission rates for downloading data and programming from the Internet is only a fraction of that.

The problem of multiple subscribers accessing the network simultaneously is potentially much worse for the cable application than it is in the case of telephone network. Internet access connections on local cable network distribution systems are shared among many subscribers, while on telephone systems the subscriber connections are mostly dedicated point-to-point transmission links. Coaxial cable systems are also subject to local interference noise from other household appliances, which are magnified as the upstream signals run through cascades of signal amplifiers. That is forcing cable systems that want to provide high-quality two-way services to upgrade their networks and eliminate network amplifiers (see Chapter 3). According to Probe Research, only 5.8 million cable TV households will be capable of using a high-speed modem by the year 2004.

This does not mean that modems connected to the public telephone network are the better investment choice. In fact, high-speed modems connected to the public telephone network have unique problems that will severely limit their usefulness. These modems will not operate properly over long distances, and the majority of households are too far from the telephone company switching office to allow for high-speed digital services. One estimate, by Probe Research, suggests that by the year 2004 only 7.4 million of an estimated 95 million households will be capable of using the new high-speed modems.

Any discussion of the Internet as a multimedia platform is also moot unless the telephone operators who provide the mass market with access to the Internet over traditional twisted-pair telephone lines choose to upgrade that infrastructure. The twisted-pair system is a painfully slow and inefficient means of transmitting graphs, charts, pictures, and in many cases even straight text. Consumers want instant access to the information they seek, and if they do not get it from the Internet, they'll find it elsewhere.

Even if the connectivity issues can be resolved, there are problems inherent to the Internet that also must be addressed. Despite the abundance of information on the Internet, much of it is useless or takes too much time to locate. A

Times Mirror survey found that PCs are not replacing the more traditional ways in which Americans get their information or conduct their business. The study stated that 12% of PC-modem owners use the device in their homes primarily for business. The PC is often used for electronic mail, but only 20% of users use the PC to access news, participate in chat groups, get travel information, or play games. In the end, it remains to be seen if the Internet will be the most commercially viable multimedia delivery system. There is no certainty that consumers will be willing to pay for information that is otherwise available for free. Likewise, it is uncertain whether advertisers will support the Internet.

Still, many computer game companies see cyberspace as a way to fight increased competition in the traditional arenas of video game console boxes and PC CD-ROMs. Sales of video game hardware and software remained stagnate at between $4 billion and $5 billion from 1992 to 1994, before dropping below $4 billion in 1995. However, sales are expected to exceed $4.5 billion in 1996. One industry source, *Computer Gaming World*, predicts that in the next decade online computer games will surpass all other forms of entertainment, including movies and TV. That projection may be a stretch, considering that online games have been around for almost 20 years (starting in the form of e-mail used to play chess) and that none of the major ventures has been a financial success. Sierra On-Line, Inc.'s online service called ImagiNation has only 70,000 of the 200,000 subscribers it needs to make a profit. Likewise, Catapult Entertainment's Xband service, which links Sega and Nintendo machines through special modems, has been a huge disappointment. Only about 50,000 of the projected hundreds of thousands of customers have bought a modem. The problem common to ImagiNation and Xband has been that they cost too much. Xband charges an average of $50 a month, but usage charges may run as high as $500 a month. That's on top of the $60 cost of the game and $20 for the modem [3].

As for the future, everyone is still searching for the "killer" application. VOD, information services, transaction services (home shopping, banking, etc.), video games, programming, and direct response advertising are often mentioned. Others that are likely, but for obvious reasons not mentioned as often, are gambling and pornography. In fact, in a recent survey, it was discovered that the most popular Web site was *Penthouse* magazine, with 54 million "hits" in one month. The other most popular sites, in order, were ESPN sports, *USA Today*, *Playboy*, and *Wired*. Netscape, by far the most popular Web browser with an estimated 80% of the market, reported that its Web site is accessed about 45 million times a day.

In the end, it remains to be seen if the rush to the Internet is really the long-term answer for the interactive multimedia business. The government's original funding of the network technology and standards along with the ubiquitous telephone network has given the commercial Internet a huge head start over other multimedia delivery systems. But while the growth in the Internet's popularity is spectacular, it still poses many unanswered questions. For example, how can several family members use the computer at once? Who will be allowed to tie up the telephone line? Will workers, many of whom spend much of their day in front of a computer, want to come home every night and sit in front of their PCs?

And what about the future of interactive TV? Will consumers want or demand more interactive services? Will VOD ever take off? Do consumers really want 500 video channels when many still watch just the three major broadcast networks? How does satellite TV fit into the picture?

It is important to note that the concept of ITV has not been entirely written off. In fact, there actually have been some ITV successes. Videoway in Canada has more than 220,000 subscribers paying an average of $15 a month for interactive services, and more than 103 ITV test markets were rolled out in 1994 and 1995. A 1994 study by Next Century Media projected that the penetration of ITV in U.S. households will top 65 million in 2003 (two out of every three households), compared with only 455,000 in 1994. It remains to be seen if that is achievable, considering the apparent shift in cable companies' priorities and that meeting that projection would require that every single cable household in America subscribe to interactive services.

In addition, digital satellite delivery systems are being deployed that could potentially serve the mass market for some interactive multimedia services. At the very least, they appear to be an attractive and cost-effective alternative for providing VOD services. They could become highly complementary to new digital CD recorders for downloading movies, news, and information services.

A lot of hype attended the introduction in late 1996 of new *high-density CDs* (HD CDs) and *digital versatile disks* (DVDs). InfoTech predicts that by 2000 American demand for DVDs and HD CDs alone will be 110 million units. That is a long way from the current numbers for regular audio CDs. There are already over 9,500 commercially available CD-ROM titles and 27 million installed CD-ROM drives worldwide. Sales of computer CD drives increased 137 percent in 1994. Sales of CD-ROMs are projected to go from $300 million in 1994 to about $1.1 billion in 1996, according to PC Data, Inc.

It seems that, for now at least, all of the aforementioned services and their associated delivery systems will coexist for some time to come, with each vying to develop more services, increase its customer base, and gain a technological edge. Consumers will benefit from more choices and lower prices. But it does not appear likely that any one system will emerge from the pack to dominate the business until it learns how to meld all the positive features of cable, computers, and telephony into an affordable and easy-to-use system, while eliminating all the downside problems associated with each system.

Microsoft, IBM, and other major industry players foresee computers evolving to fill the gap by equipping PCs with so-called plug-and-play features. That would allow PCs to be viewed by consumers as just another component in their home entertainment systems, (a much more powerful one that serves as the heart of an integrated system).

2.2 MULTIMEDIA TECHNOLOGY TRENDS

Usually in the assessment of future technology trends, a good place to start is with an examination of current and prospective demands. But there are not very many interactive multimedia services to examine. With the exceptions of the Internet, video games, and a relative handful of *interactive CDs* (CD-I), the majority of interactive multimedia applications are for specialized business services like employee training, which limits the usefulness of demand analysis. Demand for multimedia is a classic chicken-and-egg problem. Mass market consumer demand cannot materialize before multimedia networks, consumer terminals, and various other devices have been supplied, and that supply will not materialize absent consumer demand. The result is a vicious circle. The situation is similar in telecommunications and other network businesses. Mass market demand cannot take off until a critical mass of network subscribers is reached, and that cannot happen until an adequate infrastructure is in place.

Normally, market supply is demand driven, but in the early stages of a network business, the critical mass phenomenon is often the greater influence. Critical mass requires that a minimum threshold of network subscribers be achieved before market supply can be justified and sustained. The reasoning is clear: Network services are of no value to an individual consumer unless others are on the network, and value naturally rises as the number of network users and usage rise.

Of course, technological research and innovation can, and often do, occur independent of consumer demand patterns, and what we end up observing in the real world is a bunch of multimedia technologies in search of an application that would cause mass market demand to take off.

The Internet is the best example of the importance of a network infrastructure preceding mass market demand. The government funded the development of both the *Advanced Research Projects Agency network* (ARPAnet) for the *Department of Defense* (DOD) and the *National Science Foundation research and education network* (NSFnet). These publicly funded precursors to the Internet combined with the public telephone network, making multimedia on the Web possible. In the future, other mass market multimedia network infrastructures will emerge. Of the current technologies on the horizon, the one most likely to achieve mass market status is the PC-based CD-I. It remains to be seen which type of network technology will be used in conjunction with CD-Is to allow users to communicate from a distance by connecting their CD players to a network. The leading contenders to fulfill that role are new digital telephone and cable networks, but wireless technologies, including satellite to the home, are also candidates.

Unfortunately, almost all of the multimedia systems currently on the drawing board are counting on consumer demand for VOD and *near-video-on-demand* (NVOD) services to pay for the cost of the network distribution system. This is an unlikely source of funding for at least two reasons. First, it is not at all clear that one technology or delivery system will be preferred by the mass market, which makes deployment of a wired network a risky proposition. Second, even if only one preferred network delivery system does manage to evolve (which is highly unlikely), most financial business case analyses indicate that there is still barely enough potential demand in the average household to warrant the required investment.

One American trial found that VOD subscribers watched an average of two-and-a-half movies a month at about $3–$4 per movie. That yields annual revenue of $90–$120 per subscriber. But buy rates are low for early *pay per view* (PPV) cable services, because, on average, only 3% of total cable system subscribers actually buy movies or other event programs in any given month. (Some systems get considerably higher penetration, but the bulk do not). Considering that the cost of a network distribution system is anywhere from $500–$1,500 per household, it becomes clear that paying off the network investment will take a long time. Satellite firms and other companies offering NVOD services (e.g., featuring the best titles every 15 minutes or so) represent a further drain on the revenues available for VOD providers. A recent NVOD trial by TCI (the largest cable operator in the United States) and AT&T found that customers were almost as happy with NVOD as the vastly more expensive VOD [4].

It is hoped that transaction services and other new information services can make up the difference, but such hope is more a leap of faith than rigorous forecasting. The wireless network alternatives seem to hedge that risk, because a single land-based transmission antenna or satellite transponder can cover a wide area of potential subscribers, and no expense is incurred for installing subscriber equipment until a household or individual actually signs up for service.

In addition to the obvious technical difficulties, there are other equally pressing concerns. To date, there has been relatively little usage of existing interactive multimedia services, and that has industry pundits wondering if usage is low because the technology has not, as yet, become widely available, or because people are not that enamored with those services.

2.3 BROADBAND AND INTERACTIVE MULTIMEDIA INSIDE THE HOME

Assuming that mass market demand for multimedia services does take off, many issues need to be addressed or at least considered. How will a typical household configure its various multimedia terminals and devices? What type of wired or wireless distribution system will be used? Will TVs become more like computers for interactive video services, or will consumers choose computers for interactive services? Assuming a computer-based solution, might PCs also be equipped with bigger screens, replacing TV sets for more traditional video entertainment? No one currently has the answers; presumably the market itself will eventually sort it all out. Suffice it to say that there are two dominant competing models of the multimedia household: the TV set as computer model and the computer as TV model. Which is more likely: a family gathered around a large-screen PC watching TV, or that same family gathered around the TV set browsing the Web? Who knows, maybe we'll see both.

2.3.1 The TV as Computer

Although computers currently dominate the scene for interactive multimedia, it is entirely possible that TVs could catch up if manufacturers began exploiting TV's natural advantages and addressing its several disadvantages [5]. For example, it should not be forgotten that TV still reigns as the dominant household entertainment device. Still, a move should be made to equip TV sets for interactive capability and produce affordable modems for use on the network system. Also, it should not escape mention that, for years now, PCs have come equipped with CD-ROM drives, while TV sets with built-in CD players are almost unheard of. At the same time, industry decisionmakers seem willing to pump funds into functionalities and services that the viewing community has rejected. For example, TVs have evolved to offer "plug-and-play" capability with component stereo audio systems, but those features are not very popular with consumers because a TV's speaker quality is lacking, and because stereo audio systems generally are not viewed by consumers as complements to video entertainment. Since their introduction, the demand for the digital audio channels offered on cable TV systems have also languished. It seems that TV manufacturers would benefit from focusing their efforts on making a streamlined TV set that incorporates the most used functions of computers rather than offering new services for which a clear demand has never been established.

If recent announcements are any indication, manufacturers have finally adopted this logic. On realizing that CD technology is becoming a critical component of the new digital multimedia age, major TV manufacturers have announced that beginning in late 1996 they will begin to offer built-in CD players in their high-end sets. TV maker Thomson has gone a step further, announcing that by next year it will introduce TVs with built-in CD-ROM/DVD players for viewing movies and other multimedia titles. That alone will allow Thomson's TVs to become players in the new digital multimedia environment. Sony and other manufacturers have also announced plans for portable CD players that can be connected to many existing TVs sets (or computers), with target prices in the affordable $200–$300 range.

In addition, some manufacturers believe that the key to enhancing the overall functionality of traditional TV sets for multimedia services lies in the development of inexpensive digital set-top converter boxes. The key to getting the costs down is mass production of a standardized technology. The problem is that the industry has taken a long time to settle on a standard for digital signal compression, and, even now, the details are still not worked out to the point that set-top boxes from different manufacturers will work with all of the different video delivery systems planned by telephone and cable companies. Still, in early 1996, the industry's *Digital Audio Visual Council* (DAVIC) approved the first standards for ITV and digital set-top boxes, so mass production can now proceed. Even with standardization, the cost per unit, even for large orders is running in the $400–$700 range. It is expected that that figure could fall to $300–$400 per unit in 2 to 3 years.

Other firms are not relying on set-top boxes and are instead planning to build the functionality into the TV set itself. Starting next year, consumer electronics giants Thomson, Mitsubishi, and Sony are all planning to produce TV sets with built-in circuitry to support Internet access. Access to the Internet may be achieved with the click of a button, and users will be able to navigate the Web from their sofas, using only a hand-held remote control device. Increased security features will allow for inputting credit card or other payment information to support transaction services. A wireless keyboard also is planned for more involved tasks.

Building on this are IBM, Oracle Corp., and others, which have announced the introduction of inexpensive (about $500) computing devices (console and keyboard) that are really stripped-down PCs that can use a TV monitor for a screen display. Viewcall's Web PC, called Webster, is also designed to work with an existing TV screen display [6]. It is expected to become available in June 1996 at a target price of about $300. Considering that a good computer screen display alone costs about $300–$500 for a 14- to 15-inch display (more for larger displays), that could be a stroke of genius. The question is, will families sacrifice their couch potato position and tie up their beloved idiot box to browse the Web?

The proliferation of TV sets, stereos, and PCs in the home have fragmented many families, and it remains to be seen if an integrated digital multimedia entertainment center, with a large screen and better resolution and sound, will be able to bring the family together. I personally would not throw away the headphones just yet. On the other hand, perhaps costs for multimedia technology will fall so fast that everyone in the family can enjoy his or her own miniaturized personal multimedia entertainment center. Do not hold your breath, however, since it would take quite a breakthrough to make that happen.

2.3.2   The Computer as TV

Almost 40% of U.S. households own a PC (a third of which have modems). In 1994, PC sales in the United States exceeded those for TVs. Those growth rates are starting to slow, and a shift is occurring in the types of PCs being purchased. In 1995, about 90% of PCs shipped were equipped for multimedia with built-in CD-ROM drives, modems, or both.

By combining large-scale displays with their interactive multimedia capability, PCs are well positioned to invade the traditional turf of TVs. Some high-end computers costing upward of $3,000 are already set up to receive a TV signal input. As more and more video sources arrive in digital form, the PC could eventually offer the type of programming diversity and resolution now offered on cable TV systems at extremely competitive prices. New high-speed CD-ROM players capable of displaying high-resolution video images, including movies encoded with *Moving Picture Experts Group 2* (MPEG-2) video compression, are already being manufactured for PCs.

To reinvigorate the demand for PCs and expand their role in the household, Sony, Microsoft, and other manufacturers recently announced plans for software systems that, in conjunction with new plug-and-play computer hardware components, can be used to control traditional video and stereo audio entertainment systems and other household appliances. Microsoft has dubbed its new system the *Simply Interactive PC* (SIPC). It is designed to be simple to use and easy to integrate with other consumer electronic gear, such as TVs, VCRs, and stereos. Early indications are that SIPC computers will be relatively expensive at $1,500–$3,000 each, but if it makes programming the VCR possible, it may be worth it [7].

2.3.3   Set-Top Solutions for Broadband Living Rooms

Because of the huge installed base of TVs and PCs, most interactive multimedia services will be delivered via a digital set-top box, including telephone and cable modems. The $500-a-unit price tag will probably mean that most households will not purchase more than one. But if demand is to ever take off, one box very likely will not be enough. After all, it is unlikely that all of the multimedia devices found in the average household will be located in close proximity to the set-top box. That explains why the set-top box solution may not last forever. As technology progresses to further lower the cost of new digital multimedia terminals, many households are likely to prefer terminals that connect directly to a digital network pipeline coming into the home and have the requisite electronics built in. However, it seems likely that for at least the next 7 to 10 years the two types of systems will coexist. In other words, it is unlikely that a single, dominant, integrated system will emerge.

In a few years, digital broadband modems will hit the mass market by storm. The costs of high-speed modems operating in the 4- to 10-Mbps range for both cable and telephone network connections will surely fall from their initial prices of about $500–$1,000 apiece. It is important to keep in mind that a single modem located at the subscriber's premises does not a broadband multimedia system make. Similar equipment is also needed on the network end of the subscriber line, which increases the system costs well beyond those of the modem located at the subscriber premises.

In the case of telephone network modems, powerful industry alliances have requested that modems capable of accommodating real-time standard MPEG-2 compressed video be mass produced to achieve low production costs. For example, Tele-TV, a video programming consortium of some Bell Telephone Companies, is urging manufacturers to produce a digital set-top box that will work on a regular phone line and cost under $300.

Operating at only 4 Mbps, a phone line could support MPEG-2 video, download 200 pages of text in one second, or download one Web page in one-tenth of a second. There are already many *asymmetrical digital subscriber line* (ADSL) modems for telephone lines, several of which operate at speeds of 6 Mbps downstream and over 500 Kbps upstream. The short-term goal is to get the price for equipping both ends of a subscriber's dedicated phone connection down to about $500 per line. Motorola Semiconductor has announced a target cost of only $100 per end by early 1997. Telephone ADSL lines can also be interconnected to both the packet-switched Internet and Ethernet backbones [8].

With MPEG-2 video programming in short supply and DVDs just now getting into production using the MPEG-2 compression standard, the early application driver for ADSL modems will be information services obtained on the Web. Those services will be targeted at the *small office–home office* (SOHO) market. The problem with telephone modems is that the higher the modem speed, the shorter the subscriber's connection must be to use it. The modem speeds already offered exceed those that any telephone lines over 12,000 ft long can accommodate. For those subscribers whose phone lines qualify for ADSL modems, telephone companies are discussing a target price for high-speed Internet access service in the range of $35–$40 per month.

A speed of 10 Mbps is by no means the maximum that telephone lines can carry if they are short enough. *Very high bit rate digital subscriber lines* (VDSL) can, and do, operate at very high speeds. There is already a LAN service, called *copper distributed data interface* (CDDI), that operates at 100 Mbps, but only over the very short distances typical of LANs. At rates of 52 Mbps, a subscriber's access line length would be limited to about 500 feet.

Cable network modems offer even higher speeds. Partly because cable systems have more digital bandwidth to spare compared to telephone lines, many first-generation cable modems will offer speeds of 10 Mbps, which may be used for Internet access and which are compatible with the popular computer LAN Ethernet standard called 10baseT. In the first U.S. trial of that technology, Jones Intercable is charging $39.95 per month for Internet access at 10 Mbps bidirectional using an Ethernet connection to a PC.

Limited trials are one thing, and the mass market is quite another. Cable networks are shared, which means that when many households demand simultaneous access, the system can easily become overloaded. One advantage of the telephone network ADSL technology is that each subscriber's connection is usually dedicated, offering data throughput to the network equal to the nominal modem speed. Still, for both telephone and cable networks, it is not just the individual subscriber's access line connection that must support simultaneous access. The shared trunk network and information service provider nodes also must have sufficient capacity to handle multiple accesses. Network throughput for an end-to-end transmission will always be limited to the lowest-speed, or most congested, facility encountered in the network.

Forrester Research predicts that as many as 7 million households may have cable modems by the year 2000, which they will use mostly for interactive multimedia services on the Web and other online services [9]. That projection does not seem very impressive considering that a single telephone company, GTE, plans to deliver video services over telephone circuits to roughly 7 million households by the year 2005. The ability of cable network operators to convince their subscribers to use a nontelephone company for traditional telephone services is uncertain. Early trials indicated that telephone companies are firmly entrenched, and no major inroads are being predicted. Kagan Associates, a well-known cable consultancy and industry advocate, predicts that by the year 2005 cable telephony will represent a paltry $5.9 billion of the total $300-billion telephony market [10].

2.3.4   Inhome Wiring

Considering the substantial costs and potential hassles of having several modems, set-top boxes, and associated multimedia terminals in a single household, it often makes sense to wire the household as a small office with a multimedia digital LAN. This approach to home wiring has been around for a long time in a high-technology niche market, but the acceptance of "smart-home" wiring in the mass market is uncertain.

The concept of the smart home has been around for a long time, but heretofore it has been a very expensive proposition with limited usefulness. Generally, such a setup has been used for video monitoring (e.g., placing a video camera in an infant's room) and remote control of home appliances. With the convergence of technology, the same concept can be applied to distribute state-of-the-art digital broadband telecommunications throughout the home.

In an effort to promote inside wiring as compatible with a plug-and-play multimedia environment, the technology is being standardized to allow for mass production of relatively low cost, even do-it-yourself, installations. Early standards include the EIA/TIA 570A home LAN standard and the CEBus EIA IS-60 standard.[4] Most major electronic equipment manufacturers have already recognized CEBus as the standard for consumer products. The latest version of that standard offers between 5 and 890 MHz of bandwidth for multimedia applications, which is generally sufficient to satisfy interactive multimedia demand in the SOHO market segment.

A CEBus home network can have many elements and support a wide range of multimedia applications and devices, but the wiring system itself is not that complex. CEBus can use a variety of transmission media, including twisted-pair telephone wire, coaxial cable, power line, and even *radio frequency* (RF) and infrared wireless links. The system has a central controller that supports distributed control of audio, video, and data services throughout the home. Just like LANs in office buildings, these systems manage contention for system capacity from multiple devices. The CEBus system is not product specific and is easily expandable [11]. The system can be used to control everything from lawn sprinklers to security systems to entertainment centers.

In most situations, consumers may simply want to upgrade their existing home wiring to support high-speed inhome services. Such upgrades are much less expensive to do in many homes than one might think. For example, the standard telephone wire used in most homes costs about $60/1,000 ft. Upgraded category 5 data communications telephone cable for LANs costs $100–$125/1,000 ft. The cost of upgraded coaxial cable wiring is actually about the same as the old wiring. Standard coax (RG-59) for cable TV service and higher grade cable (RG-6) designed to support multiple video sources both cost about $150/1,000 ft [12]. For households that want it all, IES Technologies has announced a broadband home wiring product called FutureSmart Interactive Network, which provides for fiber optic as well as twisted pair and coax in one bundled cable [12].

For many years now, regulators have decreed that subscribers had control over the inside wiring that telephone companies had installed in homes throughout the country. Homeowners, contractors, and telecommunications equipment providers can, therefore, remove inside wire, alter it, upgrade it, or plug any device into it they wish. The same rules do not apply to cable TV wire, where regulations protect the cable operator from unauthorized use of both the

---

4. The CEBus EIA IS-60 standard specifications are available from Global Engineering Documents (tel. 800-854-7179). See also "The Residential Wiring Market: Unlocking the Potential of Home LANs," Parts I, II, and III (*Cabling Business*, June, July, and August 1995).

inside coaxial cable and the proprietary set-top converter boxes. Indeed, that protection (along with the protection provided by local franchising authorities) gives incumbent cable operators monopoly power. Even where cable competition is allowed, it is costly for competitors to rewire entire buildings. That obvious telephone company–cable company regulatory asymmetry is likely to change now that cable companies and telephone companies are allowed to compete with one another. The new telecommunications law calls for open standardized interfaces and competitive sources for consumers to purchase and install set-top boxes in the future. The FCC is actively investigating the issue with an eye toward making the rules for access to inside wire and procurement of set-top boxes similar for cable and telephone companies (new entrants and incumbents) as they move into the market for integrated multimedia services.

A host of untested legal and technical issues must be worked out, however, and because so much is at stake for competitors, years of litigation can almost be guaranteed. The sticky issues include the following:

- Determining the scope of the regulator's ability to affect the private property rights of network operators;
- Determining the scope of competitor and consumer rights of access to the private property of network operators;
- Determining the exact (legal) demarcation point where a change in ownership or control of the inside wire is to occur (e.g., the connecting block on the outside wall or the wire connection on the back of the set-top box);
- Determining whether competitors or consumers have the right to connect their own terminal gear to the network and in what manner;
- Setting standards for minimum signal levels and allowable interference, as well as other quality issues and responsibilities, for enforcement between state and federal regulatory jurisdictions.

2.4 THE INTERNET PHENOMENON

The phenomenal growth of the Internet's World Wide Web has put multimedia services front and center in the race to satisfy mass market demand for new telecommunication services. The Web has cable, telephone, and computer companies scrambling to get a piece of the action. It seems that every company in the United States, regardless of its size or line of business, has a home page, Web information servers, or both.

A brief history of the Internet might help to explain how it evolved into the hugely popular network it is today [13].[5]

---

5. See the Merit Network, Inc., home page to learn more about the history of the Internet and the recent restructuring of the NSFnet: nic.merit.edu/nsfnet/news/.release/nsfnet.retired. For more

2.4.1 History of the Internet

In 1969, ARPA within DOD commissioned the precursor to the Internet, a research network called the ARPAnet. In 1973–1974, Vint Cerf and Bob Kahn presented basic ideas for the Internet and published a paper specifying the Transmission Control Program. Also in 1974, Bolt, Beranek and Newman (BBN), who were involved in the DOD project early on, introduced Telenet, the first public packet data service network (a commercial version of the ARPAnet). From 1977 to 1981, several research and education network consortiums were formed applying the packet network technology of the ARPAnet.

In 1982, *the transmission control protocol* (TCP) and the *Internet protocol* (IP) were established as the well-known Internet protocol suite (TCP/IP). That led to one of the first definitions of the Internet as a connected set of networks. It was applied specifically to those networks using TCP/IP, which DOD declared the standard for the Internet.

In 1984, the *domain name server* (DNS) was introduced, and in 1985, the NSFnet was created, with a 56-Kbps backbone for use mainly by university researchers.

In 1987, NSF signed a cooperative agreement with Merit Network, Inc., which in turn cooperated with IBM and MCI for network management. The Merit-IBM-MCI consortium later founded *Advanced Network Services* (ANS) to administer the NSFnet backbone. In 1988, the NSFnet backbone was upgraded to T1 speed (1.544 Mbps).

In 1990, the ARPAnet ceased to exist, and the first commercial dial-up access service on the Internet (world.std.com) came online. In 1991, the World Wide Web was released by the *European Center for Nuclear Research* (CERN) along with the Gopher searching software. Also in 1991, the U.S. High Performance Computing Act, championed by then-Senator Al Gore, established the *National Research and Education Network* (NREN), and the NSFnet backbone was upgraded to T3 speed (44.736 Mbps). By that time, NSFnet data traffic was up to 1 trillion bytes/month (10 billion data packets).

In 1992, the number of host computers surpassed 1 million, and the Internet's audio and video broadcast feature, the *multicast backbone* (MBONE), was launched.

---

information on Internet statistics, see nic.merit.edu/nsfnet/statistics/INDEX.statistics. For a good reference on how the Internet operates and for some basic history, see J. MacKie-Mason and H. Varian, "Economic FAQs About the Internet," which can be found at: www.spp.umich.edu/telecom/technical - info.html. For a good, brief layperson's explanation of the Internet backbone's operational and ownership structure, see "Backbone's Connected to the..." (*Convergence*, April 1996), p. 44.

In 1993, Mosaic, a Web browser, took the Internet by storm causing usage of the WWW to skyrocket. By 1994, the NSFnet carried 10 trillion bytes/month.

During late 1994, the NSFnet backbone was undergoing a major restructuring, and on April 30, 1995, the NSFnet backbone was transitioned to a new architecture when Merit terminated its seven-year role as the administrator of the NSFnet. Today, that architecture consists of a very high speed backbone service (vBNS) and routing arbitors (RAs) operating at 155 Mbps. The vBNS network is operated by MCI on behalf of NSF. Nationwide, five regional central nodes connect the vBNS transmission links, called *network access points* (NAPs), and a large number of individual *network service providers* (NSPs). NSF directly funds the vBNS, RAs, and connections to NAPs. Ownership of the NAPs themselves were awarded to private suppliers and are used to connect the vBNS with other backbone networks, both domestic and foreign. The owners and locations of NAPs are Sprint (Pennsauken, N.J.), PacBell (San Francisco), Ameritech (Chicago), and MFS (Washington, D.C., and San Jose).

2.4.2 Size of the Internet

A survey of Internet domains was conducted in January 1996 (an estimated 34% of systems did not respond to the survey).[6] About 76,000 systems now have the domain name www, up from only 1,700 in June 1995. The number of host computers connected to the Internet worldwide has grown from 4 in 1969 to about 10 million in 1996. While no one knows how many users there actually are, most would agree that there is at least one per host computer. Other estimates put the number of users in the United States alone between 9.5 and 24 million [14].[7] Web usage represents about 26% of all Internet traffic and is increasing at 10% per month.

Table 2.3 provides the results of the latest Internet survey. The effect of the availability of user-friendly Web browser software is evidenced by the large increases recorded over the last few years.

---

6. Conducted by Network Wizards: http://www.nw.com/.

7. To look into the different estimates of the number of users, see http://etrg.findsvp.com/features/newinet.html/.

Table 2.3
Internet Growth

Date	Hosts	Networks	Domains
1969	4		
04/71	23		
06/74	62		
03/77	111		
08/81	213		
05/82	235		
08/83	562		
10/84	1,024		
10/85	1,961		
02/86	2,308		
11/86	5,089		
12/87	28,174		
07/88	33,000		
10/88	56,000		
01/89	80,000		
07/89	130,000	650	3,900
10/89	159,000	837	
10/90	313,000	2,063	9,300
01/91	376,000	2,338	
07/91	535,000	3,086	16,000
10/91	617,000	3,556	18,000
01/92	727,000	4,526	
04/92	890,000	5,291	20,000
07/92	992,000	6,569	16,300
10/92	1,136,000	7,505	18,100
01/93	1,313,000	8,258	21,000
04/93	1,486,000	9,722	22,000
07/93	1,776,000	13,767	26,000
10/93	2,056,000	16,533	28,000
01/94	2,217,000	20,539	30,000
07/94	3,212,000	25,210	46,000
10/94	3,864,000	37,022	56,000
01/95	4,852,000	39,410	71,000
07/95	6,642,000	61,538	120,000
01/96	9,472,000	93,671	240,000

The domain survey also reveals that the commercialization of the Internet is proceeding apace. As of January 1996, 45% of the domains had the commercial tag (.com). Network domains (.net) represented 24%, non-profit organizations (.org) 23%, government (.gov) 16%, followed by the U.S. military (.mil)

15%, and educational institutions (.edu) 8%. The United States accounts for about half of all Internet hosts worldwide, and about 60% of the domains.

The number of Web sites as of January 1996 was estimated to be about 100,000, about 50% of which were commercial (.com). As recently as 1993, the commercial percentage was as low as 1.5%. According to ActivMedia, Inc., the ratio of host computers per Web server fell dramatically from about 13,000 in June 1993 to fewer than 100 in January 1996. Figure 2.1 provides estimated actual usage of the Web and forecasts it through the year 2000.

According to ActivMedia, Inc., current commercial activity on the Web is estimated to generate some $436 million (1995), and the number of commercial Web sites is doubling every 3 months. That estimate may be low in light of the fact that the specialized Internet shopping network NECX, which sells computer components, reported 1995 revenues that exceeded $450 million [15]. It is anybody's guess when the phenomenal growth rates for the Internet will begin to level off, but it is not likely to happen for at least the next 3–5 years, even in the United States, which is well ahead of the rest of the world in terms of market saturation. Estimates of future commercial activity on the Web are all over the map. Some research forecasts only about $20 billion in Web commerce by the year 2000, while others predict an eye-popping $150 billion to $200 billion, with the number of users growing to over 200 million.

Hundreds of *Internet service providers* (ISPs) have emerged to provide consumers with direct access to the Web, where they can obtain online services. ISPs are quickly swamping the market for more traditional online service

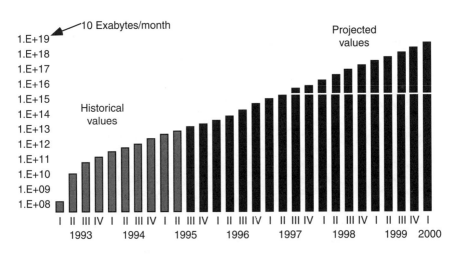

Note: 10 Exabytes = 10 thousand million bytes; roughly equivalent to 1.5 hours of full motion video per month for every citizen of the world.

Figure 2.1   Actual and forecasted Web usage. (*Source:* ActivMedia, Inc.)

vendors (e.g., WWW, America Online, CompuServe, Prodigy) that also provide Web access. By the end of 1995, ISPs had attracted almost as many Internet access accounts as America Online, the leading online services vendor. By the year 2000, Forrester Research forecasts, the total number of online service accounts from ISPs in the United States will rise above 30 million (compared to only about 13 million total for the other online service providers).

2.4.3 Video on the Internet

Video—for that matter, all high-resolution graphics and images—are difficult for the mass market consumer to obtain in a timely fashion on the Internet, both because of the lack of speed and because video has many sources. Subscribers' Internet connections or modems (not to mention subscriber terminal devices or inhome wiring) often are slow, making it very annoying to download video and graphics. Chapters 3 and 4 describe solutions for increasing the speed of subscriber connections that should mitigate the first source of frustration for video users. The second major problem is that the Internet itself is a first-come first-served conglomeration of private and public network routers and transmission links. That means the available bandwidth of the Internet is equal to the lowest bandwidth facility over which a given message might travel. Without any pricing mechanism or traffic engineering routing hierarchy, it is largely impossible to know how quickly a given video transmission will get through. Not even powerful modems and high-speed local connections can guarantee that video transmissions will get through in a timely fashion.

The Economist magazine aptly described the frustrating situation of waiting for video transmissions on the "interminablenet" [16]:

> Home viewers are still stuck waiting for video to trickle through their modem at a snooze-inducing two frames per second (compared with a film's 30 fps), with a soundtrack no better than that of an FM radio. Users of somewhat older modems receive a mere one frame of video every 2 seconds, plus squeakier AM sound.

The Web's phenomenal growth might be even greater if it could provide something akin to real-time video service to the mass market. Ideally, since so much video information will be accessed and stored in the near future on HD CDs, a good target for the Web would be the delivery of real-time MPEG-2 compressed video signals (about 4–6 Mbps would do nicely). That would allow for full motion video transmission between users' CD drives at their highest design speeds. Considering that the Web delivers video signals to the mass market at a maximum of 14.4 Kbps (the speed of most modems purchased today), there is a long way to go. (Imagine the time it would take to download a full-length movie, which is about 9.5 Gb of information, from a DVD). Even the Internet's

MBONE, a specialized multicasting service that allows for simultaneous point-to-multipoint transmissions (e.g., video conferencing) using special routers dedicated to the service, offers a maximum of only 500 Kbps throughput.

Happily, there are many new products and plans to try to speed things up. For example, a high-technology start-up, Vxtreme, has introduced a video broadcast technology with claims that it can deliver 20 fps over a 28.8-Kbps modem. VDOnet, Xing, and others claim that demonstrations have delivered 10–15 fps. While that would stand as a great achievement, it requires optimal conditions on the network (i.e., no traffic congestion) and still falls short of the standard 30 fps, which people are used to. While better subscriber line modems may be able to squeeze more video information out of the Web, there is a need for a more comprehensive solution whereby both the network itself (as in the case of specialized MBONE routers) and the subscriber connection can be optimized for video services.

For example, InterVU is developing an innovative software-based solution for accelerating access to the Internet (especially video content), based on a novel distributed, load-managed network architecture. The first implementation of the service will be available in the second half of 1996. The initial application of the technology will be on standard twisted-pair copper phone lines, but the system is designed to evolve to higher bandwidth infrastructures (e.g., digital cable and wireless networks).

Higher speed digital connections, including *integrated services digital network* (ISDN), cable modems, and ADSL modems for telephone lines also should play a major role in improving the access and transmission response times for video services on the Internet.

2.5 THE CD PHENOMENON

Since CD audio was introduced in 1982, over 400 million players and 6 billion discs have been sold [17]. Figure 2.2 shows the phenomenal growth from 1982–1995. Just like the Internet before it, the market for multimedia services using CDs is now beginning to take off. Forecasts predict that over 35 million CD-ROM drives will be sold in 1996 alone. The use of CDs for multimedia was delayed due to the industry's slowness in settling on a video compression and recording format and associated standards for manufacturing DVDs and video CD players. In late 1995, the standard was worked out by the industry, and DVDs and interactive CD-ROMs can now be mass produced. The new high-density DVDs can cram 4.7 Gb of data onto each side (about 9.5 Gb for both sides). This translates into 133 minutes of MPEG-2 compressed video. The next generation of double-sided double-density discs will hold twice as much information (17 Gb). The studio production of DVD titles is expected to explode, and some PCs are already being offered with high-speed CD players containing MPEG-2 decoders. With the new DVD standard, it will not be long before high-speed CD players are affordable for the mass market in the range of $200 to $300.

Even today's (soon to be antiquated) CD-ROMs can store 700 Mb of data compared to 3.5-in. floppy disks, which can store a maximum of less than 2 Mb. The price of CDs and recording devices are falling, and the development of read/write–erasable CDs are on the horizon for the mass market. Already, a CD recorder can be purchased for less than $1,000. A $7 recordable CD can hold about 650 Mb of data (about 10,000 pages of scanned-in text). Quantity reproductions can be made for about $2 a copy [18]. At that rate, it stands to reason that CDs will become a real force in the multimedia marketplace.

Table 2.4 provides a detailed comparison of the old CD format with that used for DVD. By the end of 1995, there were about 10,000 CD-ROM titles and about 30 million CD-ROM drives worldwide with triple-digit growth rates for two years running. It is estimated that the market for DVDs and high-density CDs will grow to 110 million by the year 2000, requiring only 11 percent of total CD-manufacturing capacity.

Table 2.4
Comparison of DVD and CD

Feature	New Format	Old Format
Disc diameter	120 mm	120 mm
Disc structure	Two substrates, each 0.6 mm thick	One substrate, 1.2 mm thick
Minimum pit length	0.4 μ	0.83 μ
Laser wavelength	635 to 650 nm	780 nm
Capacity	Two layers, one on each side, 9.4 GB total Two layers, both on one side, 8.5 GB total Four layers, two on each side, 17 GB total	One layer on one side, 0.68 GB total
Numerical aperture	0.60	0.45
Track density	34,000 tracks/in	16,000 tracks/in
Bit density	96,000 bits/in	43,000 bits/in
Data rate	11 Mbps	1.2 to 4.8 Mbps
Data density	3.28 GB/in^2	0.68 GB/in^2

Source: Scientific American, July 1996.

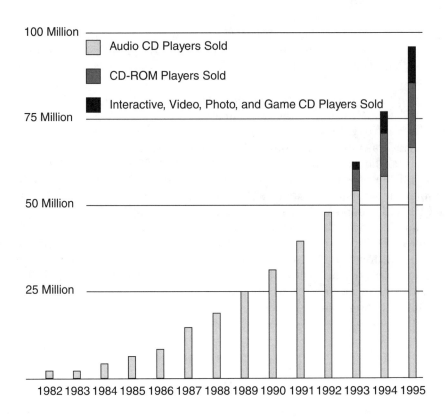

Figure 2.2   Sales of CD players worldwide. (*Source: Scientific American*, July 1996.)

PCs designed for interactive multimedia are already being offered with high-speed (9×) CD players capable of quality MPEG-2 video playback of DVDs, and most PC manufacturers are beginning to make them standard equipment. In contrast, TV manufacturers are just beginning to think about integrating CDs, instead of VCRs, into their video systems. The presence or absence of a built-in, high-speed CD player could be the deciding factor in the race between PCs and TVs for multimedia supremacy in the household. For now, PCs clearly have the edge.

2.6 DIGITAL SATELLITE SERVICE

Subscriber demand for digital satellite video services, even at introductory prices of $500–$1,500 for the reception dish and $30 for basic monthly service, has skyrocketed. Since the service was introduced to the U.S. mass market in

1994, it has grown to about 2.5 million subscribers and is still climbing. Sales of DBS dishes took off so fast that that product holds the record for early market penetration among all consumer electronic equipment, including color TVs, VCRs, and CD players. In 1995, more new satellite service subscribers were signed up than wired cable subscribers. Through the early adoption of sophisticated video signal compression technology, DBS can offer hundreds of channels of programming, dwarfing other cable systems.

Current DBS systems offer about 60 channels of continuous programming and another 60 PPV channels. The cost for each PPV movie is $3–5 and extra for sports and other packages. For example, a season's ticket for football is about $150.

That such industry giants as AT&T have recently chosen to invest in both DBS services and set-top box manufacturing should add to the momentum for DBS. The number of households subscribing to DBS service is expected to double in 1996 and grow to about 10 million–20 million by the year 2000.

Business services and subscribers are also growing rapidly, and many new consumer information and interactive multimedia services are planned.

Digital set-top box costs are falling as DBS penetrates the mass market. SGS-Thomson Microelectronics has already shipped 5 million MPEG-2 video decoder boxes and has announced integrated multimedia microchips for DVD, PC, and other applications with a target price below $65 per chip set.

2.7 BROADBAND INTO THE HOME

The jury is still out on the multi-billion-dollar question of which network delivery system will ultimately become the primary source for bringing interactive multimedia services into most households.

Figure 2.3 illustrates the known possibilities for bringing digital multimedia services into the home. The extent of the choices is overwhelming to would-be service providers who must commit substantial dollars now to a technology that may or may not be useful in the future. Both wired and wireless digital network systems will be vying for supremacy as the integrated multimedia network platform of the future. Both have their relative strengths and weaknesses, making it likely that they will actually complement one another rather than substitute for one another.

The ease and convenience of portability for consumers is undeniable, so no matter how fast or cheap wired systems become, wireless will always be preferred for many services because of its portability and mobility. On the other hand, when engaged in entertainment or work activities, users generally are indoors and stationary, so a less expensive fixed wired network connection is sufficient. (It is simply not likely that many people will want to watch TV on their wristwatches while traveling).

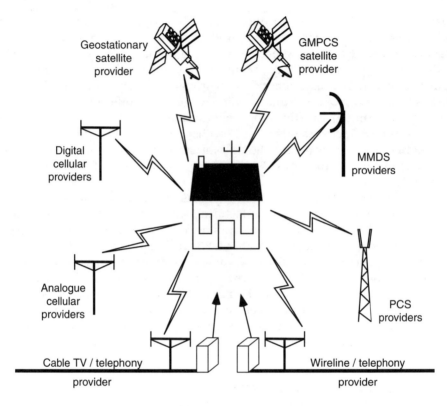

Keys: PCS = Personal Communication Services;
      MMDS = Multichannel Multipoint Distribution System;
      GMPCS = Global Mobile Personal Communications System.

Figure 2.3 Technologies competing to bring multimedia to the home.

Ultimately, the winner in the technological race to become the multime-dia platform of choice will depend largely on the cost of the network delivery system and the ability of network operators to find programming and software-based services that people want to buy. Unless an entity is planning to be a niche market provider of only one or a few services, it is probably not fruitful to pursue only one technology for the network delivery system. On the other hand, if a service provider is targeting a relatively narrow market segment or very limited set of multimedia services (e.g., video games), it is probably a good strategy to bet on only one technology, so effort can be focused on being the least-cost provider of the final product or package of products.

Programmers and information service providers who have no interest in owning or operating a network delivery system should probably hedge their technology bets and make sure their product can be delivered over many different network systems, leaving the network operators to battle it out for supremacy in the distribution business. Interestingly, that has not been the case to date as broadcasters and telephone and cable companies continue to merge with programming affiliates hoping that some important synergy or market advantage will be gained by tying programming to distribution. It remains to be seen if these strategies will work, but it is not likely, considering the fact that the government seems committed to nondiscriminatory program-access requirements. This can effectively take the advantage of vertical integration (of content and conduit) away from many companies. On the other hand, some network companies probably are motivated to venture into the programming business because they believe it necessary in order to have enough quality programming in the marketplace to make subscribing to their multimedia network distribution system attractive to consumers. That seems a questionable strategy because it can divert attention away from the highly competitive distribution part of the business.

What all this points to is software. Given the number of competing distribution systems and the amount of programming they will need, there is tremendous profit potential in developing software that allows consumers to control and manipulate their inhome network programming and information services. Consumers eventually will be engulfed by the sea of new digital gadgets and information, and they will seek out products that make it all easy and convenient to use.

The next two chapters present a detailed discussion and analysis of the underlying technology and economics of wired and wireless multimedia network platforms and try to sort out the cost and service capabilities of the major alternative network delivery systems.

References

[1] International Telecommunication Union, *World Telecommunications Development Report: Information Infrastructures*, 1995. Can be found at: http://www.itu.ch/wtdr95/.

[2] "Americans See Future and Say, 'So What?'" *New York Times*, October 7, 1993.

[3] "Wanna Play?" *Wall Street Journal*, March 28, 1996.

[4] "Tuned Out and Dropping Off," *The Economist*, November 4, 1995, pp. 65–66.

[5] Hill, G., and J. Trachtenberg, "When a TV Joins a PC, Will Anybody Be Watching?" *Wall Street Journal*, April 3, 1996, p. B1.

[6] Andrews, D., "First Web PCs Arrive," *BYTE*, April 1996, pp. 24–25.

[7] McGarvey, J., "Microsoft Pushes PC Makers To Play It 'Simple,'" *Interactive Week*, April 8, 1996, p. 8.

[8] Wilson, C., "ADSL Roaring Back as Telcos' Best Data Option," *Interactive Week*, April 8, 1996, pp. I-7–I-11.

[9] Dawson, F., "Cable Operators Examine High-Speed Options," *Interactive Week*, April 8, 1996, pp. I-14–I-15.

[10] Grover, R., and Lesly, E., "I-Way or No Way for Cable," *Business Week*, April 8, 1996, pp. 75–78.

[11] Society of Cable Telecommunications Engineers, "Consumers Guide to In-Home Wiring," 1995.

[12] Lohraff, S., "Futureproofing Your Home," *Electronic House*, October 1995.

[13] Zakon, R., "Hobbes Internet Timeline," http://info.isoc.org/guest/zakon/Internet/History/HIT.html.

[14] Gray, M., "Measuring the Growth of the Web," Mkgray@mit.edu.

[15] Colman, P., "Taking Commerce to New Heights," *Convergence*, April 1996, p. 18.

[16] "Video on the Internet," *The Economist*, January 20, 1996, pp. 82–83.

[17] Bell, A., "Next-Generation Compact Discs," *Scientific American*, July 1996.

[18] Wildstrom, S., "Cutting Your Very Own CD," *Business Week*, April 15, 1996, p. 22.

Broadband Network Deployment

3

3.1 INTRODUCTION

The NII is the U.S. government's vision of the future advanced public telecommunications network system. The international counterpart of the NII is the *Global Information Infrastructure* (GII), to which the ITU (the United Nations organization responsible for global standards and policy coordination for public telecommunications authorities worldwide) often refers. To see just how undefined both those visions are, you need only review two documents, both of which are available on the WWW:

- The Telecommunications Act of 1996 (ftp://ftp.loc.gov/pub/thomas/ c104/s652.enr.txt);
- The ITU's *World Telecommunications Development Report: Information Infrastructures World Telecommunication Indicators* (http//www.itu.ch/ WTDR95/).

While the vision of a broadband network infrastructure may not be well defined, most observers would agree that the future of the telecommunications business will be driven largely by consumer demand for multimedia service. *Convergence* is the industry buzzword that refers to the multimedia trend. On the demand side of the business equation, convergence means combining various communications media, including voice, text, data, and video, in a wide variety of consumer applications such as learning and working from the home. On the supply side, convergence refers to the delivery of various media to consumers via a common digital technology. To meet mass market demand for multimedia telecommunications, a public high-speed (i.e., broadband) digital network is required. The costs of constructing such a network and the prospects for financing it are the focus of this chapter.

Ubiquitous broadband networks are the ultimate distributors of information. As digital processing, switching, and transmission are being standardized, it is increasingly evident that basic network technology for telecommunications, computing, and electronic mass media (including broadcasting and cable television) are converging to form a homogeneous digital genre. Rapid advances in digital radio, photonic technologies, and fiber optic transmission systems represent a technological paradigm shift in the information processing and transmission industry, allowing for transmission and switching speeds in millionths (megabits), billionths (gigabits), and even trillionths (terabits) of seconds. Network hardware and basic control software eventually may provide for a standardized general-purpose information highway on which any conceivable telecommunication service can travel. The least common denominator of very high speed digital encoding and signaling will allow many simultaneous services to be integrated onto the same physical transmission link without perceptible loss of speed to the end user. That is the technological goal of would-be broadband network infrastructure providers, including telephone companies (often referred to as "telcos"), cable companies (often referred to as "cablecos"), new digital wireless network providers, and other media enterprises (e.g., broadcasting, movie, and computer companies). Although the direction of the basic technology push is becoming known, there is still uncertainty about consumer demand and unsettled institutional and public policy issues. Thus, it is not yet clear how, when, or to what extent physical network integration will ultimately occur.

There are substantial differences of opinion among academics and practitioners regarding the financial viability of residential broadband telecommunication networks. Generally speaking, most market research to date has concluded that there is insufficient demand to justify the massive investment required by telcos, cable television companies, and others to deploy broadband networks for the mass market of residential subscribers. Thus, to the extent that those players decide to deploy residential integrated broadband networks, they do so for strategic reasons, not because it is cost justified.[1]

In the future competitive telecommunications marketplace, telcos will find it extremely difficult to justify the financial investment necessary for new digital fiber optic subscriber access lines, sometimes referred to as *fiber to the*

---

1. Many analyses of the revenue and *net present value* (NPV) potential of broadband infrastructure network construction projects show that, even under a fairly optimistic demand scenario, the recovery of the initial investment would take 7–15 years. For example, see G. Kim, "The Revenues of the Infobahn," *Fiber in the Local Loop: Business, Economic, and Technical Challenges* (International Engineering Consortium, Chicago, 1995); Residential Video Dial Tone Network Service Prospectus, Bellcore, Special Report SR-TSV-002373, Issue 1, Dec. 1992; and B. Egan, "Economics of Wireless Communications Systems in the National Information Infrastructure" (U.S. Congress Office of Technology Assessment, draft, November 1994) and the references therein.

home (FTTH). FTTH systems would allow telcos to provide new services, including interactive video telephony and other high-bandwidth telecommunication services, but at an almost prohibitively high cost. Due to the difficulty in financing FTTH technology, local telcos and, to a lesser extent, cable television companies are pursuing a less expensive alternative called *fiber to the curb* (FTTC). (FTTC is also referred to as *fiber in the loop*, or FITL).[2] But even that proposition is expensive relative to the technological alternatives used by niche market players such as new digital wireless networks (e.g., *personal communication networks* or *systems*, or PCNs/PCSs; "wireless" cable; and global satellite networks). Thus, the critical question remains: Can profit opportunities for new multimedia telecommunication services justify the substantial financial investment to deploy a broadband public network infrastructure?

3.2 MULTIMEDIA NETWORK INFRASTRUCTURE TECHNOLOG AND COSTS

To evaluate the relative costs and service capabilities of alternative broadband network technologies, it is useful to benchmark them against the technologies and costs of some of the narrowband and "mediumband" technology alternatives.

3.2.1 Evolution of POTS and CATV to Multimedia Networks

Of the approximately 100 million households in the United States, about 94% of them get their *plain old telephone service* (POTS) on the *public switched telephone network* (PSTN). To provide some context for the process of upgrading traditional telephone lines for digital and multimedia capability, Figure 3.1 presents a stylized illustration of the current analog telephone network, including the proportion of total investment for each major portion of the PSTN. Figure 3.2 depicts a single residential access connection to the PSTN (loop) and indicates the various stages for upgrading that loop for digital multimedia service. The average cost for installing one additional residential POTS line is about $900 for a subscriber access line or loop, both of which are relatively

---

2. A summary tutorial of broadband network alternatives is available in presentations by P. Shumate, "Broadband Access Networks," "Broadband Subscriber Access Architectures and Technologies," "First Costs, Operations Costs, and Network Comparisons," and "First Costs, Operations Costs, and Network Comparisons—Summary" (Bellcore, 1995). Also see T. Darcie, AT&T Bell Laboratories, *Proc. Optical Fiber Conference*, San Jose, CA, 1996. A recent discussion of residential broadband network alternatives can be found in *Annual Review of Communications* (International Engineering Consortium, Chicago, 1995), especially the articles by Acosta (pp. 63–67), Davidson and Cohen (pp. 69–80), Jones and Schmania (pp. 105–110), and McConnell (pp. 203–206).

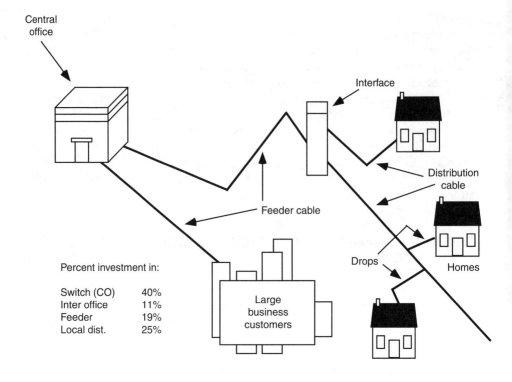

Figure 3.1 Analog telephone network and breakdown of total investment.

short (the nationwide average loop is about 10,000 ft long), and about $1,200 for longer loops (e.g., greater than 18,000 ft) because they require longer cable spans and remote electronic equipment (e.g., a *remote terminal*, or RT). Table 3.1 gives estimates of telco POTS costs by major category of investment for loops with and without remote electronics (*digital loop carriers*, or DLCs) [1].

Traditional *cable television* (CATV) operators offer *plain old cable service* (POCS), which is the analog counterpart of telco POTS service. Figure 3.3 is a stylized view of a typical CATV network for POCS and provides the proportion of investment by major category of investment. The average cost of a single POCS connection is about $300–$500 per home passed by the cable system (not all homes subscribe). Since about 60% of U.S. households subscribe to CATV service, the average cost per subscriber is in the range of $600–$900.

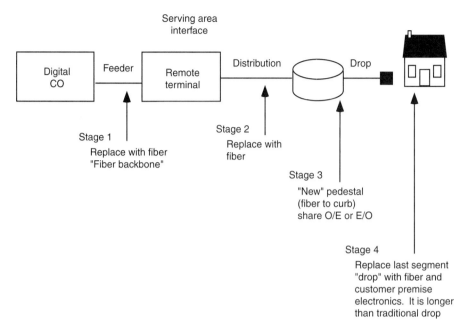

Figure 3.2 Loop network evolution.

Table 3.1
Telco POTS Benchmark Costs

Telco Distribution (Buried)		TR-8 DLC (If Needed)	
Category	*Cost ($)*	*Category*	*Cost ($)*
Distribution* cable	430	Electronics	220
Service cable	300	Enclosure	60
Pedestal	80	Battery	30
Cross-connections	25	Feeder cable	30
Protector	15	Installation and testing	100
Serving-area interface	10		
Total	860	*Total*	440

* Distribution costs include installation and testing.
Source: Bellcore.

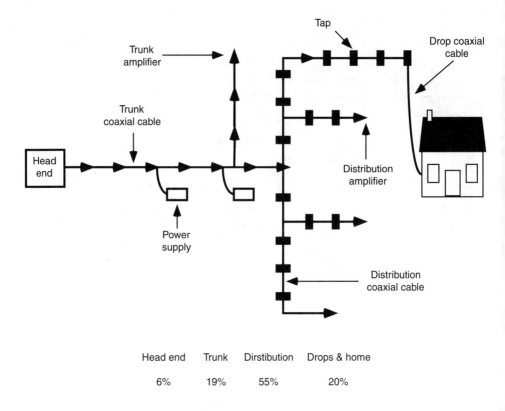

Head end	Trunk	Dirstibution	Drops & home
6%	19%	55%	20%

Figure 3.3 Typical cable network and breakdown of total investment.

Figure 3.4 depicts the network upgrade process for CATV networks to evolve to interactive multimedia service capability. Bellcore estimates that the cost of upgrading a POCS subscriber connection to provide basic telephone service would be about $400 per subscriber, for a total average system cost of about $1,125 per subscriber. Table 3.2 provides a breakdown of POCS costs for major categories of investment, and the upgrade costs for POTS to achieve POCS and POTS service capability.

As will be discussed in Section 3.3, upgrading a traditional POTS network to provide basic POCS would cost about $300–$500 per subscriber, for a total system cost of about $1,200–$1,500 per subscriber. Thus, for now, telcos are at a cost disadvantage for upgrading their systems to provide POCS relative to the cost of a CATV operator to upgrade its cable network to provide POTS.

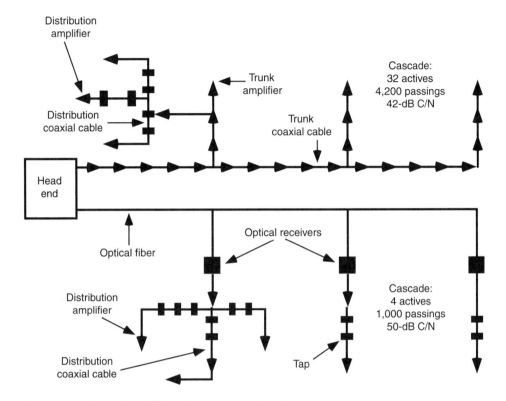

Figure 3.4 Upgrading CATV to interactive multimedia.

Table 3.2
Tree-and-Branch/HFC Costs per Subscriber With Two-Way Capability

Feeder Plant Items	Cost per Mile ($)
Feeder cable	32,000
Two-way upgrade	2,000
Uninteruptible power	1,800
Total	35,800

Cable Service	Cost ($)
100 homes passed/mile	358/home passed
65% penetration (take)	550/subscriber

Table 3.2 (continued)

Cable Service	Cost ($)
Drop cable	75/subscriber
Addressable converter	100/subscriber
Total	725/subscriber
Telephony interfaces	$400/POTS subscriber
Cable + telephony service	$1,125

Source: Bellcore.

In the future, those average system costs will tend to equalize as fully interactive multimedia systems are offered by either telcos or cablecos. The reason is that the digital technology for multimedia systems will be equally available to any would-be network operator, and cost differences among operators will be a function of a particular network architecture and technology type as opposed to a function of the embedded base of historical investments, as depicted in Figures 3.1 (for telcos) and Figure 3.3 (for cablecos).

3.2.1.1   Narrowband ISDN

The current generation of digital telecommunication technology for residential customers is called *narrowband ISDN* (N-ISDN). N-ISDN allows most local telephone lines to be used for digital information services without the use of a modem.

A single N-ISDN subscriber connection can provide at least three times the communications capability of today's analog telephone line at higher speeds and equal or better quality. Today, transmitting or receiving digital data signals on normal copper phone lines requires a modem, and reliable transmission speeds are limited to 9,600–14,400 bps (9.6–14.4 (Kbps)), and then only if the phone line is in good physical condition. That limitation will become increasingly problematic as higher speed modems become available to the mass market. Already, 28.8-Kbps modems are being mass produced and are priced at about $200; 14.4-Kbps modems can be purchased for about $100. But as data rates increase, so do error rates, so the nominal speed of a "high-speed" modem is cut down in actual use due to error corrections. Traditional copper telephone lines and the telephone network itself were never designed for high-speed digital data services, and the network will require substantial upgrades to accommodate mass market connections for two-way high-speed digital data services.

A single N-ISDN connection provides a total bandwidth of 144 Kbps and is akin to having three phone lines with only one phone number. A single copper N-ISDN phone line allows a household to use two digital channels (64 Kbps each) and one data channel (16 Kbps) simultaneously for any given service. In the context of normal voice telephone service, one of the 64 Kbps channels would be used just as a normal analog phone line is used today. The difference would be that an N-ISDN subscriber, while making a voice call, could also be using a PC online connection (e.g., surfing the Internet) and sending a fax. The two 64-Kbps channel connections are *circuit switched*, which means that the channel is dedicated to the circuit connection between the calling and the called parties that is established when the call is made. The single 16-Kbps "D" channel is *packet switched*, which means that it can be used to carry packet data traffic for any type of message being sent and is not necessarily used only for one message at a time or by one user at a time. Packet switched networks can be shared by more than one user or between more than one service at the same time (until the channel is fully utilized—then some queuing of messages will occur). The Internet is a good example of a packet switched network—it employs the statistical sharing techniques typical of computer data networks generally.

Telecommunications requirements for SOHOs are the early demand drivers for data services, as is the mass market demand for the WWW, the multimedia part of the Internet. For the SOHO market segment, narrowband ISDN service is, at the present time, a good cost-effective alternative to modem-based telecommunications.[3] However, at only 144 Kbps, basic N-ISDN service is relatively slow compared to other business LANs. Even the *primary rate* (PRI) version of N-ISDN, which operates at 1.5 Mbps, is slow compared to Ethernet (10 Mbps), the most popular LAN technology. Thus, for home office workers and others in the SOHO market who are used to working at LAN speeds back in the office, N-ISDN may not be fast enough.

By applying advanced digital signal processing techniques, digital messages can be compressed to fit on an ISDN line. In particular, slow-scan digital video signals and high-quality images can be transmitted over ISDN lines. Furthermore, it has always been known that twisted-pair copper wires used for ISDN could operate at digital signal speeds well in excess of ISDN rates. Indeed, most LAN installations use twisted-pair copper in buildings and campus environments. Copper wire LANs usually operate at speeds up to 100 Mbps. There will continue to be tremendous advances in digital signal processing technology, video signal compression, and *coder/decoder devices* (codecs) to allow higher bandwidth signals and higher resolution video signals to be

---

3. Because an ISDN line is already digital, there is no need for a modem—unless, of course, a nondigital telephone or other terminal is being used, in which case a modem will still be required.

"squeezed" onto a single ISDN line. The problem is distance, which is being addressed by other, new mediumband technological alternatives.

3.2.1.2   N-ISDN Costs

N-ISDN requires that consumers be able to convert their access lines to digital service. About 60% of the lines in the United States are served out of digital *central offices* (COs), leaving some 40%[4] to be upgraded. The per subscriber costs of converting the nationwide telephone network to ISDN is about $200–$600 per household depending on the level of deployment.[5] This figure includes the costs of upgrading both the subscriber access line and the core signaling network. It does not include the cost of the subscriber terminal(s), which must be added. For the entire United States, then, it would cost somewhat less than $30 billion to $60 billion to upgrade all households to ISDN service ($300–600 times 100M households), because only 40% of COs need to be upgraded. While the cost appears to be high, it is, in fact, only about one-sixth the cost of installing fiber optic broadband phone lines. The cost breakdown is as follows.

First, traditional analog copper telephone lines must terminate at a digital CO switch. At today's cost levels, it is estimated that all analog CO switches could be upgraded to digital for about $150–$300 per residential phone line. That translates into a total cost of between $8 billion and $16 billion for all lines not currently served by digital switches.

Second, costs associated with upgrading those embedded copper loops that will not support N-ISDN, either because they are too long (in which case, they have signal repeaters or load coils on them) or because the cable is of poor quality making the noise level higher than that tolerated by ISDN, must be considered. Generally speaking, customer lines less than 18,000 ft in length are compatible with ISDN requirements. The vast majority of existing customer lines are shorter than that and will work with ISDN. Problem lines are those that are in poor physical condition, are too long and hooked up to digital loop concentrators (remote nodes containing electronic equipment, such as multiplexers, which connect with trunk lines to a CO switch), or have been "treated" with loop electronics, such as signal repeaters or boosters. To get ISDN service, subscribers with problem loops will have to be upgraded on a location-by-location basis.

---

4. Based on 1994 FCC data, the seven RBOCs, which represent about 80% of the total United States, report that 75.3% of all switching offices are digital. However, only about 60% of access lines are served from digital switches because analog switches were originally deployed in dense urban areas and they continue to serve more subscribers per switch on average.

5. Partial deployment in a CO is more expensive per line than full conversion, because with full conversion the fixed costs are spread over more lines.

Third, electronic signaling equipment and ISDN software must be added to the digital CO to support ISDN signaling and control protocols, and electronics must be applied to the phone line itself. This situation generally refers to a CO digital termination, or *line card*, which handles the signals and conditioning for the phone line to provide a high-quality digital connection. The ISDN line connections (line cards) are about $100 per subscriber line. The frame on which the lines terminate is also different from those used for analog service, but that cost is not substantial in volume applications. The per-line software upgrade costs will vary.[6]

Fourth, as of 1994, the *regional Bell operating companies* (RBOCs) had equipped only 54% of their COs for *signaling system number 7* (SS7). In addition, only 24% of COs were ISDN capable, with a mere 1.7% of subscriber lines being equipped for ISDN. Therefore, the costs of a CO *packet handler*, an electronic device required to handle the ISDN line D channels for message control and data services, must also be added.

The telcos also need to deploy a core intelligent network to provide for end-to-end ISDN connections and to support the various information services that are contemplated. That requires the installation of network SS7, including signal transfer and control points for message routing and database memory "look-up" type functions. The installation of fiber optic backbone networks can significantly lower the operating costs of the shared interoffice network. It is estimated that the construction costs for fiber optic trunk (*backbone*) network facilities is about $100 per residential subscriber, for a total cost of about $10 billion for the nation as a whole (100 million lines × $100). The intelligent interoffice signaling and transmission network requires the placement of switching nodes, or signal transfer points, and network database nodes, or signal control points. All of these nodes are connected to the interoffice signaling network using the *intelligent network* (IN) technology known as SS7. The total cost of placing those nodes and installing SS7 is estimated to be about $5 billion.

Last, consumers would need to purchase a digital telephone or other ISDN-compatible digital terminal or an adaptive device for existing analog telephones or terminals. Little work has been done to estimate terminal design and production costs, but such purchases could easily exceed $1,000. Terminal adapters, like computer modems, are not very expensive, but they also do not provide consumers with new service capabilities. For example, the many consumer devices with built-in modems (e.g., fax machines) currently connected to the telephone network would not work with ISDN and would need to be replaced or equipped with adapters. However, prices for new ISDN-compatible terminal devices, once they are manufactured on a mass scale, are expected to fall precipitously. The future public network must be able to support whatever

---

6. In many cases, because of vendor pricing practices, the fixed costs of CO software upgrades vary with the number of lines equipped.

terminal devices are created, and the more intelligent the network infrastructure, the lower the basic terminal cost should be.

3.2.1.3 Total N-ISDN Cost

At today's cost levels, a reasonable "ballpark estimate" of the total cost of widely available N-ISDN is about $45 billion. That figure is much lower than the costs of broadband network systems, which are estimated at $200 billion to $1 trillion, depending on the capability of the network system.

Viewed in perspective, the $45 billion cost for N-ISDN is not very high. Local telcos are already spending about $22 billion annually on new construction, a significant portion of which is allocated to digital switching and fiber optic trunk cable.

3.2.1.4 N-ISDN Capital Recovery and Cost Sharing

The $45 billion cost estimate is not out of reach. In fact, the entire N-ISDN infrastructure deployment project would be paid back over 10 years at an interest rate of 12%, for only $7 per subscriber per month (in addition to current monthly charges). This figure assumes that all of the money is borrowed immediately, which of course would not be the case. If the project costs were treated in the context of an ongoing construction program over a reasonable time interval of 10 years or so, the monthly per-subscriber costs to support such a program would be only about $3.50 per subscriber in current dollar terms.

These figures assume no new revenues, which is unlikely considering projected demand for digital services. Furthermore, the average residential phone bill is about $45 a month, and the increase to cover the cost of N-ISDN is less than 10% of the total monthly bill. If you view a 10% increase as substantial, consider the fact that local telephone rates have risen more than 40% since the AT&T divestiture in 1984, mostly due to the *Federal Communications Commission* (FCC) access charge program, which added about $3.60 per line—and yet telephone penetration is up almost 2%, from 92% to about 94%. For those rate payers who cannot afford to pay more for telephone service, subsidized "lifeline" service will still be available.

There is substantial evidence that the *local exchange carriers* (LECs) will experience cost *decrements* due to operating-cost savings over the traditional analog network. Capacity costs of plant expansion are also lower with digital facilities, which generally come in larger investment "lumps" than analog equipment because of economies of scale. Advanced digital networks are more easily maintained, and realtime error detection and correction are enhanced. Furthermore, these costs include the cost of upgrading remaining analog CO switches to digital. This cost is not really an incremental cost of ISDN, since this conversion is already considered in the fundamental network plans and construction budgets of all large LECs and simply represents the next-generation switching plant.

3.2.2 Mediumband Network Systems

There is a huge difference between the international standards for N-ISDN and *broadband ISDN* (B-ISDN). N-ISDN is a circuit switched 144-Kbps network, and B-ISDN is a 155-Mbps packet switched network. In between these two low- and high-bandwidth systems are "mediumband" network systems. Because the technology of digital signal processing advances incrementally, the bandwidth performance and costs of many network systems are best viewed as evolutions of existing narrowband networks. Many next-generation network systems are upgrades to, or overlays on, existing telephone and cable networks. Furthermore, there is a huge gap in the costs and performance of narrowband and broadband systems, and some mediumband alternatives could fill that gap.

Asymmetrical digital subscriber line (ADSL) and *high-bit-rate digital subscriber line* (HDSL) are two rapidly evolving mediumband technologies.[7] Both systems require high-speed modems/transceivers on both ends of a subscriber's phone line (i.e., one at the subscriber's location and one at the telco CO switch). Figure 3.5 presents a stylized view of an ADSL system.

The first generation of ADSL systems were designed to be a short-term, relatively inexpensive method of providing one-way (i.e., asymmetrical) video service on a standard copper telephone line. ADSL systems, which enhance the POTS network with sophisticated electronics, cost about $500–$600 per subscriber as long as the subscriber's copper loop meets the minimum system requirements. In particular, the subscriber loop should be less than 12,000 ft in length. Early versions of HDSL were a symmetrical counterpart to ADSL targeted to business-market users as a substitute for high-speed private line connections (e.g., T1 1.5-Mbps two-way channels).

Through the use of advanced digital signal processing and multiplexing techniques, early ADSL systems were capable of providing both analog telephone service and one-way mediumband digital service to support video dial tone and *video on demand* VOD applications (technically a 1.5-Mbps downstream channel for single-channel VCR-quality video service). It turns out that demand for multimedia information service on the Web is far outstripping that for VOD (which is becoming available from cable and satellite broadcasters anyway), and it now appears that Web multimedia is the golden opportunity for this technology. ADSL offers an intriguing solution to increasing the bandwidth

---

7. A special volume from the IEEE provides a detailed technical discussion of the state of the art in ADSL and HDSL technology: *IEEE Journal on Selected Areas in Communications: Copper Wire Access Technologies for High Performance Networks*, Vol. 13, No. 9, December 1995. Also see *IEEE Journal on Selected Areas in Communications: High-Speed Digital Subscriber Lines* (Vol. 9, No. 6, August 1991).

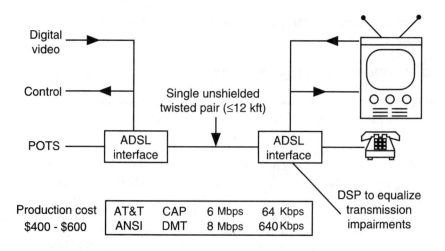

Figure 3.5 Asymmetric digital subscriber line (ADSL).

of narrowband telephone lines to meet the demand from the SOHO market and for anyone else who wants high-speed access to the Web.

Second-generation ADSL systems will offer even more downstream bandwidth, from 4 Mbps up to about 10 Mbps—an important goal considering that Ethernet operates at 10 Mbps. The lower value of 4–6 Mbps is, itself, an important performance objective, because it allows the line to accommodate high-quality digital video service using a standardized compression technique called MPEG 2.[8] But the use of MPEG 2 also requires a video decoder box, which will cost about $400 (that price is predicted to fall to $200 or less by the end of the decade).

In the upstream direction, ADSL systems will offer between 64 and 640 Kbps. As is the case for most of these systems, the primary limiting factor is the length of the phone line itself. Transmission on unshielded twisted pair copper lines will be limited to about 12,000 ft for second-generation systems operating at about 3–4 Mbps, less for future higher bandwidth systems.

The reason that ADSL is so unique and intriguing as a multimedia network alternative is that it is a modem-based technology. The beauty of this from a business and a consumer perspective is twofold: (1) modems can be quickly provisioned (relatively speaking) to subscribers wanting to purchase service; and (2) the modem equipment itself is portable and can be recovered and reused when a subscriber terminates service. It also makes network upgrades

---

8. For an excellent tutorial on this subject, see "Video on Phone Lines: Technology and Applications" (*IEEE Proc.*, Vol. 83, No. 2, February 1995).

less problematic (although planned obsolescence can shorten capital recovery times and end up being expensive).

The nature of multimedia demand, which is pretty much dictated by the Web, is that asymmetrical traffic flows are typical. Downstream traffic, or data flowing from Internet service providers to individual subscribers, is about 20 to 1. E-mail is about the only widely used symmetrical service application on the Internet, and it represents very little total bandwidth because it is used mostly just for low-bandwidth text transmissions. As more Web users' PCs are equipped with high-speed modems and connected to high-speed data lines, the asymmetry of traffic flows on the Web will become even more skewed, with downstream traffic flows equal to 100–200 times the amount of upstream bandwidth. But that is due more to the fact that consumers do not have fancy terminal devices (e.g., videophones) or personal databases and video collections to demand much upstream bandwidth. That will change in the future of video telephony. To meet the need for a more symmetrical bandwidth capability, newer ADSL and HDSL systems will offer a generous amount of bandwidth in each direction.

Future mediumband network systems, like *very-high-rate DSL* (VDSL) and *broadband digital subscriber line* (BDSL), will approach broadband system speeds, but these systems will not be available to the mass market of residential subscribers because of the short transmission distances involved. For example, phone lines equipped with newer ADSL systems operating at about 25 Mbps will be limited to about 4,000 feet, and 50-Mbps systems will only work at about 1,000 ft (well below the average line length in the United States of 10,000 ft). Planned systems with the highest bit rates (e.g., 155 Mbps) will be limited to very short distances of about 100m, while the lower rate systems (between 10 and 26 Mbps) will work for distances up to 1 km [2].

3.2.3 Fiber Backbone Networks

Both telcos and cablecos can deploy fiber backbone networks at a fraction of the cost of FTTH or even FTTC, because fiber is exceptionally well suited for high-capacity shared (nondedicated) plant and likely will be preferred to copper or coaxial cable in new construction for trunk and feeder network facilities based on cost savings alone.

Both telcos and cablecos can upgrade their local distribution networks (depicted in Figures 3.1–3.4) with a fiber optic backbone for about $50–$100 per subscriber. That, however, is where the good news ends for telcos. Even though cableco and telco fiber backbone costs are about the same per residential subscriber, the difference in quality, reliability, and future functionality leaves no comparison—cablecos win hands down. A telco fiber backbone, while perhaps more reliable and of higher quality from a network engineering and maintenance perspective, holds virtually no service advantage for consumers. Subscribers are still limited to two-way narrowband telecommunications due to the limitation of the existing copper loops. On the other hand, when cablecos deploy fiber optic backbones, the major sources of cable system failures and the major obstacle to providing telephone services (electronic signal amplifiers) are removed (see Figure 3.4).

Similar to the way in which telcos will utilize modem-based ADSL systems to upgrade their networks for video and interactive multimedia services, cable systems will utilize digital cable modems to achieve a rapid upgrade of their systems to provide telephone and interactive multimedia services. It is technically difficult to provide upstream channels to support telephone service on traditional cable systems, because they were optimized for one-way broadcast services. That means that cable system operators must first upgrade their systems with fiber optic trunk cable and reliable two-way optoelectronic signaling equipment. This can be achieved for about $200–$400 per subscriber, depending on subscriber density factors and how far downstream into the cable system the fiber optic trunk and feeder lines are extended. (Remember, this a system upgrade, not an entirely new system, which would cost about $600–$800 per subscriber). In addition, the digital cable modem, offering up to a 10-Mbps upstream channel, will itself cost about $500–$600 per subscriber (1996), but that price could fall to about $250 by the end of the decade [3].

3.3 BROADBAND NETWORK UPGRADES

On the horizon for residential phone lines is digital broadband technology, which will provide for a wide range of multimedia applications, ultimately including such services as high-resolution full-motion entertainment video, video telephony, and bandwidth on demand. Broadband technology generally requires the use of digital fiber optics, coaxial cable, or radio channels instead of the traditional copper telephone lines used for narrowband and mediumband technologies. Symmetric broadband digital service is an expensive proposition that costs five to six times that of N-ISDN and perhaps double that of mediumband alternatives.

Most broadband network architectures are based on fiber optic transmission systems, both in the shared trunk network and sometimes in the dedicated subscriber loop plant (including FITL, FTTH, and FTTC systems). Broadband systems exhibit deployment costs in the range of $1,500 to $5,000 per residential subscriber access line.[9] With about a hundred million residential access

---

9. For more detail on these types of broadband network systems, see B. Egan, *Information Superhighways: The Economics of Advanced Public Communication Networks* (Norwood, MA: Artech House, 1991); B. Egan, "Economics of Wireless Communications Systems in the National Information Infrastructure" (draft, U.S. Congress Office of Technology Assessment, November 1994); D. Reed, "Putting It All Together: The Cost Structure of Personal

lines in the United States, that amount implies a total cost for residential broadband networks in the range of $150 billion to $500 billion.

The two main incremental cost factors for broadband network systems are the initial engineering and construction costs for laying fiber and coaxial cable to the subscriber premises and the costs of optoelectronic components and devices required to allow existing *customer premises equipment* (CPE) to interface with and operate on the photonic distribution network. Fiber optic system deployment costs are substantial for all vendors (telcos, cablecos, and others) who are considering entry into the broadband infrastructure network business. However, fiber optic system costs are continually falling. Even though fiber costs have stabilized at about $.06 per meter (single-mode fiber) and about $.13 per meter (fiber optic cable), the unit costs of other fiber optic system components and optoelectronics (e.g., connectors, transceivers, amplifiers) continue to fall dramatically [4].

3.3.1 Fiber to the Home

Engineering cost estimates for telco FTTH systems yield a wide range of per-subscriber costs depending on the system architecture, functionality, and demand assumptions. None of the cost estimates presented herein includes CPE or other types of costs incurred on or inside a customer's premises. The most often quoted forward looking (1998–2000) average cost numbers are in the range of $1,500–$3,000 per subscriber line. The cable industry has not shown any significant interest in FTTH since they view their broadband coaxial cable subscriber loops as being adequate for two-way residential broadband service when fiber optics is deployed in their trunk networks.

Figure 3.6 shows a stylized FTTH access line architecture in a residential setting. Figure 3.7 shows various subscriber loop architectures for both business and residential customers and indicates the data rates that loop optoelectronics may provide. The distinguishing characteristic of a full-fledged FTTH system is that the fiber optic loop is dedicated to a single subscriber, and the optoelectronic interface is located at the subscriber premises. This Cadillac of multimedia systems has the potential capacity to provide virtually limitless bandwidth on demand. It is also the most expensive type of system because the loop and loop electronics are dedicated to a single subscriber. The most popular type of FTTH system is a *passive optical network* (PON), which means that it is a nonswitched network architecture. Depending on the loop plant design,

Communications Services" (Working Paper No. 28, FCC Office of Plans and Policy, Washington, D.C., 1991); and D. Reed, "The Prospects for Competition in the Subscriber Loop: The Fiber-to-the-Neighborhood Approach" (FCC Office of Plans and Policy, presented at 21st Annual Telecommunications Policy Research Conf., September 1993) and the references therein.

it is possible to have a host digital terminal for extended systems to allow for some sharing of trunk cable plant in the feeder portion of the network (Figure 3.6). Sharing of common feeder trunk lines, distribution cables, and optoelectronic equipment among subscribers can allow considerable cost savings in FTTH systems. It is possible in such systems to modularize the construction, so that as more subscribers sign up for service, system capacity can be expanded by adding optoelectronic components.

The estimated per-subscriber cost of a POTS fiber access line is about $3,000 for this stylized suburban residence network service configuration. Notice that, for this particular loop architecture, the feeder portion of per-subscriber costs is only about $100—about 3% of the total costs of subscriber loop plant, (which is about two-thirds the total per-subscriber costs) and associated electronics (which are about one-third the total per subscriber costs).

Notice that the underlying cost structure of FTTH systems is fundamentally different from that for current analog telco networks (shown in Figure 3.1). In fact, it is very nearly the opposite, which serves to stand traditional engineering economics on its head. In traditional telephone networks, 75% of the total system investment cost is in the outside-plant facilities portion of the loop (the feeder and distribution cable). In FTTH systems, it is not uncommon for 75% of the costs to be in the inside-plant portion (switching and electronics).

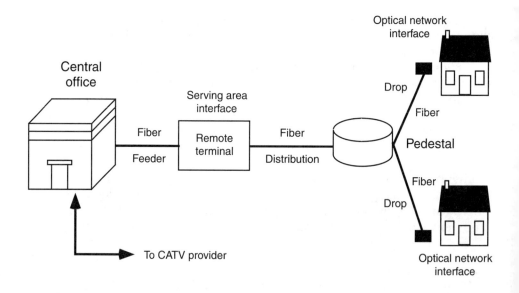

Figure 3.6 FTTH network for residential setting.

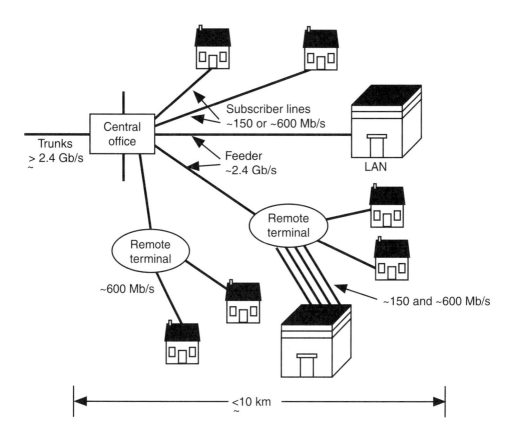

Figure 3.7 Business and residential subscriber loop architectures and associated data rates. (*Source:* T.E. Darcie, AT&T Bell Labs, OFC 1996.)

With continued progress in lasers and specialized microchip design, this system cost could be reduced considerably. The target cost of such systems is about $1,000 per subscriber after the year 2000.

3.3.2 Fiber to the Curb

There is a big difference in the cost of fiber backbone networks and FTTC (also called FITL and *fiber to the node*, or FTTN), whether for cablecos or telcos. In fact, most available estimates of per-subscriber costs of telco FTTC for two-way broadband networks are in the range of $1,000–$1,500—over 10 times those of per-subscriber fiber backbone costs (though some newer vendor systems do claim lower costs for mass deployment).

Figure 3.8 is a stylized view of a telco FTTC network. FTTC loop network architectures come in many flavors, so many that there are now several different classifications in the vernacular of telecommunications managers. The distinguishing feature of FTTC systems is that they utilize both fiber optic and metallic (i.e., coaxial and copper) cables. In that manner, FTTC systems use fiber optic cables in the high-capacity trunk backbone portion of the network system connected to lower capacity metallic feeder and distribution cables, which themselves are connected directly to subscribers. Compared to an all–fiber optic system, FTTC allows a system operator to take advantage of the relative cost efficiencies of optical fiber (the required optoelectronic components for high-capacity shared plant), while using metallic cable facilities and the simple electronics for the lower capacity distribution cables and dedicated subscriber connections.

The combination of increased sharing of common plant facilities and the use of relatively inexpensive metallic facilities for dedicated subscriber connections makes the per-subscriber costs of FTTC systems considerably lower than those for FTTH. In FTTC systems, about 40% of the per-subscriber system costs are for inside plant, including optoelectronic components, as opposed to 75% in FTTH systems. Figure 3.9 is a broad-gauge average cost comparison for most popular types of FTTC systems [1].

Hybrid fiber coax (HFC) (see Figure 3.9) is the most popular current FTTC system. The reason for HFC's near-term popularity is that this system maintains analog video capability, which, in its current analog form, is very convenient and relatively inexpensive to provide because the costly digitization process of existing broadcast signals is avoided. The deployment of HFC systems also minimizes up-front investment costs among other residential broadband sys-

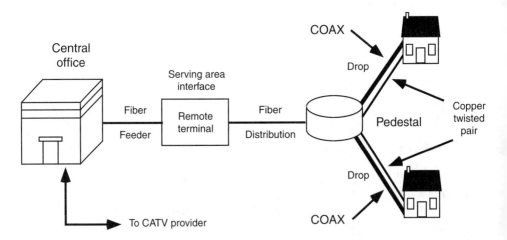

Figure 3.8 FTTC network.

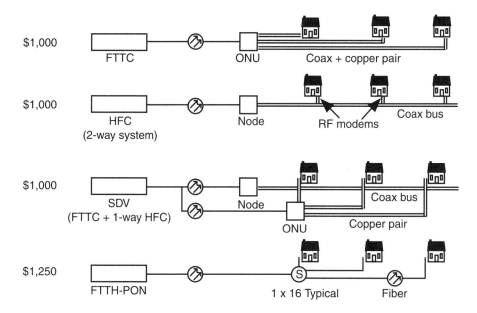

Figure 3.9 Cost breakdown for most popular FTTC systems.

tems because it allows for a pay-as-you-go approach to system deployment. In other words, system investment costs rise as more homes that are passed by the network choose to subscribe. This allows for a better matching of revenues and cost streams. Figure 3.10 presents estimates of the costs per home passed of deploying an HFC system and shows how that cost changes as subscription rates rise. Figure 3.11 illustrates how corresponding monthly revenues per home passed also rise as system subscription rates rise.[10]

Switched digital video (SDV) systems (see Figure 3.9) exhibit the same types of service capabilities as HFC systems but usually deliver video signals in digital format, potentially allowing for a more flexible video services format for such things as *video dial tone* (VDT),[11] interactive video, virtual VCR, and video signal editing. SDV systems are also better suited to providing interactive

---

10. These HFC costs and revenues are from the vendor of a branded system called Homeworx, by ADC Telecommunications, Inc. See also J. Cadogan, "Hybrid Fiber Coax: Today's Broadband Solution" (*1993–1994 Annual Review of Communications*, International Engineering Consortium, Chicago); also see the Homeworx presentation by J. Lehar, of ADC at the OFC '96, San Jose, CA, February 28, 1996.

11. VDT refers to a unique type of video system. It connotes a "common carrier" type video system in which third parties may provide whatever programming they wish by purchasing capacity from the VDT network operator. Such a system is similar to the Internet's WWW, in which third-party information and programming providers give users access to a variety of information over phone lines. In the case of VDT, the desired information is video.

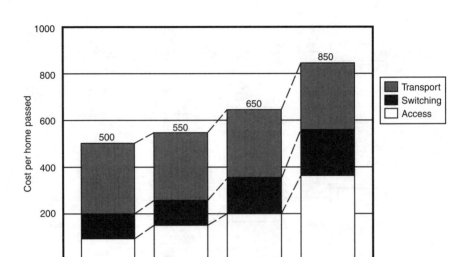

Figure 3.10 Estimates of the costs per home passed of deploying an HFC system.

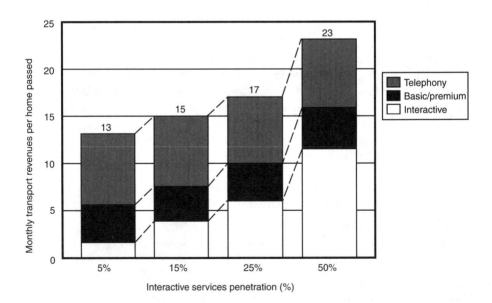

Figure 3.11 Corresponding monthly revenues per home passed.

switched broadband services like video telephony. In the short term, SDV systems are more expensive to deploy than HFC systems, especially if the system is to be used primarily for broadcast video services like traditional cable networks. HFC systems are also easier to deploy and cost less than SDV systems if the system is used as an upgrade or as an overlay to an existing state-of-the-art telephone or cable network. In the case of new system construction, the target per-subscriber costs of both HFC and SDV systems is similar, at about $1,000. If, however, the network operator sees the most revenue potential coming from VDT, video on demand, and other point-to-point video services, then an SDV system may be the best investment, since this is its relative service advantage.[12]

Tables 3.3 and 3.4 provide a comparison of FTTC and FTTH systems based on technical features, service capability, and service demands. The different hybrid fiber/metallic network delivery systems present so many engineering possibilities that it would be presumptuous to assume that any category of systems (i.e., FTTC, HFC, SDV) could or could not be designed to support any particular type of service (e.g., analog video, video telephony). That is especially true considering that, given the right combination of optoelectronic components, any of these systems can be transformed to perform whatever network function is required and to deliver any given service that subscribers want [3,5]. Suffice it to say that both HFC and SDV systems are upgradable to fully interactive digital broadband networks.

Table 3.3
Comparison of Broadband Access Networks

Criterion	HFC	SDV	FTTC	FTTH
Analog capability	High	High	None	High
Video reliability	High	High	Higher	Highest
Voice/data compatibility	High	High	Higher	Higher
Voice/data compatibility	Lower	High	High	High
Video quality	High	High	High	Highest
Voice/date blocking	Possible	No	No	No
Maintenance	Higher	Higher	Lower	Lowest
First cost per subscriber	(SDV)	(HFC)	Lower*	Higher
Installation Time	Fast/Slow[†]	Slow	Slow	Slow

* Depends on set-top cost and number required.
† Fast if coaxial plant is already in place.

---

12. For a comparison of the per-subscriber costs and revenues between HFC and SDV network systems and, for that matter, a good background discussion of issues, see J. Jones, "Video Dialtone: Choosing the Right Network Architecture" (Broadband Technologies, Inc., April 1993).

Table 3.4
Network Choice Depends on Service Plans and Customers

Criterion	FTTC	HFC	SDV	FTTH
Voice telephony service	Yes	No	No	No
Broadcast television service	No	Yes	No	No
Broadcast + interactive TV	No	Yes	Yes	No
Telephony + interactive TV	Yes	Yes	Yes	No
Broadband ISDN services	Yes	No	Yes	Yes
Operator with coaxial plant	No	Yes	Yes	No
Business + residential services	Yes	No	Yes	Yes

Source: [1].

Most of the discussion until now has focused on telco network upgrades. Some cable network operators have already deployed fiber optic backbone networks for only $30–$40 per subscriber in early applications. Extending the cable network fiber trunks into the feeder portion of cable plant can add substantially to that cost (about $150–$250 per subscriber). Figures 3.3 and 3.4 illustrated the difference between traditional coaxial cable systems and systems that have been upgraded with fiber optic trunks. It should be noted that fiber optic cable trunk and feeder networks generally do not imply extension of the fiber cable as far downstream in the network to the subscriber as is implied by telco FTTC. The typical telco fiber/copper "node" in FTTC systems may be placed at the point of existing copper drop lines. This point, called the *pedestal*, will often serve only 4 to 16 homes. In the case of cable television, current hybrid network designs allow the fiber/coax "node" to serve about 250 to 500 homes. Thus, the functionality of telco FTTC systems and the cable hybrid networks may be quite different, making per-subscriber costs difficult to compare.

3.4 COMPETITIVE ADVANTAGE

FTTC systems cost only about one-third to one-half per subscriber what an FTTH would. In the case of telcos, at least, FTTC may offer a significant increase in network functionality for subscribers that fiber feeder backbones do not. As a result, there is substantial interest in telco deployment of FTTC. A host of telco FTTC network architectures have been proposed by a number of vendors and no doubt more will be announced.

Reliable cost data are sparse, although some major industry sources put the total per-subscriber costs of FTTC at nearly equal to the costs of new residential copper access lines. This is not surprising considering the cost advantages of fiber in shared network facilities. It is, after all, the dedicated subscriber portion of the loop and optoelectronic equipment that make FTTH so expensive. FTTC allows for sharing of optoelectronic network interface devices and components among many subscribers.

Telcos must consider deployment of FTTC since fiber feeder backbones simply do not offer much in terms of value added for customers. With FTTC, a high-quality broadband capability may be achieved by interconnecting to coaxial cable for the final subscriber loop segment. Initially, only POTS and one-way video will be likely in telco FTTC networks, but that is the minimum configuration necessary to match the potential functionality of advanced two-way cable television networks. It is not that they would not rather provide real-time two-way broadband functionality. Rather, the costs of doing so are so high that this capability will be deferred until the second (or third) generation of local networks is contemplated. In fact, as of this writing, almost all major LECs have endorsed a policy of FTTC instead of FTTH for their next generation of broadband-capable subscriber loop plant.

Cablecos also need to deploy FTTC to be able to match the potential subscriber network functionality of telco FTTC. However, it is somewhat easier for cablecos since the critical (and relatively expensive) last network segment—the subscriber connection—is already broadband coaxial cable. There may be cost-effective ways to connect cable fiber optic backbones to telco switched network facilities to achieve a high-quality two-way telecommunication capability. The cost data available to date, however, indicate that cableco FTTC deployment costs could also be quite high ($1,000 per subscriber or more) if switched two-way digital telecommunication functionality is required.

If the demand for interactive multimedia ever lives up to the industry hype, there are two potentially important shortfalls in many of the telephone and cableco FTTC systems: (1) a lack of switching for point-to-point connectivity; and (2) the capacity to handle real-time interactivity (including the capacity to handle increased demand for symmetric broadband services like video telephony). Most vendor FTTC systems being considered for the next generation of residential broadband networks do not use switching for some, or even most, network services because of the relatively high up-front costs. The result is a relatively passive network architecture for one-way video distribution that is limited in terms of functionality compared to an all-switched video system. The bandwidth offered on the coaxial cable portion of the network in FTTC systems may not be able to support real-time subscriber interaction with the communication media, especially in an integrated environment.

3.5   LONG-TERM INFRASTRUCTURE NETWORK OPTIONS

There is still the possibility that the deployment of a fully switched broadband network infrastructure would be a serious mistake from a cost perspective. Some experts remain convinced that the cost advantage in a future broadband multimedia marketplace will accrue to those network operators who rely on transmission capacity to provide bandwidth on demand and that the preoccupation of most telecommunications engineers is misguided. That preoccupation is partly due to inertia caused by the fact that telephone networks have historically evolved as circuit switched point-to-point and point-to-multipoint, operations. It is also partly due to the historical telephone monopoly's preoccupation, reinforced by regulators nationwide, with the ultrahigh quality service that only a circuit switched network can provide. Circuit switching means that the entire bandwidth of any given telecommunications channel is surrendered to only a single application (e.g., voice phone call). Regardless of the actual bandwidth necessary for the call, the entire end-to-end circuit connection between the calling and the called party is dedicated to that single use for the entire duration of the call. That design is essentially the opposite of statistically shared computer networks and, in particular, is not consistent with fast packet *asynchronous transfer mode* (ATM) networks, which are expected to become the PSTN of the future.[13]

The coexistence of the various types of networks may not be as inconsistent as it seems. Most networks will be able to interconnect and "interwork" with one another (of course there is a cost associated with such interworking). In fact, one can take comfort in the fact that, even now, the circuit switched PSTN is providing end-to-end connectivity for the leading public packet network protocols used by the Internet (TCP/IP) and is working well with the leading LAN protocol used by the Ethernet's token ring network (10baseT). That is not to say that things could not be moving in a more effective direction.

Never having been a monopoly or even regulated, computer networks typically have not employed switched point-to-point architectures and the dedicated circuits typical of the PSTN circuit switched network architectures. But computer networks are for data communications either between machines or between people and machines. In the world of voice telephony, where delay times and busy signals are annoying (at least to humans), it is understandable that telcos and their regulators did not deploy facilities in the manner used by computer networks. Nevertheless, in the future world of multimedia, where public network communications increasingly involve machines, there is cer-

---

13. For a tutorial on network fundamentals that summarizes the basic distinctions between different types of network architectures, see J. Berthold, of Bellcore, "Fundamentals of Networks" (OFC '96, San Jose, CA, February 27, 1996).

tainly something to be said for the computer network model and the cost efficiency of statistical sharing.

That raises the question of whether it even makes sense to combine voice telephony or any other type of telephony on the same network with machine telecommunications. Perhaps (just perhaps), it makes no sense from a network cost and service provisioning perspective to integrate human-sensitive services with computer data services.

One thing is certain, if it is desirable to integrate all services on the digital network information superhighway, then the engineering solution must lie in a fast packet network solution like ATM, which still leaves open the question as to whether the ATM network infrastructure itself should be switched or non-switched. That remains a question of the relative costs of switching versus transmission. If and when switched point-to-point networks like the PSTN begin to employ superfast electronic or optical switching devices so that signal speeds become as fast (in terms of data throughput) as passive fiber optic transmission network systems employing fiber ring and bus architectures, it does appear that a network infrastructure based on the latter is more cost effective, both in terms of initial deployment costs and in terms of the costs of system capacity expansion. In any event, it will be a very expensive proposition to make ATM access available to all households the way that the PSTN is now.

The risk for telcos and cablecos of not investing in an efficient shared network infrastructure to support their high digital data traffic is that new players in the market for high-speed public network services, including computer companies, may eventually deploy a more efficient network infrastructure, one with which circuit switched networks will have trouble competing.

3.5.1 The Social Costs of Infrastructure

The long transition path from today's public telecommunications network infrastructure to an advanced switched digital broadband infrastructure capable of delivering interactive multimedia to the mass market raises some important cost issues. While the advancement of fundamental network technology is rapidly changing, it is not going to be possible, in practice, to minimize the total social cost of technology adoption (due to the long-lived nature of the investments made in the short to medium terms of the transition phase). The economics of technology adoption, especially in a market-driven environment, are partly a process of discovery. Fixing one optimal plan is not possible. Nevertheless, it is incumbent on the research community to try to identify those generations of technologies that represent socially efficient expenditures. In theory, if one were starting from scratch with what economists call a "scorched Earth," there is a dollar investment amount that represents the *total incremental cost* of deploying a broadband communication network infrastructure based on a least-cost long-run network optimization model. However, because we are not starting from scratch, the relevant cost issue is the expenditure to upgrade the existing infrastructure based on the total incremental cost of the least cost network design.

In practice, many short-run market-driven strategies for network upgrades will not be consistent with the social long-term optimum network; therefore, the sum of the various incremental upgrade costs will exceed, perhaps considerably, the theoretical total incremental cost based on a scorched-Earth situation. What short-term strategies fall into this category? The most popular next-generation network vendor systems utilize passive delivery of video communications in FTTC local network architectures. Based on the work of Reed, among others, it is clear that the use of passive delivery of video signals in hybrid broadband networks will increase the total cost of ultimately deploying a fully switched system [6]. There are many reasons for this, not the least of which is the inertia inherent in the existing video technology utilized by current analog video networks and associated CPE. Deployment of such network systems may raise substantially the total cost of ultimately achieving an advanced two-way switched digital broadband network infrastructure (assuming, for the purposes of this discussion, that this is the way to go). In all probability, this scenario could be avoided only by rapid—and significant—advances in switched network system designs, which are not currently cost competitive with passive systems for the provision of distributive video service. Unfortunately, that is not likely.

Reed and other researchers have shown that little is to be gained in terms of engineering economies by combining distributive video service delivery with two-way switched telecommunications on a single fiber optic or coaxial cable access line. If it is true that the economies of scope from physical integration of those two services is very small, deployment of such networks will not be justifiable on the basis of engineering cost savings alone.[14]

So why do telcos still pursue this short-term strategy? The reason is twofold. First, even if the preferred long-term vision is a fully switched integrated broadband system, the market for traditional narrowband services is a sluggish one. Everyone who wants a phone has one, and annual growth rates for per-household telephone usage are only slightly higher than population growth rates. On the other hand, new digital video services and digital data services exhibit high growth rates, and forecasts continue to be optimistic. If telcos are going to be players in the high-speed data and video services market, they cannot wait for the technology that is best suited for achieving the long-term vision. They must move soon or risk losing their dominant position in the telecommunications marketplace. Thus, even though there may not be much to

---

14. Indeed, to this day in the United Kingdom, which has a nascent and vibrant market in which U.S. telcos are allowed to jointly offer both cable television and telephone services, these companies have chosen to keep each service on a separate distribution cable.

gain in terms of network cost savings by integrating passive delivery of downstream video services with the telephone network, it at least represents a step in the direction of network integration, which has become something of a holy grail to telecommunications network engineers.

Second, the only alternative to network integration is to continue the status quo of separate services (e.g., voice, data, video) provided on separate access lines. To telco management, this is not an exciting proposition, because it would force telcos to continue to serve the less than glamorous basic telephone service market. Furthermore, without integration with the PSTN, regulators might not allow telcos to invest proceeds of the telephone business in other closely related lines of businesses like cable television. Many regulatory agencies consider this type of cross-subsidy to be against the public interest they are trying to protect. Regardless of the regulatory treatment of the costs of new construction, it is probably true that, in the very long term, demand for real-time two-way multimedia telecommunications will grow to the point that it will be worthwhile for telcos to pursue. It follows that it is best to enter the market early by beginning the network integration process now.

Thus, it is likely that FTTC systems will be widely deployed by telcos. What is, perhaps, most disconcerting about this for the long term is not that passive video delivery will be employed, but that it calls for deployment of "plain old coax" for the subscriber line connection. After all, there is already a coax line for cable television now, and if the long-term vision calls for FTTH systems, is it socially efficient to deploy FTTC for the next generation of residential local networks? In the case of the shared fiber trunk network investments, cost savings alone will justify rapid deployment of fiber optics, which is consistent with the long-term vision for a switched broadband infrastructure. But that is not the case for the coaxial cable access lines.[15]

Regulators are caught in the middle. They would like to see a strong and progressive telco network, but the costs of the network upgrades, even for FTTC, are substantial. If regulators allow FTTC investments to go into the regulated cost base of telcos, subscriber rates may have to rise.

One obvious, albeit partial, solution to some of these problems is to allow the telcos and cablecos to cooperate locally to provide new broadband telecommunication services. Right now, this is against the law. Neither Bell operating companies nor cable network operators are allowed to have a significant financial or managerial stake in one another's operations (the legal limit is 10% equity share). They are not even allowed to engage in joint service arrangements,

---

15. Shumate in particular makes the point that it may be time to reconsider the potential for fiber optic access lines. See P. Shumate, "Broadband Access Networks," "Broadband Subscriber Access Architectures and Technologies," "First Costs, Operations Costs, and Network Comparisons," and "First Costs, Operations Costs, and Network Comparisons—Summary" (Bellcore, 1995).

thereby sharing the risk of major investment projects like broadband multimedia infrastructure networks.

The bottom line is that the institutional and political environment itself is largely responsible for what we observe in terms of suboptimal network technology adoption. Even with the new law that "deregulates" the industry, outdated regulations will continue to take their toll on deployment of network technology.

In summary, we are moving toward a long-term vision of a fully switched digital broadband network infrastructure, but with suboptimal short- and medium-term investments in network upgrades. Market processes are dynamic and full of risk and uncertainty and are therefore a process of discovery. No doubt suboptimal network investment decisions are natural in the short and medium terms, but aggravating the situation with public policies that provide the wrong investment incentives for network operators is shortsighted.

3.5.2 Timing

The hybrid fiber/metallic alternatives for upgrading telephone and cable network infrastructures could definitely be available to the mass market within a decade. That may not be enough time for ubiquitous deployment, but it would be enough to make on-demand access to a nearby fiber node affordable for most households.

Widespread deployment of fiber optic interactive two-way broadband networks (the "second generation") would probably take another full decade to complete *if* a public policy imperative developed. Left to the demand-pull of the marketplace, ubiquitous residential broadband network access could be a full two decades beyond the next generation FTTC networks.

3.6 CAPITAL BUDGETING CONSIDERATIONS

When demand is uncertain and, in any event, significantly lags network construction, the decision to invest in capital-intensive construction projects, like fixed-wireline telecommunication networks (whether for initial investment, upgrade, or enhancement), is an extremely difficult one. That is especially true of telecommunications relative to other capital-intensive industries (e.g., automobile manufacturing and petrochemical processing) because many telecommunication services are public in the sense that demand directly depends on others also being hooked up to the network. Therefore, until the network itself is widely deployed, demand cannot take off.

Demand uncertainties notwithstanding, there are some advantages to public communication network investments compared to other capital-intensive industries. Multiservice communication network firms feature production processes characterized by very high joint and common costs. Telephone network operators are somewhat unique in that even though capital assets are fixed and immobile ("sunk"), they are also widely dispersed geographically in the network. In many situations, local subscriber distribution plant comprises half, or more, of the total network investment. That provides unique opportunities for flexible technology-adoption strategies. In other capital-intensive industries, such as auto, steel, and energy production, adoption of a new technological paradigm involves major lump-sum investments for large centralized production facilities. Contrast that with telecommunications, for which the spatial distribution of network nodes and transmission facilities is relatively decentralized, allowing for selective upgrading and modernizing of facilities without major production interruptions. However, the slower the rate of network technology adoption in response to such fundamental changes as substituting analog copper and coaxial cable with digital fiber optics, the more it will raise the cost of interworking the two technologies during the transition phase. Nevertheless, it is useful from a capital-budgeting perspective to have the flexibility.

It is much easier and less expensive to alter, postpone, or cancel a telecommunication network construction project in response to market conditions than it is to do so for, say, a nuclear power plant. The primary reason is that portions of the network are useful for providing advanced communication services to customers in a certain geographic area without having to make those services available everywhere. It is not necessary to have the entire network converted to be able to provide new services. New telecommunications services often are phased in gradually, even though, from a demand perspective, it would be preferable to offer new communication services to everyone at the same time.

The risk associated with future demand is lower for telcos than for other capital-intensive firms because of much shorter lead times and greater flexibility in construction. Construction time horizons for telecommunication network facilities are only 1 to 3 years for trunk and switching equipment. Likewise, the financial implications of reaching a point of no return in construction are less treacherous. Half a network node or trunk group, in terms of fungible plant capacity, is better than none at all, but half a nuclear power plant is not. That provides a certain amount of comfort for telco management. Should future cash flows fall short of expectations, heavy borrowing from external sources is not necessarily required. The bottom line is that no investment costs are sunk until actually committed.

Many possible engineering tradeoffs in broadband network design and construction serve to complicate costing out each physical network alternative in a business case analysis. Most important are the tradeoffs among fixed and variable network cost components. For example, two classic alternatives for capacity additions are: (1) to lay more fibers in an initial cable installation; or (2) to plan for increases down the road by adding more electronic and photonic equipment as needed, thereby reducing up-front fixed investment in favor of future incremental investment in electronic devices. Such considerations alter the mix of fixed, sunk, incremental, avoidable, and variable costs for any given network architecture over the planning horizon. In turn, the regulatory cost allocations of telcos are affected since (mostly fixed) *non traffic sensitive* (NTS) costs (e.g., fiber optic trunk cable) are allocated to state and interstate regulatory jurisdictions in different ways than the *traffic sensitive* (TS) costs for electronic devices. Even in the case of unregulated cable television firms, such tradeoffs among fixed and variable costs can alter pricing and capital recovery strategies prospectively, thus affecting the demand side of a business case.

The embedded base of network assets also may affect marginal investment decisions. Any given mix of embedded, fixed, and variable network facilities costs alters the decision parameters for future capacity additions. That is true even though, in theory at least, the embedded base of fixed costs became sunk and nonavoidable (irrelevant) for future decisions (or future business case analyses) immediately after they were incurred. The past selection of network technology and architecture can (and likely does) affect future marginal technology and capacity decisions forever.

It is, therefore, important to anticipate the impact of current business decisions, such as choosing the next-generation network architecture, on future decisions. In a world of rapidly changing technology it is important not to simply adopt the least expensive short-run network architecture alternative (especially if it might not be robust enough to accommodate anticipated future demand and technology changes). In addition, how robust current investment decisions are to unanticipated demand changes also must be considered. In the case of residential broadband networks, the unknown factors of new digital wireless loop technology and satellite technology may come into play and make the land-based wireline loops prematurely obsolete. No doubt other new alternatives for local telecommunication will appear on the scene. It is important to avoid the potentially large financial mistake of deploying dedicated fiber optic subscriber loops if, indeed, alternative technologies around the corner can provide similar service at lower cost.

Another important strategic consideration on the supply side of the broadband network equation is to accurately identify the life-cycle stage of each customer service that is contemplated. Forecasts of demand growth and construction of network capacity requirements must be consistent with the place on the S-shaped life-cycle curve for each service. This obviously can affect the target for capital recovery or payback period. For example, even though many entertainment video services represent a brand new growth business for telcos, they are well along in their respective life cycles, and full market penetration may be around the corner. Growth in traditional mobile cellular service is also going to slow down at some point, and it is critical to accurately forecast that point. Network investment strategies for both network hardware and software must be geared toward capturing new customer service markets that are early in their growth phase.

Currently, only two major players are anticipated to have a significant role in the future game of network deployment for advanced two-way residential broadband communication networks: cablecos and local telcos. On the other hand, as might be expected, there is a host of potential players in the business broadband communications market, including cablecos, telcos, private satellite and radio network vendors, private fiber optic networks, LANs, *wide area networks* (WANs), and *metropolitan area networks* (MANs). Due to relatively early demand drivers, such as those required for high-resolution imaging to support health care, high-speed computing, *computer-aided design* (CAD*), computer-aided manufacturing* (CAM), *computer-aided engineering* (CAE), and other business applications, the development of the market for business broadband services will proceed quite differently from that for residential services. The differences could be minimized if government authorities chose to promote an infrastructure approach to broadband network technology deployment. Indeed, unless an infrastructure approach is taken, in which government authorities promote early retirement of old technology, it may take several decades for residential switched broadband networks to be widely deployed (because of the long-lived embedded base of traditional copper phone lines).

No one knows for certain what path is socially optimal for residential adoption of broadband network technology. Predictions, to date, vary from one or two decades to never. Whatever the predictions, business decisions must be made in both the public and private sectors to determine the next step. The world does not stand still; telcos and cablecos have to continue to upgrade their networks both to achieve network efficiency and to compete in the near term for new—and advanced—communication services. Furthermore, from the perspective of the nation as a whole, our performance for domestic economic and social productivity and international competitiveness may not allow us the luxury of simply waiting for the existing base of public telephone network technology to physically wear out before replacing it with a broadband alternative.

3.6.1 Cash Flow Analysis

The results of standard calculations of the *net present value* (NPV) of future cash flows, properly applied to the life-cycle costs of network investments can be used to evaluate the financial viability of both short- and long-term network upgrade alternatives for telephone and cable network companies.[16] Life-cycle

---

16. For a description of the capital budgeting decision process and the role of NPV analysis, see R. Brealey, S. Meyers, and A. Marcus, *Fundamentals of Corporate Finance* (New York: McGraw-Hill, 1995), chaps. 6–11.

costs include both the upfront network construction costs (*installed first costs*, or IFC) and the annual ongoing expenses associated with operating the network expressed as an annual or monthly annuity amount.

For a very high level analysis, some rules of thumb exist for evaluating the costs of telecommunication network investments. For example, in the telephone industry, an annual charge factor of 40% is used to reflect the annual costs (sometimes referred to as carrying charges) of new network investment. There are four components of the 40%:

- The annual depreciation charge of 10% (indicating an average service life of 10 years);
- Annual interest charges, or "cost of money," of 10% to reflect that funds tied up in network investments could be earning interest (or that interest has to be paid on the borrowed funds for the network investment);
- Annual income tax expenses of about 4% (composite corporate tax rate times the 10% return);
- Annual operating expenses and allocated overhead of 16%.

Using those four components, for every additional $1 of network investment cost, a telco would need to increase its annual revenue (immediately in the first year and every year thereafter) by $.40 to recover its original capital investment over 10 years' time. Thus, for every $1,000 of broadband/multimedia network upgrade costs per subscriber, a company would need to start collecting $400 more a year ($33 monthly) from every subscriber just to recover its original investment costs.

Very few comprehensive capital budgeting analyses of broadband network upgrades have been made public, but the results of quite a few general analyses have, by now, been published without revealing the detailed (often proprietary) underlying analysis.[17] One of the most comprehensive studies to attempt a standardized financial comparison of the NPV of both telco and cableco broadband upgrade alternatives was performed by a consortium of European researchers employing a host of technical and financial evaluation tools developed by the European RACE project [7]. That study juxtaposed the likely per-subscriber costs for both telcos and cablecos for: (1) new construction;

---

17. For example, see P. Shumate, "Broadband Access Networks," "Broadband Subscriber Access Architectures and Technologies," "First Costs, Operations Costs, and Network Comparisons," and "First Costs, Operations Costs, and Network Comparisons—Summary" (Bellcore, 1995); B. Egan, "The Case for Residential Broadband Telecommunication Networks: Supply-Side Considerations" (Columbia Institute for Tele-Information, Research Working Paper Series, Columbia Business School, February 1991); and "Overview of New Technology Deployment Model: Broadband and Associated Depreciation and Overheads" (Telecommunications Industry Analysis Project, University of Florida, Gainesville, FL, *Quarterly Bulletin*, Vol. 16, No. 4), pp. 569–573.

(2) relatively modest broadband network upgrades using ADSL and HFC alternatives; and (3) relatively aggressive upgrade scenarios for switched integrated broadband service. The study also projected demand for each scenario and calculated the NPV of cash flows for both telcos and cablecos for the 15-year period 1996–2010.

The analysis indicates that the broadband network upgrade costs for both telcos and cablecos are roughly comparable to the investments already made in their current networks (actually somewhat higher for telcos and somewhat lower for cablecos). The NPV and the payback period of the upgraded networks in the modest upgrade scenario were similar to that for the base case scenario of no network upgrade at all (i.e., present method of operations to the year 2010). In fact, viewed in isolation, when the incremental costs of the network upgrades themselves were compared to the incremental revenues from new interactive broadband services that the upgrade made possible, the investments never paid for themselves.

Another comprehensive analysis was conducted by Bellcore to evaluate the NPV of cash flows associated with various alternatives for upgrading the telco POTS network to provide VDT service [8]. The results of the Bellcore study are similarly pessimistic. The average discounted payback period for recovery of the initial upgrade investment costs was 6–7 years for ADSL type systems and 11–15 years for HFC and FITL systems, with most of those being closer to 15 years. Even under some fairly aggressive demand assumptions, the NPVs (and corresponding internal rates of return) are never very high.

From the viewpoint of telco management, in an age of increasing competitive alternatives for multimedia distribution systems, 15 years is a long time to wait to recover an initial investment. Despite the early pessimistic financial forecasts, given that the future high-growth telecommunications markets will be for digital data and interactive multimedia services, broadband network upgrades remain a strategic imperative for most incumbent telcos and cablecos. After all, the alternatives—gradually losing market share[18] or abandoning the public network business by selling out—are not so rosy.

References

[1] Shumate, P., "Broadband Access Networks," "Broadband Subscriber Access Architectures and Technologies," "First Costs, Operations Costs, and Network Comparisons," and "First Costs, Operations Costs, and Network Comparisons—Summary," Bellcore, 1995.

[2] Salamone, S., "Higher Data Speeds Coming for Plain Phone Lines," *BYTE*, January 1996, p. 30; Shumate, P., "Broadband Access Networks," Bellcore, 1995.

[3] Darcie, T., AT&T Bell Laboratories, "Broadband Subscriber Access Architecture and Technologies," OFC '96, San Jose, CA, Feb. 28, 1996.

---

18. Of course, milking the cash cow in the process can be a financially attractive option.

[4]   Hobbs, J. (Kessler Marketing Intelligence), "Cost Factors Affecting Fiberoptic Deployments in Communications Networks," *Proc. OFC '96*, San Jose, CA, February 28, 1996.

[5]   Jones, J., "Video Dialtone: Choosing the Right Network Architecture," Broadband Technologies, Inc., April 1993.

[6]   Reed, D., *Residential Fiber Optic Networks: An Engineering and Economic Analysis*, Norwood, MA: Artech House, 1992. Also see Reed, D., "The Prospects for Competition in the Subscriber Loop: The Fiber-to-the-Neighborhood Approach," FCC Office of Plans and Policy, presented at 21st Annual Telecommunications Policy Research Conference, Solomons, MD, September 1993.

[7]   Olsen, B., et al., "PNO and CATV Operator Broadband Upgrade Technology Alternatives: A Technoeconomic Analysis," *Proc. OFC '96*, San Jose, CA, February 28, 1996.

[8]   "Residential Video Dial Tone Network Service Prospectus," Special Report SR-TSV-002373, Issue 1, Bellcore, Livingston, NJ, December 1992.

The Economics of Wireless Communications Systems

4

4.1 WIRELESS TECHNOLOGY IN THE NII

Wireless telecommunications is receiving a lot of attention these days. Some would say an inordinate amount. Why? After all, wireless telecommunications has been around a very long time, and by now nearly everyone knows about the convenience of cordless and mobile telephones. The key to what is going on lies in that, for the first time ever, wireless technology is progressing to the point that it will not only be convenient to use but also affordable for the mass market of American consumers.

Due to the obvious advantage of portability, it is a given that wireless telecommunications will be very successful with consumers. However, any meaningful discussion of the role that wireless communications might play in the grand scheme of things cannot occur in a vacuum. While it is a foregone conclusion that the convenience inherent to portable wireless telecommunications ensures that it will always be an integral part of the telecommunications business, the real public policy issue for governments around the world is whether wireless technology will be an important part of the public network infrastructure for the 21st century. This raises some obvious questions:

- Can wireless technology meet the service demands placed on a modern network infrastructure?
- Can this be done at a cost lower than that for wired alternatives?
- If the answers to the first two questions are yes, then what could (or should) the government do to stimulate investment in a public wireless network infrastructure?

A critical roadblock to obtaining a market-based answer to those questions has been the government's historical policy of compartmentalizing wireless network operators by restricting the use of their particular slice of spectrum to one specific type of service(s) and expressly prohibiting entry into the market for traditional telephone services. Finally, however, the FCC has adopted a new "flexible-use" philosophy, and it is considering allowing wireless network operators into the market for traditional telephone services. This is a huge step in the right direction, but it is only a beginning. As of this writing, the FCC has not removed this barrier to entry, but it is considering the matter.[1] This chapter begins an investigation of the potential for wireless technology to augment or even replace the traditional wired telecommunications network infrastructure.

The Clinton administration's characterization of the NII provides the context for the discussion of wireless technology herein. But the administration has said very little about what the form or substance of that infrastructure would be. For example, will the NII use wireline or wireless technology or some combination of the two? Will it be digital, analog, or both? Will it be capable of narrowband voice and data services applications, or will it support multimedia and broadband services such as TV or even video telephony? The answers to those questions simply are not addressed in the current version of the administration's vision of the NII. Therefore, each of those possibilities will have to be considered.

Some aspects of the future NII are clearer. There are three basic pillars of the administration's vision of the NII as an advanced network infrastructure: (1) investment for infrastructure will be privately funded; (2) services will be widely available and accessible to the public (i.e., "affordable"); and (3) the various networks that the NII comprises will be compatible or standardized.

Given those principals, what current or future wireless network systems are likely candidates for the NII? Since virtually every household in America already has affordable access to analog broadcast radio, television, and *cordless telephone* (CT) service, the NII must represent more. Maybe the NII is best described in terms of new digital television, ITV, new digital voice and data mobile telephone systems, satellite radio paging, and messaging systems.

Such questions must be answered before any meaningful progress can be made in deciding what characteristics make wireless systems candidates for use in the NII. The key is to narrow the field of likely candidates without eliminating the possibility of novel systems that may not even be on the drawing board yet. The way to do that is to focus on the generic role and capabilities of the technology itself rather than analyze specific wireless network systems or try to guess the uses to which the technology will ultimately be put.

For analytical convenience, it is useful to view wireless technology as potentially playing a role in three major aspects of the physical NII: (1) consumer

---

1. In FCC WT Docket No. 96-6, the Commission has proposed to eliminate the restriction on cellular radio network operators (CMRS) that barred them from providing service in competition with the fixed wired local networks of the telephone companies. However, the proposal still does not recommend lifting restrictions on all types of enhanced and multimedia services, only on basic voice and low-speed data services.

terminal devices or cordless "handsets"; (2) the over-the-air transmission network; and (3) the interconnection, or wireless access, arrangement connecting a consumer terminal or private network to the nationwide public network infrastructure.

Wireless access, the focus of this chapter, implies the use of a handset and is, therefore, the term that best captures the essence of the role of wireless technology in the NII; it represents the wireless counterpart of the all important on-off ramps to the information superhighway. Of much less importance for the NII is the role of wireless transmission for internodal transport—the "open road" part of the information superhighway. Besides representing only a small fraction of the total cost of an advanced network infrastructure, experts agree that there will be many private industry players and much investment flowing into this segment of the NII using a mix of routing and transmission technologies.

The great potential for wireless telecommunications in the NII is obvious once one considers the basic economics of supply and demand. On the demand side, the raw convenience offered by truly portable, personal, and private telecommunications in the everyday activities of all Americans makes it an unambiguous winner in the marketplace. On the supply side, wireless telecommunications technology is progressing to the point where it meets or exceeds the cost-performance and quality characteristics of wireline alternatives for traditional telecommunication services.

4.2　WHAT EXACTLY IS "WIRELESS" TELECOMMUNICATIONS?

The answer to that question is invariably a matter of scope and context. Broadly speaking, the term *wireless* simply refers to telecommunications that does not involve a tether. But this definition is too broad to be meaningful in the context of the NII, which places certain minimum requirements on infrastructure network technology (e.g., that it feature capacity and cost characteristics to make it publicly accessible and affordable for most Americans).

Many popular forms of electronic communication already rely on wireless technology but would not be suitable for the NII. Mass market demand for wireless communications has been rapidly expanding for decades, but it generally has been limited to use in the immediate household area (e.g., broadcast and satellite TV, radio, and cordless phones). In most cases, noninteractive wireless "networks" using truly portable wireless media have been featured (e.g., PCs and diskettes, video tapes and VCRs, and compact discs and CD players).

Wireless telecommunications in the context of the NII is different. The technology must support two-way real-time (i.e., interactive) digital telecommunications. In common parlance, it simply means a phone without wires, but technically that's what a cordless phone is. So, to be more precise, the wireless phone of the future would be connected via a digital wireless interface to the PSTN. Just where that point of interconnection would be is one of the most critical—yet fluid—issues being considered by would-be wireless network operators.

Thus, *wireless* refers to the ability to engage in real-time, private, two-way voice and data communications at a distance without the use of wires.

Note that the issue of whether wireless systems may support only narrrowband voice and data services or include multimedia and broadband services (e.g., video telephony) is not presupposed, nor should it be at this early stage. Any requirement that the future NII include broadband service will depend partly on the government's public policy objective and partly on the costs of achieving it, the latter being the focus of the discussion that follows.

4.3 PORTABILITY ASPECTS OF WIRELESS ACCESS

On the demand side of the wireless market, the convenience aspect of portability poses a substantial inherent advantage over wireline service. Holding constant the quality and price of wireless transmissions, the combination of portability and reliability will be paramount in determining the winners and losers in the marketplace. However, supply-side considerations dictate that portability itself is a matter of degree and is directly related to wireless system cost. From a wireless network operator's perspective, meeting consumer demand for a wide range of service capabilities and portability features is potentially very costly.

It is one thing to say that a given consumer application of wireless telecommunications is "portable," but quite another to claim that it is possible to use a portable phone to call anyone anywhere, anytime—the all-important three As of portability. The availability of portability anytime is a given; the ability to call anyone is not and will depend on the connectivity of wireless and wireline networks. Portability anywhere is yet another matter.

Anywhere means that a telephone handset will work at home, in the office, while the user is walking or driving. More formally, three possible "modes" are associated with calling anywhere: (1) using wireless access while at the home base station location; (2) moving about (e.g., walking) at a distance from the home base station in a given home base station area; and (3) roaming (e.g., driving) at a distance from the home base station area. The economics of wireless access systems critically depends on the relative costs and demand for each of the three modes of portability. The overall cost of building wireless access systems varies dramatically as portability is expanded from modes 1 through 3.

These modes of portability should not be confused with the lingo in the cellular phone business today that characterizes as multimode "smart" handsets, which are capable of changing frequencies. In the lingo of wireless marketing, so-called dual-mode handsets are contemplated for switching between paging and cellular service, between home area and roaming, or between different types of new digital cellular systems. For example, Nextel is introducing a triple-mode phone capable of performing dispatch, paging, and cellular functions.

To place the three portability modes in context, the stationary, or fixed, wireless mode is akin to using a cordless telephone in the home, in the office (including wireless LANs and wireless *private branch exchanges*, or PBXs), or at another base station location (e.g., shopping mall, airport, bus or train terminal). The fixed mode of operation is the least expensive to provide.

The second portable mode for wireless access, moving about in a given geographic area at a limited distance from the home base unit (e.g., walking, driving in town), requires that base station unit(s) provide continuous coverage throughout the local geographic area. Personal and portable phones, pagers, two-way radios, and car phones will work in such situations, using radio, cellular, broadcast, or satellite technology. This mode of operation is considerably more expensive to provide than the fixed-location mode, but it is much less expensive to provide than complete portability.

The third portable mode works even when the user is driving fast in a car potentially far away from home base. Mobile cellular systems are one example, but satellite-based systems would work as well. This mode of operation is the most costly, so would-be wireless system operators must carefully evaluate the value to consumers of this type of system relative to other more limited systems to ensure that the substantially increased costs associated with offering *fast hand-off* mobile capability and *roaming* service capability are worth it.

A single multimode handset may potentially function on one or more wireless systems simultaneously for all three modes of portability, providing a full range of portable narrowband voice and digital data services. In addition, all three modes could support a host of other digital services, including non-real-time messaging (e.g., paging and locator services, data, computing, and fax services) and other transaction services (e.g., *smart-card* debit/credit financial services, electronic databases, and information services). Other, more exotic portable applications could include satellite "briefcase" phones (actually a portable Earth station) for communicating from isolated and remote locations, air-to-ground phones, and ship-to-shore phones. However, none of these niche market applications is important for the mass market contemplated in the NII, which connotes a publicly available infrastructure.

One of the most crucial decisions that prospective wireless network operators face is the tradeoff of consumer demand for the added convenience of multimode operation versus the additional cost of providing the sophisticated hybrid network hardware, electronics, and handsets necessary to make it all work. Through the use of sophisticated network electronics, digital signal processors, and intelligent control software, any of the three modes of portability may be used in conjunction with one another in hybrid wireless network systems via interconnection arrangements and so-called overlay networks, used to combine features of different wireless systems.

The network design and cost structure of these systems will be examined in more detail later in this chapter, but first, a basic description of known wireless access alternatives and service capabilities will be discussed.

4.4 WIRELESS ACCESS ALTERNATIVES

There are many wireless network alternatives. Four major categories of wireless systems that will be discussed here include cellular radio, noncellular radio, broadcast radio, and satellite. Within each of these categories are any number of alternative methods for network access, transmission, and routing functions. The basic technical and economic aspects of the most popular wireless access systems are briefly described next.

4.4.1 Cellular Radio Systems

Any given geographic market area is segmented into "cells," each with its own radio base station. This arrangement is often cost effective relative to noncellular arrangements because it allows for the possibility that different users could share the same radio frequencies within a given cell and allows for reuse of the same frequency spectrum in different cells across the entire geographic coverage area. Depending on network system design criteria, a network operator may utilize relatively large cells (*macrocells*), relatively smaller ones (*microcells*), or even smaller ones (*picocells*). All other things being equal, smaller individual cell areas allow for higher system traffic capacity, but the overall system cost is higher due to the larger number of network nodes (antenna sites) per total system coverage area.[2] Current-generation mobile cellular systems use macrocells; future ones will use microcells or picocells (e.g., PCN/PCS).[3]

There are three primary modes of operation for cellular networks: *frequency division multiple access* (FDMA), *time division multiple access* (TDMA), and *code division multiple access* (CDMA). The terms refer to the method by which network operators provide, and by which individual users access, particular communications channels. The relative costs and perform-

---

2. For a good primer on the cellular structure and basic engineering criteria for cellular radio networks see A. Hac, "Wireless and Cellular Architecture and Services" (*IEEE Communications Magazine*, November 1995), pp. 98–104.

3. For a brief discussion of the evolution of PCS from the perspective of a major vendor, see P. Petersen, "Motorola's Wireless Vision for PCS" (*1993–1994 Annual Review of Communications*, International Engineering Consortium, Chicago, 1994).

ance of these three access methods will be discussed in Section 4.5. FDMA and TDMA are well-known technologies that work. CDMA systems have yet to be deployed, which has given TDMA digital systems a huge head start in the market. However, some major industry players are betting that CDMA is imminent and could even become the predominant technological choice for digital wireless access systems.[4]

Even within a given cell area, system capacity and unit cost performance may be improved through sectorization of the cell into smaller geographic coverage areas via the use of directional antennas and variable powering schemes for handsets or base stations to reduce interference. Cell sectorization comes at a cost, but it is usually a less expensive method of adding system capacity compared to adding entirely new cells or to splitting the older, larger cells into newer, relatively smaller cells.

4.4.2 Noncellular Radio Systems

The wireless access system of noncellular radio is conceptually similar to traditional analog radio networks with a single radio station transmitter serving a given geographic area (e.g., mobile phones, taxi dispatch, and emergency services). Such systems typically feature very limited capacity compared to cellular systems but may be less expensive to develop and operate, depending on the type of service being offered (e.g., dispatch and paging services). By using the same methods for frequency sharing as cellular systems (i.e., FDMA, TDMA, CDMA) and sectorization of the radio coverage area, it has become possible to dramatically improve system capacity and performance. The leading applications for this technology are *specialized mobile radio* (SMR) and *enhanced SMR* (ESMR). It is possible to expand the capacity of those systems by migrating to a cellular or cellular-type network structure and setting up many relatively low power radio signal repeater sites that may reuse the same frequencies as other such sites within the same geographic coverage area.

From a network service perspective, it is not yet clear whether the cost of migrating an existing SMR system toward a cellular structure is a less expensive proposition than building and operating a digital cellular network, even assuming that the service quality of the former could equal the latter.

4.4.3 Broadcast Radio Systems

Traditionally used for one-way analog video and audio service, broadcast radio is being reborn using digital technology to expand system capacity and allow for two-way transmissions. Digital signal processing and compression of video

---

4. For a brief discussion of the TDMA/CDMA technology race, see "Cellular Phones: Short Circuited" (*The Economist*, July 29, 1995), p. 45.

signals have dramatically improved channel capacity and functionality of broadcast systems, ultimately providing for two-way interactive communications, including basic telephone services. The most popular systems being considered for the NII are two-way "wireless cable" systems called *multipoint multichannel distribution systems* and *local multichannel distribution systems* (MMDS and LDMS). A number of other system designs feature more limited capacity, geographic coverage, and functionality, including some that are single-channel broadcast systems, for example, *single master antenna television* (SMATV), ITV, and *low-power television* (LPTV).

The economics of migrating one-way broadcast networks toward two-way digital service capability are unclear, since the technology is in the very early stages of development. To date, it appears that, for a relatively small incremental investment in network electronics, such systems soon will be able to support a digital narrowband upstream data channel for limited consumer interactivity for such services as PPV video, NVOD, and, ultimately, virtual VCR video service. Most of the consumer cost of subscribing to ITV services will be for sophisticated electronics to perform the required signal decoder and memory functions in the television set-top boxes. Voice and video telephony are another matter, however, and the incremental network investments required to provide those services over traditional broadcast networks may be substantial. Whether it will prove to be cost effective to upgrade digital radio broadcasting systems to provide for interactive multimedia service, the FCC has at least started to remove a major regulatory barrier to entry into this market—but only for LMDS systems, not MMDS. In July 1995, the FCC issued rules that allow LMDS operators to provide services that compete with the fixed wired local service of local telephone companies. The wording of the applicable provision effectively limits LMDS operators to providing basic voice services. It does not permit LMDS operators to provide other enhanced telecommunications services, but it is certainly an important step in that direction.

4.4.4 Satellite Systems

Satellite wireless systems rely on orbital satellite transmissions, as opposed to terrestrial or land-based wireless alternatives. The most popular systems being considered for possible application in the NII are *low Earth orbit* (LEO), *medium Earth orbit* (MEO), and *geosynchronous Earth orbit* (GEO) satellites using high-frequency (e.g., Ku- and Ka-band) *radio frequency* (RF) spectrum. So many new digital satellite systems have been announced that it is difficult to sort out the similarities and the differences.[5]

---

5. George Gilder provides a detailed discussion of proposed systems and evaluates their market positions in "Telecosm: Ethersphere" (*Forbes* ASAP, October 10, 1994).

GEOs have historically dominated the scene for satellite telecommunications both for two-way telephone service and for one-way broadcasting services using C-band RF spectrum. Because of their high orbital altitude (36,000 km) and correspondingly slow orbit period (24 hours), GEO telecommunication systems may achieve effective global coverage with only three satellites. Traditional GEO systems use relatively low frequencies and therefore require lower operating power levels. Such low-powered, low-frequency systems required system users to install rather unwieldy (and unsightly) large signal receiver dishes. The first applications of this technology for telephony were made by the traditional C-band public satellite telecommunication network systems such as COMSAT, INTELSAT, and, more recently, PanAmSat.

The newer GEO systems use higher power levels and frequency bands (Ku and Ka band) to provide one- and two-way video and data communications. These satellite systems support popular data networks called *very small aperture terminal* (VSAT) telecommunications systems, which use small, unobtrusive, and relatively inexpensive signal receiver dishes.

Using the same high-frequency bands, new digital GEO systems providing DBS services are being deployed to provide digital video broadcasting direct to the home. As in the case of traditional land-based broadcasting systems, new digital signal processing techniques will continually expand the capacity and functionality of many types of satellite broadcast systems, ultimately allowing for two-way voice and data transmission. The FCC's flexible-use spectrum policies may be extended to allow DBS networks of the future (or, for that matter, any other broadcast satellite networks) to provide two-way voice and data services. One persistent problem faced by GEO broadcast video and data systems that want to migrate toward two-way voice service capability is the inherent transmission delay time associated with geosynchronous orbit uplink and downlink, which causes annoying echo and cross-talk in voice telephone calls.[6] While such problems have always existed with traditional C-band long distance telephone systems, consumers will be less willing to tolerate those problems when good substitutes are available for global transmissions like transoceanic cables.

LEOs and MEOs are low-flying satellites that reduce considerably the problems of signal delay in GEO systems, making them more acceptable for voice services.[7] Such systems require that more satellites be launched and maintained or replaced. Whereas GEO systems may achieve effective global coverage with only a few satellites, MEO systems require 10–15 satellites with an orbit altitude of 10,000 km and orbit period of 6–12 hours. LEO systems re-

---

6. Round-trip signal propagation delay for GEO systems is 280 ms.

7. Round-trip signal propagation delay for MEO systems is 80–120 ms; for LEO systems, the delay is 20–60 ms. However, some systems, like Iridium, that may utilize satellite-to-satellite links in lieu of satellite-to-Earth station links will increase propagation delay.

quire more than 48 satellites at an altitude of only 700 km and orbit periods of about 1.5 hours.[8] One proposed LEO system, Teledesic, uses hundreds of satellites. In the near future (pending final approval from the FCC), new global and domestic *mobile satellite systems* (MSSs) using the higher frequency Ku and Ka bands will begin providing two-way voice and data services. These systems are being planned and financed by many major communications companies. Pending other FCC decisions regarding spectrum allocation, still other future satellite systems are targeting not only mobile but also otherwise fixed telecommunication service markets which could ultimately compete directly with traditional wired voice and data telephone networks.

Given the level of business activity that is already occurring, it is safe to say that all of the four categories of wireless services discussed here will be players in some portion of the NII. Some of them, perhaps even most of them, will be only very small players. The relative cost and service advantages of each will dictate which ones ultimately become viable for mass market deployment.

4.5 FUNCTIONALITY OF WIRELESS

There is considerable dispute within the telecommunications industry as to the service functionality of wireless access. In other words, what functionalities are possible using wireless access (e.g., analog/digital switching and transmission, narrowband and broadband signal speeds, circuit and packet switching and transmission, routing, network control), over what types of networks (e.g., cellular, SMR, broadcast, satellite), to provide what range of services (voice, video, text, data), capable of satisfying what types of end-user applications? The answer to those questions is: All of the above. It is simply a matter of network cost and the cost and availability of RF spectrum.

The functionality of wireless transmission is potentially universal, just as it is for wireline transmission, with one obvious benefit on both the supply and demand sides of the market equation: it is portable ("untethered") and does not require costly installation because there are no physical transmission cables or wires. The actual functionality of wireless access (as opposed to wireline access) is directly related to the government's willingness to provide sufficient usable frequency spectrum so as not to limit its capacity and, in turn, its service capabilities. The more relevant questions then become: In light of what we know or can forecast about the government's spectrum allocation policies, what functionalities will various types of wireless access systems be able to pro-

---

8. There are also *highly inclined elliptical orbit* (HEO) systems, like the proposed Ellipsat system, which fly in an elliptical orbit as far away from the Earth as 42,000 km and as close as 1,500 km. These satellites are actively providing service only in their high-orbit phase, giving them a coverage area, or *footprint*, that is about the same as that for GEO satellites. HEO systems require 5–12 satellites for global coverage and exhibit delay times of 200–310 milliseconds.

vide, and what types could be cost effectively provided vis-à-vis wireline alternatives?

Among the technical improvements in digital telecommunications technology, advances in wireless signal processing techniques to enhance network functions, such as access, routing, transmission, control, and message encoding and encryption, will continue to improve system functionality and performance. It is already known that any of the four types of wireless access can now or will soon be able to utilize digital technology to perform both circuit and packet network functions capable of providing two-way voice and data telecommunications. However, some of the wireless network alternatives have a huge head start (e.g., cellular mobile service started back in 1984), while others have just begun to be developed or have yet to be conceived (e.g., local lightwave telecommunications using so-called photonic phones are on the horizon).

The following section on wireless access network cost structures and cost estimates includes underlying assumptions regarding the amount of frequency spectrum available to each individual network operator in a given stylized geographic area and will therefore indirectly address the issue of system functionality. Of course, it is dangerous at this early stage of technology development and deployment to presuppose the functional uses to which wireless access systems will be put, especially in business markets with highly specialized applications and system requirements.

It remains to be seen whether portable wireless access systems will be capable of providing for two-way or interactive real-time video telephony. Some of the systems could potentially do so, but only at a significant incremental cost above narrowband digital service. Most experts agree that such cost is at least high enough to preclude it from being universally available to the mass market of POTS subscribers, making it ineligible for the NII. What's more, from a practical perspective, it is not likely that most people will be watching TV while on the move using digital wristwatches or pocket phones. It is most likely that people will be stationary when participating in activities involving broadband telecommunications.

Upgrading a two-way narrowband voice and data cellular telecommunication system for broadband telephony or "bandwidth-on-demand" service requires enormous capacity additions and financial investments to current and even planned systems. Expansion of capacity in wireless systems, especially those capable of handling both mobile and fixed services, implies additional expenses for support structure and scarce public rights-of-way (e.g., light poles and rooftops) as well as for additional antennas, transceivers, and associated electronics. For wireline systems, broadband capacity expansion generally can be accomplished by adding digital processing equipment within the existing support structure and rights-of-way.

Picture a fixed fiber optic or coaxial cable phone connection from a subscriber location to a network node that, in turn, is connected to a fiber optic backbone network. Compare that to expanding capacity on wireless connections requiring line-of-sight connections in a mobile environment. Once the initial network system is constructed, it is generally less expensive to expand capacity on the fixed wireline network connections, which do not require line of sight. Frequency spectrum limitations of wireless systems notwithstanding, the additional electronics and new cell sites required to significantly expand wireless network system capacity is relatively expensive compared to similar incremental capacity expansion of an existing wireline network system.

Unless there are radical and, as yet, unanticipated advances in both wireless access technology and the FCC's spectrum allocations, the future vision of integrated broadband access offering end-user bandwidth-on-demand-type service will likely be reserved to the province of wireline technology.

Rather than thinking of integrated bandwidth-on-demand service as an extension of next-generation narrowband cellular systems, it is much more likely that wireless technology, should it become the vehicle for the information "superpipe" of the future, will develop as an extension of next-generation digital broadcast and satellite systems, such as two-way MMDS, LMDS, and DBS systems. Like wireline systems, these wireless systems are primarily designed to serve fixed-service demand. Therefore, line-of-sight and support structures are significant issues only for the initial deployment of the network, not for capacity additions to serve increased demand for bandwidth. Still, the additional costs associated with electronics for providing two-way bandwidth-on-demand service, combined with existing limitations on frequency allocations effectively make this scenario an unlikely alternative to the wireline solution.

There is no doubt that the specific areas of the frequency spectrum that are or will be assigned to digital wireless access systems of all types will be nominally capable of providing multimedia and broadband telecommunications (including video telephony). The significant issue is the bandwidth of the particular slice of spectrum licensed to any one network operator, which may easily be less than that required to support mass market multimedia and broadband services featuring simultaneous (and random) access by network subscribers. Given the relatively small slice of spectrum that the FCC has licensed to individual wireless PCS network operators (no more than 40 MHz in a geographic market area), it is clear that those operators are not going to be in the multimedia or broadband business for the mass market.

That is not necessarily true, however, in the case of satellite and other broadcast radio network systems, which may be licensed with sufficient spectrum to provide for any type of digital multimedia service, including broadband video telephony. While technological advances on these types of wireless access systems will eventually make it possible to provide an integrated bandwidth-on-demand service capability, it is not probable that that scenario will ever be realized for the mass market unless the FCC allocates even more frequency spectrum to this industry segment.

4.5.1 Frequency Spectrum and Wireless System Functionality

Cost and service characteristics inherent to a given wireless access system notwithstanding, all types of wireless access facilities require RF spectrum to function. The potential for success depends on the FCC's procedures governing the allocation and licensing of spectrum. Allocation refers to the amount of spectrum and the specific uses to which it may be applied. Licensing refers to the FCC's assignment of the exclusive rights to use a portion of the allocated spectrum to a company in a particular geographic area. Usually the right to use the spectrum is granted subject to provisions specifying the types of services that may be provided and stating that it is granted only for a temporary (but not always specified) time period. Thus, quite apart from the issue of technical cost and service advantages inherent to any particular technology or network design, the scales of competitive advantage may be tipped in favor of one type of wireless access system or another, depending on FCC spectrum policy.

The FCC is in the process of implementing an entirely different regime of flexible spectrum allocations and market-based licensing procedures (i.e., auctions), while grandfathering its past non-market-based decisions.[9] Beginning with the spectrum allocations and licenses for PCS services, the FCC has allowed licensees to provide whatever services they wish (except broadcast and point-to-point microwave services). This new flexible-use policy, which the FCC has adopted as a pillar of its new market-based spectrum allocation policies for the future, should be expanded to include other portions of the RF spectrum. For example, if broadcasters were allowed to use (or offer to others for use) their spectrum endowments under current and future licenses, then such spectrum could possibly serve as a platform for two-way digital telephony in the NII.[10]

Specifically, in the future environment, traditional analog broadcast video channels may be digitally compressed, resulting in dramatic increases in spectrum efficiency to support many more channels per unit of radio bandwidth, including upstream voice and data channels. That would argue for the FCC to expand its flexible-use policies so that broadcasters and wireless cable systems may become full players in the NII by providing two-way digital voice and data services. Once the FCC's service restrictions are removed from current broad-

---

9. See the FCC's March 8, 1994, and June 9, 1994, decisions regarding PCS frequency spectrum. For an economic analysis of the FCC's spectrum policies, see T. Hazlett, "Regulating Cable Television Rates: An Economic Analysis" (Working Paper Series No. 3 (revised), Institute of Government Affairs, University of California, Davis, CA, September 1995); and E. Kwerel and J. Williams, "Changing Channels: Voluntary Reallocation of UHF Television Spectrum" (FCC OPP Working Paper No. 27, November 1992).

10. In July 1995, the FCC in CC Docket No. 92-297 took a small step in this direction by allowing LDMS wireless cable network operators to provide services that compete with the fixed wired local service of local telephone companies.

cast spectrum licenses, the playing field among competing wireless access alternatives is leveled to the point where the least-cost network systems could emerge to compete with the wireline systems of cablecos and telcos.

Expanding the FCC's flexible-use policies puts a complex twist into any analysis of potential winners and losers in the market for wireless access alternatives. But this is a market risk that must be accepted by the wireless players as they vie for market position. That is not to say that the FCC should be empowered to randomly, and without notice, change its spectrum policies thereby devaluing existing or future licenses. That would create tremendous uncertainty among prospective wireless system network operators and could seriously dilute the value of (and the monetary bids for) new spectrum licenses. What it does mean is that the FCC must make clear its long-term intentions regarding spectrum policy: that it will gradually and deliberately expand both the spectrum allocations and the flexible-use rules.

In any event, the FCC's spectrum policies going forward must try to balance the business risk associated with investing in new communication networks with the interests of consumers seeking more market choices and lower costs. The FCC should generally opt for those policy options that favor the latter over the former. The old spectrum policies did as much to protect the business interests of competitors as it did to promote the interests of consumers. The new policies are pointed in the direction of reversing that situation and should be pursued aggressively.

For purposes of the technical analysis to follow, to the extent possible, the impact of FCC spectrum allocation policy on any given wireless access alternative will be considered neutral as among alternatives within the four categories of the technologies listed here.

However, between and among wireless access alternatives, this assumption is problematic for two reasons. First, the total bandwidth allocated within any geographic area between competing wireless access systems will potentially affect the per-unit costs of providing service and, in turn, could dictate winners and losers in the marketplace. Second, regardless of the absolute amount of spectrum associated with a given license, the old service restrictions, which were conditional with the granting of the license, severely limit the market opportunities available for any type of wireless system and, in turn, may be enough to dictate winners and losers in the marketplace. If the FCC is serious about extending its new flexible-use policies beyond those for the relatively narrow PCS bands, they should begin the process as soon as possible, to allow all types of wireless access systems to achieve their full potential in the NII.

Invariably, spectrum allocation rules for a given technology or service type will, in practice, directly affect its relative cost performance. The primary reason is that the total amount of spectrum allocated to a wireless service or that portion granted to one licensee (e.g., 30 MHz out of 200 MHz total allocation), along with its corresponding underlying network or technology (e.g., PCS/TDMA) and position in the range of electromagnetic spectrum (e.g., 1.8–2.0 GHz), in large part determines its cost, performance, and market viability. A full discussion of such issues is beyond the scope of this analysis, but the implications of known spectrum licensing rules for wireless system economics will be evident in the cost and service evaluations to follow.

The implication of the FCC's current spectrum policy for new digital wireless access systems is that, while these systems can fare well in the NII (as a narrowband service platform in the case of PCS or as a one-way broadband service platform in the case of wireless cable), they will not fare well if the vision of the NII includes broadband telephony. The reason is that when the FCC licenses spectrum, it has traditionally done so with strict spectrum usage limitations. For example, when the FCC allocated SMR spectrum, it issued licenses to individual applicants with a strict proviso that it be used only for local radio dispatch services. Recently, that restriction has been relaxed to allow for provision of new two-way narrowband telephony.

The same practice is true, although somewhat less so, under the new flexible-use rules that the FCC is applying to its recent spectrum allocations for PCS services. While a licensee is allowed to use the PCS spectrum for any service it wants (except broadcast and point-to-point microwave service), it is assigned so little spectrum that there is still an implied or effective service limitation. By limiting any given service provider to 40 MHz of PCS spectrum in a given geographic area, the FCC is effectively precluding them from mass market broadband service applications like video telephony. In other words, if very many system subscribers chose at any point in time to access multimedia and broadband telephone services, the system capacity would quickly exhaust, leaving no room for other users to sign on.

Thus, under the FCC's current spectrum policy, the market for two-way broadband services will be the province of wireline access alternatives, primarily fiber optic and coaxial cable, perhaps in conjunction with satellite and other land-based broadcast networks. The irony of this may be that the FCC, in the name of promoting digital wireless technology, has simply not allowed for wireless access to be the technology of choice in the race to develop a fully integrated broadband network system, leaving the winner's circle to wireline access alternatives.

This is not a criticism of the current administration or the FCC, both of which favor changing the rules to liberalize spectrum usage restrictions. Indeed, the current policy has been constrained by historical practices, which can only be changed gradually over time. The past restrictive spectrum use policies were not a serious problem in the days when digital radio services were nonexistent or nascent and the public demand for spectrum was relatively low. Nevertheless, it remains important for the government to aggressively pursue the new policy direction toward flexible use so that the NII may develop unencumbered by obsolete spectrum policies.

In any event, it still remains to be seen if, in the very long run, the market for a wireline information superpipe to the home will ever become financially viable in the presence of cheaper, nonintegrated alternatives, including digital wireless access and digital broadcasting systems.

4.5.2 The Cost of Frequency Spectrum

Since 1993, the FCC has raised about $9 billion from auctions of RF spectrum for new digital services. The first auctions were for relatively low valued spectrum for ITV and regional and national paging services. Since then, ITV has gone nowhere as a service, while paging has grown rapidly. The early auctions generated only about $2 billion, compared to frequency spectrum auctions for PCS, which concluded on March 19, 1995, and generated about $7 billion. The top bidders were established companies with deep pockets, including, in the top three, the telco-cableco consortium called Wireless Co. ($2.1B—Sprint, TCI, Cox, and Comcast), AT&T ($1.8B), and PCS Primeco ($1.1B—AirTouch, Bell Atlantic, NYNEX, U S West). In response to the future development of digital television broadcasting, the FCC is considering extending the auctions to new (or even existing) broadcast frequency spectrum. If it does, it could bring in another $40–$80 billion in bid revenues.[11]

The up-front cost of purchasing the rights to use the RF spectrum, either via FCC auction or by purchasing them from an incumbent, is substantial. This situation causes many bidders to complain that auctions, rather than posing a market opportunity, are actually a barrier to entry. However, bidding merely implies that there is a perceived financial payoff from owning the license, and that perceived benefit is at least as high as the bid price. Thus, the cost of the license for the rights to use a slice of RF spectrum is straightforward for a prospective system operator to incorporate into a business case analysis. It simply represents a (potentially huge) startup cost, which is amortized over the system life or other planning horizon. Presumably, there is also a (potentially huge) salvage value of spectrum rights as well. The overall effect may be just like having money in the bank. Indeed, spectrum itself can be banked; just like oil in

---

11. This may not happen, though, if the government follows through on its announced intention to give away huge amounts of RF spectrum allocated to digital television service. The whole issue of broadcast spectrum auctions for land-based systems has become politicized to the point where it almost held up passage of the new telecommunications law. The issue concerns the death of so-called "free" over-the-air television, a very emotional and political issue. Yet the government would not hesitate to charge for new satellite and wireless cable broadcasting spectrum, which is also "over the air"—it has just never been free. See M. Lewyn, "The Great Airwave Robbery" (*Wired*, March 1996), p. 115.

the ground and fallow land held for future use, there is latent value inherent to some types of stored assets, and spectrum happens to be one of them.

As an up-front, fixed cost, once incurred, spectrum cost has little real impact on future competitive market outcomes. In a financial model, spectrum license fees are simply rents assigned at the outset to either the government or incumbent private interests, depending on which has the spectrum rights. Therefore, any would-be market entrant must offset that amount against the NPV of cash flows from network operations.

This is not to say that the startup costs of spectrum cannot be so substantial as to give a would-be wireless access operator serious pause to enter the market. It is simply an observation that, no matter what the up-front cost of spectrum, an incumbent or entrant firm will make a bid for it unless there is simply no profit to be made by entering the wireless access business. In that case, no one would come forward to bid for spectrum, and the government would have to reevaluate its spectrum allocation and licensing policies.

There is also the risk of overbidding for an FCC license, especially if the FCC does not guarantee that additional spectrum would not be allocated in the future, thereby diluting the value of a license purchased today. However, this is hardly a legitimate complaint against the government's auction. In a world of uncertainty, the risk of overbidding is always there.

This point is important because many observers believed that the auctioning of spectrum would discourage market entry or otherwise distort market outcomes, due to the apparent asymmetry of requiring new wireless access network operators to purchase spectrum when incumbent firms (or future entrants) may be endowed with "free" spectrum. That would be true except that incumbents, too, have an opportunity cost and market value associated with their own spectrum endowments. Therefore, the spectrum auction fees are really just a one-time assignment of the rents associated with spectrum rights to the government instead of the private sector. The net effect on market entry and network operations of either the new wireless network operators or the incumbents should be neutral, with one important caveat: that either is free to compete with the other if it so chooses and is willing to pay for the privilege.

This is not a trivial caveat. If the government were to regulate and limit the uses to which spectrum could be put, the possibility exists that competitive market outcomes would be precluded and that monopoly quasi-rents associated with spectrum rights would exist. In the case of wireless access for PCS services, the government has so far followed the advice of economists and has not restricted the uses to which the new spectrum allocated for wireless access could be put.[12] In turn, the government has not precluded incumbent cellular

---

12. At least in principle. In actuality, even the new "flexibly licensed" PCS spectrum has limits placed on its actual use. However, progress is progress even if only incremental; in the political economy of spectrum rights, the FCC is a world market leader.

operators, previously endowed with spectrum via so-called set-asides of half of the spectrum to PSTN operators and the other half, via lottery, allocated to non-PSTN operators, from using it for new wireless access services.

The FCC's flexible-use policy for spectrum allocated to PCS is totally new and represents a true sea change in spectrum licensing policy.[13] However, the FCC still retains strict rules limiting how the vast majority of licensed RF spectrum is used. For example, broadcast spectrum cannot be used for nonbroadcast services, and spectrum allocated to wireless telecommunications cannot be used for broadcast services or point-to-point microwave services.

Critical to business case analysis of wireless access operators are the uncertainty and risk associated with changing FCC spectrum allocation policies and whether more total spectrum will be assigned to wireless telecommunications. For example, in the future, the FCC could, in further pursuit of its new flexible-use rules, allow UHF spectrum, which lies adjacent to cellular spectrum, to be used for wireless telecommunications. Or the FCC could allow wireless cable spectrum, which lies adjacent to PCS spectrum, to be used for telephony. That would represent a veritable flood of additional spectrum into a competitive wireless telecommunications market, diluting the value of the licenses of the early licensees.[14]

In pursuit of its objective to allocate spectrum for use by wireless service entrepreneurs and innovators, Congress has ordered that another 200 MHz of RF spectrum below 5 GHz be reassigned from government to private sector use. While the process of reallocation of all 200 MHz may take up to 15 years, the FCC has since requested that the government immediately specify and transfer 50 MHz to be allocated for unlicensed private wireless telecommunication services and has invited comment on who should be able to use it [1].

4.6 MASS MARKET DEMAND AND SUPPLY

When the market efficiency and desirability of various technologies for the NII are being considered, the ex ante market supply conditions do not really matter if ex post market acceptance never materializes. Successful market entry will hinge on issues of service choice, quality, convenience, and low prices. The market cannot be ignored during the development of government technology

---

13. The FCC's flexibility started with SMR spectrum and has become manifest with its landmark decision on PCS spectrum use.

14. For a rather interesting and enlightened view of the entire spectrum "scarcity" debate, see P. Baron, "Is the UHF Frequency Shortage a Self Made Problem?" (Marconi Centennial Symposium, June 23, 1995).

and competition policy. Successful market entry into the wireless access business will require that a system feature portability, that it is interconnected to those not on the local system, and that its (quality adjusted) price is affordable.

The costs of making a call on wireless networks is—and will be, for some time to come—more expensive than wireline network calling, partly because it actually costs more to provide the service and partly because there is a willingness to pay a premium for the convenience of portability. The clearest evidence of this for the mass market is the explosive growth of cordless telephone units in spite of their relatively high price (on average about four times that for a wired telephone). The cost of usage is the same for both, so this is a good metric for evaluating the value to consumers of portability, at least for fixed-base-station service.

While it may cost only a dollar to make a cellular phone call from a car, that same call from a remote location or a cruise ship can easily cost 10 times that. Of course, part of the reason is the lack of alternatives in captive markets, just like pricing food services in a ballpark, but part is also due to underlying costs of providing the service. Communicating while on the move has always cost more than doing so while standing still; calling long distance has always cost more than calling locally; and so on. Even for a satellite service, which, due to the nature of the technology, tends to be distance insensitive, there are still significant issues of the technology and cost of transporting the call to the exact location of the called party, who may be on the move.

It is obvious that consumers accessing the NII would like their portable phones to work in all three portability modes, providing a full range of services—but not if the price is too high. There is always the option of having two or three different phones and putting up with the hassle of having to remember to always have the right phone in the right place. Many Americans already have a cordless and a cellular phone in addition to their normal wireline service, and inexpensive pagers are now rapidly being added to the mix.

Nevertheless, a huge debate is raging among experts on the demand side of the wireless future as to whether people will continue to buy so many different phones and at what price, even if each one is relatively cheap compared to a triple-mode phone. Indeed, as cellular phone operators have discovered, the price that consumers face for handsets, as well as the cost of making a call, is an important determinant of mass market demand. Regardless of the assumption that many consumers will pay a premium to avoid the hassle of owning more than one phone for each mode of operation, the mass market will remain very price sensitive. Therefore, to ensure a high level of residential demand and mass market penetration, the incremental cost to consumers for handsets featuring multimode operation had better be somewhere close to the total cost of owning different handsets. Consumers today seem to be able to put up with the hassle of owning a separate pager, cell phone, and CT unit without too much complaint.

Residential local phone calls are provided "free" almost everywhere in the United States. Only a few states charge for local PSTN calls, and even then the charge is quite low at $.01 to $.02 per minute. That means that whatever the cost of providing for local phone calls, the telephone companies recover it from the monthly charges on other services, especially business phone lines, or from long distance services. This fact makes it difficult, if not impossible, for independent wireless network operators to enter and compete in the mass market for local telephony because it is hard to compete against a zero price for usage when there is no source for subsidizing such entry (e.g., toll calling revenues). This sets the mass market entry price bogey for wireless companies to be in the range of monthly charges that incumbent wireline telephone companies charge for local service, which currently runs about $18 per month per household nationwide.

This also helps to explain why interconnecting carriers, especially long distance companies interested in becoming the beneficiary of their own payments to local telcos, are at the forefront of those clamoring to get into the wireless access business. For example, AT&T's takeover of McCaw Cellular allows for the possibility that McCaw customers will be saving on payment of subsidies associated with their calls using AT&T, which, in turn, saves by reducing the amount it must pay out to the local telcos for access to the PSTN.

Cable television companies are the next most logical entrant into the market for local telephony. They see their use of wireless access as a two-way voice and data channel that allows them potentially to become a full service multimedia communications provider. Cable companies and other independent wireless network operators however, face the daunting prospect of paying high prices for interconnection to the telco's PSTN facilities to guarantee nationwide service capability to their subscribers. No wireless access system can become a viable market player unless ubiquitous call terminations anywhere in the country can be achieved. Current local telco interconnection charges are very high, at an average $.07 per minute. This is so high as to be the single highest nonnetwork operating expense of potential wireless access service providers.[15]

The FCC has often stated that it believes that wireless access services are the best hope for introducing competition into local telephone service markets. If that is to be the case, then federal and state regulators need to level the playing field of market entry by reducing toll and interconnection charges and business service cross-subsidies by deregulating the local and toll charges of incumbent wireline carriers.[16] If that is not possible, then the administration

---

15. T. McGarty analyzes the relative importance of PSTN interconnection charges to the cash flow of wireless access systems in "A Precis on PCS Economics."

16. The FCC is actively investigating the issue of PSTN access charges for wireless operators in CC Docket 95-185. It is not clear, however, what the resolution of that docket will be or whether the individual state regulators will go along with FCC recommendations governing access

and the FCC had better plan on seeing some familiar faces on the wireless scene as incumbent suppliers jockey for position to bypass one another (or even themselves) using the new wireless network alternatives to save on paying cross-subsidies.

Unfortunately, the new telecommunications law is not much help in this regard. While the new law does contain suggestions for reforming PSTN access charges, it also recommends that all service providers interconnecting to the PSTN (which, by definition, includes new wireless network operators) share in the burden of cross-subsidizing the ongoing costs of funding the universal availability of advanced wired networks. In years past, this cost burden was largely borne by long distance service providers in the form of PSTN access charges. Now it may be applied to wireless operators as well. This would be sure to substantially increase the costs of interconnection for wireless network operators. On a more positive note, the new telecommunications law does eliminate requirements for *commercial mobile radio service* (CMRS) providers, which encompasses all cellular carriers, to provide so-called equal access to their systems. That allows for cellular carriers to join together in exclusive dealing and interconnnection arrangements with long distance service providers or others, increasing the opportunities for cellular carriers to bypass the networks and local access charges of traditional local exchange carriers. However, for terminating cellular calls to wired public network subscribers, it will always be difficult to bypass the high access charges of local telephone companies.

For now and for the foreseeable future, local telephone companies cross-subsidize a portion of the costs of providing basic local exchange service from profits on business services and access charges paid by interconnecting toll carriers. That artificially raises the price of interconnection to the public telephone network for toll carriers and lowers it for the interconnected local networks. Eliminating all or a portion of the artificial cost burden this places on interconnecting toll carriers and, in turn, the cost benefit it confers on local telephone companies will cause interconnection charges among local telephone companies to rise, which will dramatically reduce barriers to entry in markets for local access and transport services.

While it is entirely possible that, under deregulation, the same telcos and cablecos would eventually dominate the new wireless markets anyway, it would be preferred for the FCC to allow entry on an equal footing to new entrants if, as the FCC has stated more than once, its new wireless policy is to "let a thousand flowers bloom."

---

charges for interstate services. One proposal the FCC is considering is making wireless operators pay the same (high) access charge tariff rates that long distance companies pay to local telephone companies for access to the local PSTN. That would surely cut the demand growth of wireless services and cut the profit margins of wireless network operators.

4.7 COST STRUCTURE OF WIRELESS COMMUNICATIONS

The conceptual model of a wireless access network system is simple. Just like all radio communication systems, wireless access is fundamentally a line-of-sight technology. The basic characteristics of wireless network systems are illustrated in the stylized network in Figure 4.1. The simple generic system includes the essential aspects of all digital land-based systems now being considered for the NII, some of which are up and running in actual test market applications and most of which are still in the prototype testing or development phase.

Wireless access systems in the NII will be open networks, allowing for public access on demand for both call originations and terminations (assuming system capacity and spectrum utilization are engineered to meet demand in a given market area). That is not to say that the handsets or other consumer terminal devices required to access the wireless network are themselves open. While most wireless access system network operators in the NII will need to conform to generic *network-network interface* (NNI) requirements, that is not necessarily the case for the *user-network interface* (UNI), which connects user terminals to the network. Many local wireless network operators, especially very large ones, may use proprietary signaling protocols for transmissions between handsets

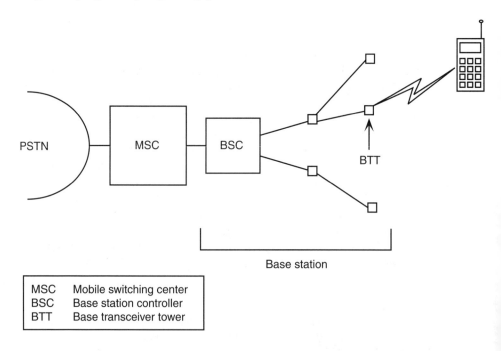

Figure 4.1 Basic characteristics of wireless access system.

and base stations, depending on the particular choice of technology and network control software.

In Figure 4.1, a base station tower is connected to a subscriber's handset for two-way digital transmission. The connection may or may not pass through other network node points between the tower location and the handset, depending on the type of wireless access system. Each base station is potentially also connected to another base station tower in the network or through a network switching center, which is itself connected to the PSTN so that calls from the subscriber handset can terminate anywhere. The *mobile switching center* (MSC) is a primary network node that represents the control point of the wireless access system. The MSC is the brains of the network and performs complex network operation and control functions, including, in cellular systems, call hand-off. For roaming functions and other future intelligent network functions (e.g., call waiting, three-way calling), the MSC communicates via a packet data link with a *home location register* (HLR), not shown in Figure 4.1. The HLR is a computerized database that keeps track of the locations of mobile units and performs other functions yet to be determined.

The counterpart of the MSC in analog systems is the *mobile telephone switching office* (MTSO), which serves as the network host node for existing mobile cellular systems in North America (*advanced mobile phone service*, AMPS). In a network system, the MSC node will be interconnected, usually by high-capacity wireline or point-to-point microwave radio trunks, to the PSTN. In certain types of single-coverage-area wireless access systems (e.g., SMR), the MSC node location may also serve as a base station connected via RF links directly to subscriber handsets. In cellular network systems, the MSC serves as a digital host network controller connected via microwave or fiber optic links to one or more base stations, also called *base station systems* (BSS). Figure 4.2 illustrates a BSS. A *base station controller* (BSC) is the host node of a BSS. The BSC performs basic network functions such as channel allocation, link supervision, transmitted power level control, and transmission of network signaling information. The BSC serves remote nodes called *base transceiver stations* (BTS). In cellular systems, the BSC could be connected to BTSs via either wireline or wireless trunk connections.

Conceptually, newer land-based wireless access systems are no different from the way an old-fashioned *mobile telephone system* (MTS) works. But that is where the similarities end. The poor signal quality, lack of privacy, small coverage area, short distance, and congestion typical of old analog two-way radio systems would never have developed into full mass market penetration because nearly every household already has ready access to a regular phone line to obtain high-quality telephone service.

To overcome the list of problems with traditional analog two-way radio services, digital wireless access systems are immensely more complex. Through the use of sophisticated microelectronics, digital wireless access sys-

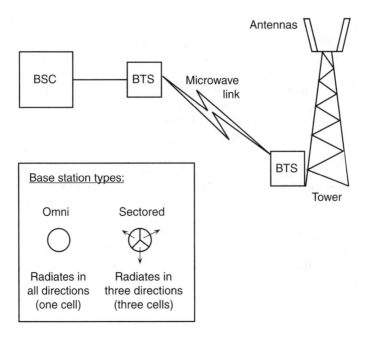

Figure 4.2　Base station.

tems are potentially able to meet or even exceed current wireline network quality and reliability for voice and data services. Several wireless network systems featuring unique network design characteristics and cost structures currently are contending for prominence in the NII.

4.8　WIRELESS ACCESS NETWORK CHARACTERISTICS AND COSTS

This section describes the basic network design for the four types of digital wireless access systems: cellular, noncellular, wireless cable, and satellite.

4.8.1　Cellular Network Design

The cellular category of digital wireless access systems includes all types of cellular configurations regardless of the size of the individual radio cell (macrocell, microcell, or picocell). For purposes of discussion, digital CT technology will also be discussed even if it does not conform to the cellular radio model, because it is likely to be used in conjunction with some cellular systems. The basic wireless access network depicted in Figure 4.1 and discussed earlier features all the basic building blocks of digital cellular networks. The primary distinction between different cellular network configurations is the size and the structure of the cells and, in turn, corresponding differences in system signaling, handset power levels, channelization schemes, co-channel interference, and reuse factors.

4.8.1.1 Digital AMPS (AMPS-D)

Digital signal processing techniques will ultimately allow for substantial capacity gains over AMPS. In the United States, the purpose of first-generation digital cellular systems is primarily for upgrading analog AMPS systems to expand network capacity. That is not true in other developed countries where digital cellular systems are separate from older analog mobile systems or in less developed countries where analog systems were never deployed.

AMPS utilizes FDMA techniques, and AMPS-D systems use the more efficient TDMA techniques. To allow for a smooth migration of subscribers from old- to new-technology AMPS-D systems are designed to operate in dual mode with current AMPS systems (i.e., using a portion of the same 25-MHz spectrum licensed to AMPS operators). Now that the FCC has allocated an additional 120 MHz of radio spectrum to PCS, new TDMA network operators will soon be on the scene.

Just as in the application of time division multiplexing in the PSTN, digital radio TDMA techniques allow AMPS operators to expand capacity by sharing the same communication channel among users. Digitally enhanced versions of AMPS (e.g., *narrowband AMPS* or NAMPS) provide an effective short-term method of expanding the capacity of AMPS cellular systems and provide a bridge to deployment of fully digital systems.[17] Under the IS-54 North American TDMA standard, NAMPS uses the same bandwidth per carrier channel as AMPS (30 kHz), but by allowing three users to share it, the bandwidth per voice channel is only 10 kHz (20 kHz duplex) instead of the full 30 kHz (60 kHz duplex), for a 3-to-1 capacity gain.

Tables 4.1, 4.2, and 4.3, taken from Uddenfeldt [2], are valuable for gaining a basic understanding of some distinguishing characteristics of leading alternative digital cellular systems. Table 4.1 compares digital cellular standards for European, American, and Japanese systems. Table 4.2 compares those standards to the capacity of current North American AMPS systems. Table 4.3 provides the basic distinguishing characteristics of macrocell, microcell, and picocell systems.

---

17. For a detailed description of a digitally enhanced AMPS system called NAMPS, see M. Kotzin and J. Kay, "Cellular Systems Technology: Narrow Band Development and Digitally Enhanced Cellular Services" (*1993–94 Annual Review of Communications*, International Engineering Consortium, Chicago, 1994), pp. 632–636.

Table 4.1
Comparison of Digital Cellular Standards for European, American, and Japanese Systems

Parameter	GSM	ADC	JDC
Access method	TDMA	TDMA	TDMA
Carrier spacing	200 kHz	30 kHz	25 kHz
Users per carrier	8 (16)	3	3
Voice bit rate	13 Kbps (6.5 Kbps)	8 Kbps	8 Kbps
Total bit rate	270 Kbps	48 Kbps	42 Kbps
Diversity methods	Interleaving, frequency hopping	Interleaving	Interleaving, antenna diversity
Bandwidth per voice channel	25 kHz (12.5 kHz)	10 kHz	8.3 kHz
Required C/I	9 dB	16 dB	13 dB

Note: GSM = European standard; ADC = North American standard; JDC = Japanese standard.
Source: [2].

Table 4.2
Comparison of Capacities of European, American, Japanese, and
Current North American AMPS Systems

Parameter	Analog	GSM		ADC	JCD
	AMPS	Full rate	Half rate		
Total bandwidth (Bt), MHz	25	25	25	25	25
Bandwidth per voice channel (Bc), kHz	30	25	12.5	10	8.33
Number of voice channels (Bt/Bc)	833	1,000	2,000	2,500	3,000
Reuse factor (N)	7	3	3	7	4
Voice channels per site (M)	119	333	666	357	750
Erlang/sq km (3-km site-to-site distance)	12	40	84	41	91
Capacity gain	1.0 (ref)	3.4	7.1	3.5	7.6

Source: [2].

Table 4.3
Basic Distinguishing Characteristics of Macrocell, Microcell, and Picocell Systems

Parameter	Macrocells	Microcells	Picocells
Bandwidth allocation	11.34 MHz (11,134 channels)	1.26 MHz (126 channels)	1.26 MHz (126 channels)
Channel allocation	Fixed	Adaptive	Adaptive
Transmit peak power per voice channel	6W	0.6W	0.03W
Antenna configuration per site	120° sector	Omni	Omni
Erlangs/site	148	6	2
Site-to-site distance	3 km (hexagonal)	0.3 km (rectangular)	0.06 km (rectangular)
Erlang/sq km and Mhz	1.6	52	2,300/floor
Erlang/sq km	18.2	66	3,000/floor

Source: [2].

In the current macrocell environment, the capacity of a cellular system is normally determined by calculating the number of simultaneous users, M, per base station cell site for a given amount of RF spectrum, Bt. The system capacity therefore is: $M = 1/N(B_t/B_c)$, where Bc is the equivalent bandwidth of a voice channel and N is the RF reuse factor. Table 4.2 shows that current-generation digital cellular systems using TDMA offer three to eight times the capacity of AMPS systems without adding new cell sites or resorting to microcell deployment. AMPS-D is at the low end of the range.

Lee also has estimated the capacity gains in a comparison of new digital cellular systems with AMPS. Using a fixed amount of spectrum for a radio carrier channel (1.25 MHz), AMPS FDMA systems feature a capacity of 6 radio channels per cell, while TDMA features a capacity of 31 channels (5 · FDMA). CDMA techniques, which are relatively new in commercial applications, offer system capacities of 120 channels (20 × FDMA) [3].

The per-subscriber capital costs of current AMPS systems are about $700–$1,000 [4]. The per-subscriber capital costs of AMPS-D systems (TDMA) are much lower, at about $300–$500 [5].

4.8.1.2 Global System for Mobile Communications

The earliest and most prevalent global standard for digital cellular service is the European *Global System for Mobile Communications* (GSM) (TDMA) standard.

GSM, like North American cellular radio telephone systems, operates in two distinct frequency bands which the government has allocated to cellular mobile (900 MHz) and PCS services (1800 MHz GSM and 1900 MHz U.S.). For obvious reasons, it is important that these two systems are able to interwork with one another, and GSM has proved that they can. Many GSM systems are already operating or are in the deployment phase throughout the world. The newer-version 1800-MHz GSM systems are called DCS 1800.

GSM TDMA techniques can achieve considerable capacity gains over AMPS (about 7 to 1; see Table 4.2). While the United States has already adopted the interim IS-54 (TDMA) standard for AMPS-D, it is still possible for new U.S. wireless access network operators (or incumbents, for that matter) to adopt GSM techniques.[18] Indeed, large GSM system equipment vendors (e.g., Ericsson) will be targeting U.S. markets to compete with North American standards. In allocating cellular spectrum for PCS, the FCC has left wide open the choice of wireless access scheme. Carrier channels in GSM have considerably more bandwidth than those in AMPS-D and therefore may handle more voice channels per carrier. The real advantage of GSM's wider carrier channel bandwidth (200 kHz), however, may be the migration from supporting voice to multimedia and high-speed data services. The per-subscriber costs of GSM systems are in the range of those found for AMPS-D.

Antenna diversity and cell sectorization techniques can expand the capacity of digital macrocell networks like AMPS-D and GSM. For example, the transceivers and associated omnidirectional antennas at BTS cell sites can be reconfigured by the use of directional antennas to split the cell into sectors, like slices of a pie. Altering the power output to distinguish handsets according to near/far conditions is another technique that may be used to gain capacity within the cell area. In that case, the cell is split into concentric zones based on distance from the antenna location, rather than like slices of a pie.[19] It is also possible for cellular system operators to adjust cell sizes and cell coverage areas using combinations of directional antennas and powering schemes.[20]

Increasing capacity to handle increased demand in wireless access systems often simply involves the placement of more transceivers on an existing BTS tower. For example, assume that service begins by placing a single omnidirectional antenna on a tower serving a single carrier radio channel. In a GSM system, a single radio channel is time multiplexed into eight virtual channels,

---

18. "There is every chance that they [U.S. operators] will choose the European standard—which is based on GSM—because it is a proven technology," *Financial Times*, October 17, 1994.

19. See W. Lee, *Mobile Communications Design Fundamentals* (New York: Wiley, 1993), pp. 116–119 and pp. 207–213 for a discussion of antenna diversity schemes and pp. 184–196 for a discussion of cell sectoring schemes.

20. See the discussion in Hac, "Wireless and Cellular Architecture," regarding mixing cell sizes to customize cellular coverage to meet local conditions and requirements.

seven of which may be accessed by subscribers and one of which is reserved for network functions. To increase capacity, a BTS cell may be split into three sectors by the placement of additional transceivers on the tower and the use of directional antennas, each serving one carrier channel, like a pie sliced into thirds. This situation can be characterized as a 1-by-1-by-1 antenna configuration (i.e., one antenna facing each direction). When capacity at that BTS site needs to be expanded further, additional directional antennas can be placed on the same tower and added for the particular cell sector that needs capacity relief (e.g., $1 \times 2 \times 1$, $2 \times 2 \times 1$, up to a $3 \times 3 \times 3$), or when no more carrier channels are available.

Other methods can increase system capacity while holding constant the available RF spectrum. Digital signal processing techniques can be used for adaptive channel allocation and lowering the bit rate for digital voice coding.[21] Incremental changes in per-subscriber or per-minute system costs associated with the adoption of these innovations in voice coding are not yet available.

4.8.1.3 GSM Evolution

By now, GSM systems have moved beyond their initial phase of deployment to add new functionality and services. Because new system capacity constraints were experienced and to save on system expansion costs, GSM systems turned to half-rate (8 Kb) voice coding schemes to achieve a nominal 2-to-1 gain in the number of subscribers the system can support. This decision, however, involves a trade-off in voice quality, especially when calls are made between two cellular subscribers, each with half-rate digital voice coding. This is also problematic if GSM is to evolve toward a universal wireless service for the mass market and as a potential substitute (or at least a not too inferior complement) for fixed wired telephone service operating with voice channels of 64Kbps each (even though high-quality voice does not require that much bandwidth) and with digital cordless systems offering 32-Kb voice channels. New GSM *enhanced full-rate* (EFR) voice coders operating at about 13 Kb increase voice quality and still fit within the 16-Kb full-rate GSM voice channel. This EFR coder may also be used in the U.S. 1900 MHz systems.

Other new GSM network services on the horizon include conferencing and related group calling services, enhanced and intelligent network services such as call forwarding and call blocking, packet data, and even high-speed data services.[22] Compared to fixed wired networks, the use of enhanced and

---

21. See, for example, W. Lee, *Mobile Communications*, and R. Steele, *Digital Mobile Communications* (New York: IEEE Press, 1992). However, this may degrade voice transmission quality.

22. For an update on GSM's technical advances and system evolution, see M. Mouly and M. Pautet, "Current Evolution of the GSM Systems" (*IEEE Personal Communications*, October 1995), pp. 9–19.

intelligent network service features and functions in cellular radio networks presents some special problems for the core network system, which always has the difficult task of keeping track of where subscribers are located to preserve the integrity of individual messages and connections across cells or even neighboring systems. This would call for substantial expansion of the functionality of the HLR, which can be viewed as the future wireless counterpart of the *network control point* (NCP) component of the traditional IN used to provide network routing and number translations for the public telephone network.[23] GSM network designers are also considering ways to allow GSM and advanced digital cordless systems like the *digital European cordless telephone* (DECT) to work together. Similarly, an effective interface to allow interworking between GSM systems and new digital global satellite systems is being investigated.

One very attractive future GSM system development involves globalization. As more and more countries adopt the technology, extended intercountry roaming becomes possible. A standardized *subscriber identity module* (SIM), an electronic card that plugs into a cellular handset, could provide the necessary functions required for system compatability, fraud protection, and billing accuracy across totally different cellular systems.

4.8.1.4   Macrocell Mobile Systems and PCS

The beauty of cell sectoring in cellular radio systems is that, using essentially the same type of network equipment, the cost of increasing individual cell capacity may grow incrementally over time as demand grows. Even in a microcell environment, it is possible to employ cell sectoring schemes to increase capacity and transmission quality.

The use of cell sectorization techniques in a macrocell environment to improve RF reuse in a given market area has the same effect, but at less cost, as implementing microcells. By piggybacking early PCS service demand on the macrocell network, mobile system operators believe that they can compete against the capacity and performance of new microcell wireless access systems. In fact, this pronouncement has been made by most major cellular operators in the United States, which contend that they have a significant head start and market advantage over new microcell PCS operators. Some of these pronouncements are suspect, because the FCC has restricted incumbent cellular network operators to acquiring a total of 15 MHz of PCS spectrum per market area.[24] This places incumbents at a competitive disadvantage to other new PCS opera-

---

23. In addition to expanding the functionality of the GSM HLR, the use of sophisticated high-technology "itinerant agents" to perform enhanced cellular network functions are being investigated. See D. Chess et al., "Itinerant Agents for Mobile Computing" (*IEEE Personal Communications*, October 1995), pp. 34–49

24. Incumbents already have 25 MHz for their existing cellular mobile systems.

tors, which are allowed a total of 120 MHz (40 MHz each) per market area. Thus, it behooves existing operators to announce early on their intention to compete in the PCS market so they can gain customers and signal new entrants of their intentions to compete using their existing cellular system and RF endowments.

Once the demand for PCS grows to the capacity limitations in the sectored macrocell environment, the mobile network operator still has the opportunity to split the coverage area into smaller cells, which further expands system capacity and begins to mimic the network design of the microcell system operator. That should give microcell network operators pause if they believe that their choice of technology is somehow unique in serving the market for PCS. In fact, recent research suggests that both TDMA and CDMA may be cost effectively applied in a macrocell environment until such time as capacity constraints require adopting a microcell system structure.[25]

Other things being equal (e.g., system demand), it is always more expensive to deploy microcells than macrocell systems because doing so involves more radio tower sites and associated transceiver equipment costs. Microcell systems require the placement of many more nodes (BTSs) per coverage area. To the extent that such placements may be delayed by macrocell network system operators without sacrificing tapping into the early PCS market potential, it behooves mobile network operators to squeeze as much capacity out of their macrocell network as possible. If, however, PCS service demand skyrockets, mobile operators will have to worry about system capacity shortages.

4.8.1.5 CDMA

CDMA macrocell cellular systems can use essentially the same architecture as that for TDMA, AMPS-D, and GSM systems. The primary difference is the considerable gain in system capacity by a reduction of the spectrum reuse factor from 7 to 1. The gain in spectrum efficiency (i.e., system capacity) for a given radio coverage area and fixed amount of radio spectrum is inversely related to the numerical value of the spectrum reuse factor. In CDMA spread-spectrum systems, the bandwidth of the radio carrier channel is much greater and is shared among many more subscribers in the same cell. Cell sectoring techniques are also used to expand capacity in CDMA systems.

CDMA macrocell systems are not yet deployed, and a number of possibilities exist for channelization schemes. Qualcomm, a major supplier of CDMA systems, has proposed a 1.25-MHz carrier channel bandwidth that can accommodate 25 voice channels. With cell sectoring (three sectors per BTS), CDMA

---

25. For a discussion of the technical performance of competing TDMA and CDMA cellular systems, see G. Pottie, "System Design Choices in Personal Communications" (*IEEE Personal Communications*, October 1995), pp. 50–67.

carrier channels have a capacity of 75 voice channels. For urban CDMA systems employing this sectored cell network configuration with over 50,000 subscribers, McGarty reports a per-subscriber capital cost of about $350 [5]. The per-subscriber costs for urban systems are sensitive to subscriber density within a cell and the size of the coverage area. Holding constant the total radio coverage area, the per-subscriber system costs increase rapidly for subscriber levels below 50,000 and could easily be two to three times the $350 number for very low penetration (e.g., 10,000 subscribers). The per-subscriber costs slowly decrease as demand expands beyond 50,000 subscribers but flattens out quickly. The same would be expected to be true for TDMA and even AMPS-D cellular systems.

4.8.1.6 PCS Microcell

Microcell TDMA and CDMA wireless access systems use fundamentally similar radio technology compared to their macrocell counterparts, but with reduced cell sizes (e.g., 3-km radius versus 0.3-km radius). Reed studied microcell PCS network costs and reported the per-subscriber capital cost to be about $500 for both TDMA and CDMA systems [4].[26] Interestingly, McGarty reports fairly similar per-subscriber costs (considering the rough level of the analysis) for large urban macrocell systems using either CDMA ($373) or TDMA (GSM) ($453) [5].[27] These per-subscriber system cost estimates are derived from static calculations of total construction costs, including startup, divided by a target level of subscribers (e.g., 50,000). Using a different approach, once the initial system is built and operational, the estimated incremental capacity cost per minute for growth in network usage multiplied by the average system usage per subscriber (180 minutes per month) yields a TDMA per subscriber cost of about $200 [6].

In a mobile environment assuming fast hand-off capability, the implication from the available data is that microcell network structures have no inherent unit cost advantages over macrocell ones and that a network operator should delay the conversion from macrocells to microcells until capacity constraints require it. This reactive mode of operations could backfire, however, if the early microcell system operator is better positioned to fill (unanticipated) PCS demand. It is also possible that if, in the near future, it is perceived by the macrocell system operator that capacity constraints in the macrocell system

---

26. Reed states that it is too early to try to distinguish cost differentials for TDMA and CDMA systems. For the technical parameters associated with two PCS CDMA systems, see the table on p. 68 in D. Reed, "Putting It All Together: The Cost of Personal Communications Services" (FCC OPP Working Paper No. 28, November 1992).

27. However, the CDMA architecture incorporated a three-sector cell, while the TDMA cell was not sectored.

would create the need to reduce cell sizes, squeezing as much capacity out of a macrocell system design before converting to microcells to relieve capacity might actually end up raising the total long-run cost of operations. That would be especially so if the costs incurred for macrocell system capacity expansion were nonrecoverable before it became necessary to eventually convert to microcells to improve capacity to levels required by rising demand.

4.8.1.7 Expense Factors in Cellular Networks

There exists a wide range of estimates of marketing and operating expenses associated with new digital cellular wireless access systems [4,7,8]. Since the fundamental operations among competing carriers for stand-alone cellular systems are homogeneous (system administration, service provisioning, repair, maintenance, etc.), the ongoing expenses for network operations are likely to be similar, or at least that is a reasonable assumption. Because competing carriers operate in the same markets to attract the same customers, marketing expenses could also be expected to be similar across carriers in the same market area. In the case of incumbent cellular carriers, especially vertically integrated ones, there may be some economies of scale and scope from reduced interconnection, operating, and marketing costs. However, little is to be gained at this early stage in comparing expense estimates, since it is not likely to be the determining factor *ex ante* in selecting one type of network system over another.

4.8.1.8 Cordless Telephone Technology

Mobile multimedia is the ultimate concept in digital cordless technology. The international vision of this concept has been called the *universal mobile telecommunications system* (UMTS); in the United States, it is referred to as the *future public land mobile telecommunication system* (FPLMTS). In 1992, the *World Administrative Radio Conference* (WARC) assigned 230 MHz of spectrum around the 2000 MHz RF band to UMTS. It remains to be seen if that bandwidth is enough to ever support a true mass market wireless multimedia network infrastructure. Most likely, it is not, and more bandwidth in higher-frequency bands will be needed.

The goal of UMTS is to provide high-quality high-speed services with unlimited mobility and global coverage. Needless to say, achieving that goal will take a lot of work in the research and development community, and there will no doubt be some serious setbacks. However, it is a valuable goal from a social infrastructure perspective and a useful vision to keep in mind for guiding wireless technology developments. UMTS is based on *personal telephone numbers* (PTNs). A PTN is like a *personal identification number* (PIN) in that it will follow an individual wherever he or she goes and to whatever terminal that individual uses. International standards bodies are investigating how to keep track of incoming and outgoing calls in this new environment so that network and equipment standards may be developed. The reason that CT service, which is normally associated with a very limited and fixed coverage area, may be the key to the beginning design of a cost-effective global mobile system is that most people most of the time are close to their home or workplace. Recent polls in the United States (one of the most, if not *the* most, mobile society in the world) indicate that 90% of the time that individuals are out of the office they are either in the same building or nearby. To achieve global coverage, UMTS may be interconnected with emerging global digital satellite networks. The many proposals to launch satellite personal communication networks may represent the forerunner of the satellite portion of the global UMTS system [9].

Compared to cellular service, CT technology generally features very low power, slow (or no) hand-off, and a limited base station coverage area. For very short distances from a base station unit, CT handsets may handle the network control functions, which in cellular roaming modes would have been handled within the network. For example, a CT handset should be capable of automatic selection of an open channel from those available at the base station, the way some cordless phones already do today.

For obvious reasons, most individuals use the telephone while at home or in an otherwise stationary situation (e.g., office, shopping mall). This simple fact of life is what allows CT technology, which costs a lot less than a stand-alone cellular system, to become a potential market winner. The market for digital cordless phones in 1994 alone was $1 billion, doubled in 1995 to $2 billion, and is estimated to grow to $32 billion by the year 2000.

The network and handset costs associated with fast hand-off and roaming features offered in a mobile environment are very high compared to CT systems offering only fixed location or slow hand-off capability. However, such supply-side cost advantages may mean little in terms of market success if consumers truly desire and are willing to pay for the added convenience of total portability in a mobile environment.

The United States has no significant players planning to deploy CT tech-nology except in conjunction with other plans for wireless infrastructure. CT's role in the NII will be as a complementary service offered in conjunction with or interconnected to other wireless networks or as a cheap substitute for more expensive wireless network systems for those consumers that either do not want or cannot afford such access.

It is not that there will not be a demand for CT service. Indeed, the explo-sion in the demand for cordless handsets in American households makes that a given. In fact, we should anticipate the day when infrared light is used in addi-tion to or as a substitute for current CT radio frequencies in the home. Using photonic phone technology, the numerous remote control devices for televi-sions and stereos could double as portable phones, pagers, and intercoms. As the futurists have put it, "We'll be watching our phones and answering the TV."

Around the globe, notably in the United Kingdom, Japan, and (soon) Canada, CT technology is beginning to be deployed in various forms. Compared to those countries, there is much less excitement and anticipation in the United States among consumers or major players in the wireless access industry. The United States has adopted no standard for advanced CT technology. Still, to maximize both the functionality and capacity of planned wireless access systems, U.S. cellular carriers are considering CT technology for near-base-station communications. GTE's proposed TELEGO wireless network system is one early example. TELEGO is touted as a fully functional portable phone service that switches from fixed-location CT mode in the home base station area, to on-the-move mode when outside the home base area, then to mobile roaming mode when the user is driving a vehicle far from the home base area.

The question of whether there is a market for CT-type networks depends on the nature of demand. In particular, is the mass market characterized by a dense and not very mobile population, both in terms of the speed of movement (slow) and the proximity of subscribers (close) to the base station? If the majority of the urban population tend to congregate in very limited areas (e.g., downtown rail and subway stations) and are not very mobile (usually on foot or on a bicycle), then CT networks may be the best market alternative among wireless access systems. Many Asian countries (among others) meet these criteria and are likely to be early adopters of this technology.

Even in the United States, there are potential mass market applications for advanced forms of CT technology, especially those that may become a good substitute for digital wired phone service, even in rural areas. For example, a standard has been specified for a *personal access communications system* (PACS) suitable for PCS and fixed wireless loop applications.[28] PACS employs a very low power microcell TDMA technology featuring a relatively low cost infrastructure capable of providing high-quality digital service. However, no vendor in the United States is actively pursuing deployment of such a system because the FCC has not licensed suitable spectrum for that purpose. Even with sufficient spectrum, for CT systems to be financially viable in rural applications, high power levels would be required, thereby increasing the coverage area of a single antenna site. The FCC's power restrictions associated with spectrum licenses (to avoid interference), often designed with dense urban areas in mind, becomes a limiting factor for the market viability of rural CT systems.

---

28. See American National Standards Institute (ANSI), "Personal Access Communications System Air Interface Standard" (SP-3418, JTC(Air)/95.04.20-033R2, June 5, 1995). PACS is also known as *wireless access communications system* (WACS). For a description and technical parameters, see S. Lin (Bellcore 1995) and Bellcore Technical Reference, TR-INS-001313, "Generic Criteria for Version 0.1 Wireless Access Communications Systems (WACS)" (Issue 1 (revised), November 1994).

The network cost of CT technology deployment can range from nearly zero, in the case of the vastly popular household units, to very expensive, depending on the sophistication of the technology, power level, distance capability, functionality of the handset (e.g., paging, intercom), and the number and spatial distribution of base station locations and remote nodes. Advancements in the technology include increasing the practical operating distance between the base station unit and the handset and increasing the number and locations of base station units and remote electronics (e.g., signal repeaters and amplifiers, trunks). The CT mode of operation is relatively cheap to provide compared to mobile radio service and is almost strictly a function of the number of base station units, subunits, and electronics. The handsets are small and relatively inexpensive because they can operate on very low power. Capital costs associated with a CT network for trunking and interconnection are minimized because the phones work only near a base station and because relatively unsophisticated plug-in connections to the PSTN can be used. As in any other portable communications network system, usual operating costs include marketing, sales, network operations, administration, billing, and so on.

Cell sizes for CT technology are very small (e.g., 100m to 500m radius). There are many versions of CT technology. The first-generation cordless telephones (CT1) were simple single-base-station phones on a single fixed RF connected to the PSTN. Beginning in 1985, the *Conference of European Posts and Telecommunications* (CEPT) initiated a standard for second-generation cordless digital systems called *Cordless Telephone 2* (CT2) service, also called Telepoint. CT2 is the first cordless technology to use digital voice coding (FDMA) and multiple base stations in a limited coverage area. CT2 functions like a normal cordless phone in non-Telepoint mode. When away from the home base, CT2 allows for only originating calls. CT2+, a second-generation standard, allows for slow hand-off between cells. In 1988, Telepoint was introduced in the United Kingdom with much fanfare and dubbed "the poor man's mobile phone." By now, most CT2 service providers have given up and are being displaced in favor of newer digital cellular systems.

Also in 1988, CEPT decided on a new cordless system operating at a different frequency. Introduced in 1992, the new system was DECT (European standard) and CT3 (Ericsson), a third-generation CT technology that employs TDMA GSM techniques allowing for send and receive capability and adaptive channel allocation. Compared to CT2 phones, DECT doubled the transmission range (up to 300m outdoors) and permitted hand-off between base stations. DECT, like CT2, uses 32-Kbps voice channels, but DECT may allow for combining channels for high-speed data services. While DECT does allow for hand-off and complete coverage in the area where the system is located, its geographical coverage area is usually restricted to a campus environment. The system design makes it an expensive proposition to cover a very wide area and still allow for roaming. The very popular worldwide standard for TDMA cellular networks,

GSM, is basically the same as the European CT standard, DECT. The DECT standard defines compliant protocols for interworking with both ISDN and GSM.[29] Such compatibility promotes the deployment of the technologies, since they may grow in tandem due to network compatibility and interconnection.

Early applications of public CT network technology were championed in the United Kingdom. Mercury has already launched the first PCN system (dubbed One-2-One), which now competes in certain market segments with macrocell mobile carriers. The consumer markets served by these two types of network access systems may not overlap as much as one might think. So far, the demand for the CT alternative in the United Kingdom has not substantially slowed the demand growth in cellular systems. In just two years, Mercury has signed up over 300,000 subscribers, two-thirds of whom never used a cellphone [10].

In Japan, DDI has introduced the *personal handy phone* (PHP), which, because of widespread deployment of base station units, will feature wide area coverage and two-way capability. The PHP will not, however, allow for mobile communications because of a lack of fast hand-off capability [11]. In Canada, the government has adopted a CT2+ technology standard, allocated spectrum, and licensed several CT networks (e.g., Popfone, Telezone, Personacom).

4.8.2 SMR

SMR systems are the only wireless access technology being considered for the NII, which is based on the traditional (noncellular) model of two-way mobile radio. SMR systems use RF frequencies located adjacent to mobile cellular service frequencies. When the FCC allocated them, they were single (paired) channel frequencies intended for high-power, single-antenna, large coverage areas for two-way radio and dispatch-type services.

Beginning in 1987, Fleetcall (now Nextel) and other companies began purchasing and aggregating thousands of SMR frequencies in cities throughout America to achieve scale economies. With the help of FCC rulings that allow for different radio system configurations, Nextel was authorized in 1991 to construct digital radio networks using SMR frequencies. Today, a handful of players have pieced together coast-to-coast service capability.

With the assistance of *Motorola's integrated radio system* (MIRS) technology, the enhanced version of SMR, ESMR, relies on the same advances in digital signal processing that have opened up the future for all the land-based wireless access companies. In ESMR systems, the old familiar scratchy and haphazard transmissions of taxi and emergency dispatch systems will be digi-

---

29. See P. Olanders, "DECT Standardization—Status and Future Activities" (*Proc. 5th Symposium on Personal, Indoor and Mobile Radio Communication*, The Hague, 1994), pp. 1,064–1,069.

tally enhanced to the point where they may compete with newer cellular systems.

ESMR systems using MIRS technology operate in a TDMA cellular-like environment. Such systems may expand the capacity of a single SMR radio channel sixfold, allowing ESMR wireless access systems to have enough capacity to compete for the customers of cellular network systems. However, as is the case with current cellular networks, the capacity of ESMR systems for serving mass market demand still may become limited if PCS demand takes off.

The ESMR system cost per subscriber for wireless access is difficult to estimate because some of the system infrastructure is already in place for existing lines of business, including dispatch and radio paging services. It is reasonable to assume that the per-subscriber costs of upgrading SMR systems to ESMR using MIRS is lower than the system startup and build-out costs of PCS competitors and are probably less than digital cellular upgrade costs on shared AMPS/AMPS-D systems.

Because of the historical use of SMR radio frequencies for two-way radio dispatch and paging-type services and the installed base of subscribers to those services, ESMR wireless access system handsets will be among the first to offer multimode service. In fact, because ESMR systems will be built in market areas where a radio network infrastructure is already in place, they will be bringing the service to market potentially two to five years ahead of PCS systems, which cannot even begin the network build-out until some time in 1996. That head start could represent a huge marketing and service advantage. However, as is often the case with being the first to bring a new technology to trial, ESMR is having early service problems. As one ESMR business customer in Los Angeles put it, calls on the network "sound like you're under water" [12].

4.8.3 Wireless Cable Systems (MMDS/LMDS)

Originally planned as a wireless broadcast alternative to cable television service, wireless cable systems are potentially capable of two-way digital access services. Originally, the FCC allocated spectrum (2.596–2.644 GHz) for the new wireless cable services. Thirteen video channels called *multipoint distribution service* (MDS) and *multichannel multipoint distribution service* (MMDS) were allocated for use by licensees. Additional spectrum (20 channels) using frequencies originally set aside for educational programming has been made available to MMDS operators so that a total of 33 channels could be offered. The FCC has since set aside certain RF spectrum for response bands for upstream signaling for interactive video services.

More recently, the FCC has proposed allocating another 2 GHz in the 27.5–29.5 GHz band to a new service dubbed LMDS for uses similar to MMDS, but it has not yet granted standard operating licenses. Recently, the FCC announced plans to allocate and auction more spectrum for this service in the 6-GHz band.

Both MMDS and LMDS plan to use digital technology to increase broadcast channel capacity and to provide for limited two-way interactive service. If the FCC gives the go-ahead under its new flexible-use policy, wireless cable systems could use two-way digital channels for telephony.

The basic cost structure of wireless cable technology is illustrated in Figure 4.3. The systems will consist of a head end for combining video signals from terrestrial network and satellite feeds for transmission directly to subscribers. Subscribers to wireless cable systems receive the signals via a small antenna and signal downconverter and television set-top box for channel selection.

The primary distinguishing characteristic of wireless cable systems' cost structure is their substantial up-front fixed and startup costs and, in turn, the low incremental capital cost of adding subscribers. Almost all of the incremental investment associated with subscriber additions is CPE, including the

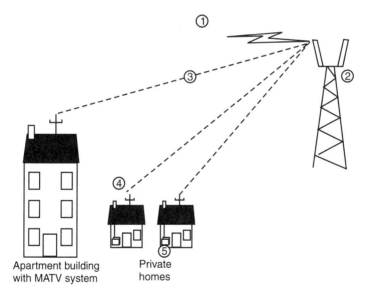

Apartment building with MATV system

Private homes

1. Programming is beamed via satellite to a receive/transmit station.
2. Signal is converted for multichannel, multipoint distribution.
3. Signal is transmitted via microwave.
4. Signal is picked up by antenna and converter and reformed for TV reception.
5. Signal travels to a decoder box atop television, where it is unscrambled for viewing.

Figure 4.3 Basic cost structure of wireless cable network.

installation of the receiving antenna, signal downconverter, and television set-top box. For that reason, such systems are especially well suited for high-density urban applications. Due to line-of-sight requirements for clear television reception, wireless cable systems will have area coverage problems when adverse weather, terrain, and man-made interference factors are present.

As with any video delivery system, use of digital technology by wireless cable networks is brand new. Significant advances in the network application of digital signal processing and compression techniques to support VOD and interactive services are still largely on the drawing board. It is a forgone conclusion, however, that digital signal processing technology will be applied and that two-way capability eventually will be a reality. Since the original purpose of these wireless access systems was to provide television service at fixed locations, the portability aspects associated with roaming have not been investigated. If roaming capability is ever to be in the cards for these systems, it will likely have to come from interconnection to other mobile systems that are interconnected to the PSTN.

While industry observers have mixed opinions regarding the ultimate capability of wireless cable systems to provide wireless access services as part of the NII, it is generally agreed that they will be potentially significant players in the digital video business. Therefore, wireless cable systems may be used by subscribers as a platform for broadband service in conjunction with other narrowband wireless access networks (e.g., PCS) to provide a totally wireless mass market service platform in the NII.[30] Relative advantages of digital wireless cable systems include the rapid deployment feature of the technology and its ability to fill in the gaps for areas not otherwise served by wireline alternatives. Wireless cable's relative disadvantage is the adverse effects of bad weather and terrain (especially trees) on the quality of the signal.

4.8.3.1 System Upgrades and Costs

Due to the very large coverage area from a single antenna site (e.g., 3,000+ sq mi, 30-mi radius) and the high subscriber densities offered by urban areas, the fixed network capital costs on a per-subscriber basis for wireless cable (MMDS) systems are very competitive, lower than those for traditional wired cable systems. Average per-subscriber system costs are about $500 [13]. Variable costs for existing analog wireless cable systems are the dominant cost factor at about $350–$450 per subscriber, about half of which is CPE and half installation.[31]

---

30. For example, G. Berzins (Kidder Peabody, "Wireless Cable: A Strong Competitive Threat to Cable Television," research report, New York City, July 20, 1994, p. 5) does not think that MMDS has potential for two-way telephony.

31. See W. Vivian and A. Kreig, "Wireless Cable: Today's Low Cost, High Capacity Pathway Worldwide to Residences, Schools, and Businesses" (Wireless Cable Association International,

People's Choice TV in Tucson reports an incremental per-subscriber system capital cost of $525, $380 of which is fully reusable if a subscriber discontinues service [14].[32]

On the near-term horizon is the digitization of wireless cable signals, which will allow for video compression and a dramatic increase in channel capacity (250 channels) and system functionality (e.g., VOD, NVOD). The system costs on a per-subscriber basis will remain steady in a digital environment, although the cost of set-top boxes will rise somewhat at first, but the increased system capacity will cause the per-channel cost to fall dramatically. The price of digital signal converter boxes (by late 1996) is estimated to be about $300–$350 [15].

Upgrading a wireless cable system that already has digital broadcast video capability to provide two-way digital wireless access service capability should not be too difficult, but very little hard data are available on the cost of doing so. One of the main reasons for that is that the FCC has not licensed the spectrum for telephony. Assuming that network operators are already planning to digitize their networks and use digital compression technology to expand channel capacity, the incremental fixed network costs to provide digital wireless access for two-way telephone services on a wireless cable system should be low. All that is required is that a portion of the broadcast radio links be assigned to upstream signal carriage.

There may also be a network cost incurred to aggregate upstream traffic in a cellular-like or sectored environment similar to the way other narrowband wireless access systems plan to backhaul subscriber traffic. For example, wireless cable operators might employ remote antenna sites and signal repeater/amplifier stations for traffic aggregation, allowing for shared use of upstream channels. To conserve broadcasting spectrum and make efficient use of that portion of the total available spectrum (about 200 MHz) band that must be dedicated to upstream communications channels, MMDS systems could employ the same shared access techniques used in PCS systems such as CDMA. Another possibility is the use of wireless LAN access techniques. Like other wireless access systems, wireless cable systems could team up with local wireline network providers (telcos, cable television companies, etc.) for backhauling and terminating upstream traffic originating on the wireless system.

The variable per-subscriber cost required to upgrade a wireless cable customer for two-way digital service, however, will not be nearly as low as the costs required for the network portion of the system. In any event, it should not be any higher than that which a wired cableco would have to incur, since both

Washington, D.C., draft, July 1995), p. 8; and Nordberg Capital, "Third Generation Television" (New York, October 21, 1993), p. 10.

32. Additional cost will be associated with subscriber "churn," estimated to be $125–$250. See Gerard, Klauer, Mattison & Co., "Wireless Cable Industry Report" (New York, 1995), p. 11.

require sophisticated set-top boxes to separate, combine, modulate, and de-modulate the incoming and outgoing signals. A set-top "transverter" unit (a combination radio signal transceiver, codec, and up/down signal frequency converter) would be required to make the system work on a customer's prem-ises. Network equipment manufacturers have not yet announced the availabil-ity of digital equipment for wireless cable applications; therefore, reliable cost data for upgrading the systems for digital wireless access service are not available.

LMDS systems differ from MMDS in network design and operation. Oper-ating in a very high frequency band, the LMDS head-end location will be capa-ble only of serving a much smaller coverage area compared to MMDS due to the higher frequency signal propagation. Serving an entire city, metropolitan area, or remote locations will require the system head end to feed signals to remote signal repeater/amplifier antenna sites designed for smaller coverage areas in a cellular design. To date, the FCC has issued only one license for LMDS service, to CellularVision, which operates a single experimental system in New York, but which plans to license its technology for many more systems throughout the United States. CellularVision's provisional license provides it with over five times the nominal spectrum available for use by an MMDS operator. This obviously is an advantage as long as the costs required to cover an entire metro-politan area in a cellular arrangement are low enough to compete with single-tower two-way cable systems. The costs of these systems still should be lower than wired cable, since, like MMDS systems, the cost of laying cables and main-taining the wired system with all its signal amplifiers is avoided.

LMDS cells sizes will vary, but they may be as large as a 12-mi radius (for very flat areas with no trees and dry climate) or as small as a 1-mi radius or even less in areas with varied terrain, like large urban centers. CellularVision claims its transmitter provides excellent service for a coverage area of 48 sq mi (4-mi radius). Thus, to serve a major city of, say, 1,000 to 2,000 sq mi would require 20 to 40 transmitter sites [16]. LMDS technology has the capability and, if the FCC licenses it, the available spectrum, to provide two-way services in-cluding video telephony.

LMDS system capital costs per subscriber will be somewhat higher than for MMDS systems and will depend on the number of cells and remote trans-mitters required. In time, the costs of production CPE (e.g., antenna, downcon-verter, set-top box) will likely be the same as for MMDS. The same is probably true for the cost of upgrading subscribers for two-way telephony using digital signal transverters located on the subscribers' premises.

But for the greater spectrum bandwidth allocated to LMDS service, these systems face many of the same problems of MMDS operators in establishing two-way mobile and roaming services and would probably have to consider in-terconnection with another mobile system operator to become a full-service wireless communications company.

One example of a prototype LMDS video system using today's technology and subscriber equipment with cell sizes of 1- to 3-mi radius (urban) estimates network system investment costs at about $40 per home passed. For larger cells requiring repeater/amplifier nodes (e.g., 12-mi radius suburban system), the cost is about $110 per home passed. On a per-subscriber basis, the cost would be much higher depending on the penetration rate assumed. Associated CPE costs are estimated at about $700. Adding two-way narrowband telephone service adds substantially to those costs. The system capital costs per subscriber quadruples to about $200, and associated CPE costs are about $1,200. Both the network equipment and CPE costs should fall dramatically once manufacturers begin to provide production quantities. In a production mode, it has been estimated that equipment costs will fall such that a two-way LMDS system may be installed for about $700 per subscriber.

4.8.4 Satellite

Due to high upfront investment costs and the wide area coverage, the cost structure of satellite network systems is similar to that for wireless cable systems. The greater signal coverage area of the satellite system compared to land-based wireless systems make the potential per-subscriber costs of satellite network systems very competitive. As with wireless cable systems, most of the variable cost per home passed or per subscriber will be for CPE. Many types of new high-powered, high-frequency satellite networks and services are on the horizon, including LEOs, MEOs, and GEOs. The FCC licenses providers of MSSs operating in the 1.5–2.5 GHz band. Already the FCC has approved five licenses for so-called "Big LEO" systems.

The initial applications of the technology will be in niche markets for locator services, mobile roaming, and remote telecommunications, where it eventually could dominate the scene. While it is technically possible that MSS networks may be used as a mass market substitute for fixed wired access, this seems unlikely in developed countries with nearly universal access already available. While not a substitute for land-based wired networks, new digital satellite systems will potentially become important complements, providing for worldwide connectivity. Many systems will offer dual-mode handsets capable of using either the satellite or interconnected land-based systems, whichever offers the more convenient or lower-priced service.

In addition to proposed digital satellite systems offering voice and data services, it is anticipated, because of desirable cost characteristics, that DBS networks will be the dominant technology for distributing broadcast video signals worldwide for use by other land-based video distribution networks or directly to end users themselves, especially in remote locations or in locations not otherwise served by terrestrial networks. But that is not the only possibility. Assuming that the FCC continues to liberalize the uses to which spectrum may be put, DBS systems may be able to profitably expand into the two-way telephony business.

Table 4.4 is a summary of proposed MSS systems.[33] Table 4.4 makes it clear that major players are vying for a share of the MSS service market. Within the satellite services market, so-called small LEOs operating at lower frequencies will primarily serve niche markets for data and locator services (e.g., global positioning), while big LEO operators (e.g., Iridium, Globalstar, Teledesic) will target the market for worldwide two-way mobile voice and data services, including rural and remote locations and less developed countries.

Table 4.4

Proposed MSS systems

Organization	Investors	Cost to Build	Service Date	Description
Iridium, Inc., Washington, D.C.	Include Motorola, Sprint, STET, Bell Canada Enterprises, and Daina Denden	$3.4 billion (includes launch)	1998	66 LEO satellites to link handheld wireless phones with global reach
Odyssey	TRW, Teleglobe	$2.5 billion	1998	12 MEO satellites
Inmarsat, London	Inmarsat, a treaty-based co-op of telecommunications operators from 73 countries, or a privatized Inmarsat spin-off	$2.6 billion	2000	"Inmarsat-P" system still undefined but leaning toward 12 MEO satellites to link handheld wireless terminals
American Mobil Satellite Corp., Reston, VA	Include Hughes, McCaw Cellular, Mtel, and Singapore Telecom	$1.2 billion	1994	3 GEO satellites to link handheld phones in North America
Ellipsat International, Inc., Washington, D.C.	Include Mobil Communications Holdings, Inc., Fairchild Space & Defense, and Israeli Aircraft Ltd.	$700 million	1997	16 elliptically orbiting LEO satellites for service to handheld terminals

---

33. For a discussion of these systems, see G. Staple and R. Frieden, "The New Space Race" (*Infrastructure Finance*, June/July 1994), especially the table on p. 54.

Organization	Investors	Cost to Build	Service Date	Description
Globalstar, Palo Alto, CA	Include Loral, Qualcomm, Alcatel Deutsche, Aerospace, Air Touch, Vodafone, and Dacom	$1.8 billion	1998	48 LEO satellites to provide worldwide voice, data, paging, and facsimile
Teledesic, Kirk-land, WA	Include Craig McCaw Development Co. and William Gates	$6.3 billion	2001	840 LEO satellites to provide global coverage for broadband data, video, and voice service

Motorola, a major player in wireless network equipment and consumer terminals, is the driving force behind the Iridium LEO system and will be in a good position to link up to other ground-based wireless access systems or the PSTN whenever complementary joint service opportunities arise. However, at a preannounced price of $3 per minute, it is clear that Iridium is not a mass market substitute for land-based wireless access service in the NII. The Iridium handset itself is very expensive, at an estimated $3,000 [17]. Planned for service in 1998, the Iridium system includes about 66 LEO satellites orbiting about 500 miles above the Earth and operating in the 1.5–2.5 GHz band, at a launch cost of $13 million each. Motorola's Iridium system is unique in that it will utilize satellite-to-satellite links to transit traffic between user locations, bypassing terrestrial networks in transiting countries. Iridium will even be able to transmit directly to user handsets, but it will usually make use of its domestic "gateway" cellular providers' terrestrial network for call terminations or originations.

Most other proposed systems have somewhat less ambitious plans than Iridium and plan to utilize the existing facilities of terrestrial carriers. For example, Globalstar's LEO system is planning to augment land-based wireless access systems using 48 satellites and 200 Earth station gateways. Globalstar's handset costs are estimated at $700. American Mobile Satellite Corporation, a GEO system, plans to operate a relatively inexpensive system of dual-mode satellite/cellular mobile service covering only North America and has announced target prices that are among the lowest. Based on the early price announcements, many MSS firms will be much more price competitive than Iridium. Target per-minute usage charges for these MSS systems range anywhere from about $0.25 to $2.00. Competition should force those prices to come closer together, probably somewhere in the middle of the estimates. Handset prices also will vary at first, but competition should also force some convergence.

Among planned MEO systems, Inmarsat, the international satellite consortium providing telecommunications service for shipping and airlines, has announced the introduction of a new personal satellite phone service called Inmarsat-P, available in the 1998–2000 time frame. Odyssey, a satellite system backed by TRW and Teleglobe, recently announced a two-way global MEO network consisting of 12 satellites orbiting about 6,000 miles above the Earth. This system would also plan to compete for mass market telephony services as well as niche market applications.

The Teledesic network backed by McCaw and Microsoft is an even more ambitious technological effort than Iridium. Operating in the very high frequency Ka band (20–30 GHz), these "birds" would be capable of providing global coverage for two-way broadband services, including video telephony and multimedia. Such projects are hugely expensive, however, and, while the potential telecommunications capabilities and applications of these superbirds is impressive, it is also still very experimental.

Recently, even more global satellite systems besides those listed in Table 4.4 have been announced. Spaceway, a new all-digital satellite system proposal before the FCC and made by the Hughes Communications division of General Motors, is an MEO Ka-band wireless access system. The proposed system, consisting of 17 satellites, would be designed to provide bandwidth on demand for all types of narrowband and broadband telecommunications services in competition with land-based alternatives. A novel feature of this system, which allows for spectrum reuse, is that an individual satellite will use transponder "spot beams" to segment the very large signal coverage area (or footprint) normally provided by geosynchronous orbit birds. Subscribers would be connected with so-called *ultra small aperture terminals* (USATs), measuring only 66 cm across and costing less than $1,000.

With so many grandiose announcements from so many deep-pocket investors, it is safe to assume that some, perhaps most, of these global satellite communications systems eventually will become operational (though some industry consolidation is likely). The investment community views the future as risky, so attracting external financing has not been easy. Two of the leading contenders in the race to deploy satellite systems, Globalstar and Iridium, have both failed recently to attract investor interest in recent bond offerings, even at fairly high coupon rates [18].

In addition to approximately 320 communication satellites already operating, satellite networks providing a wide variety of services will become a ubiquitous public infrastructure. Pelton [19] provides estimates of revenues for global satellite service markets, which are forecast to more than triple by the year 2002. Nevertheless, even after considering the pronouncements of the major industry players, satellite services will be relegated to serving niche market applications; therefore, their role in the American NII will be limited. Perhaps the greatest potential for the new global satellite systems would be to take advantage of their relative cost performance and coverage capability to provide modern digital telecommunications service in rural, remote, or otherwise undeveloped parts of the world.

4.9 EVALUATING NETWORK COSTS

A comparison of the economics of various alternatives for wireless access systems requires an examination of the time path of the expenditure stream against the anticipated revenues. The focus here is on that portion of the expenditure stream that reflects the capital costs of building a wireless access network system. These costs come in several different flavors: (1) so-called first costs, or the total installed costs of the initial wireless access system upon activation; (2) build-out costs, or the costs incurred over time to expand the system coverage area to its long-term target; and (3) system growth and maturation costs, or the variable costs that result from rising system usage. In the case of existing wireless access systems, there is also a difference in the costs to upgrade or otherwise modernize the system to handle new service capabilities compared to the costs associated with building a system from scratch.

The third item, the variable costs of operating the system to handle increased demand, is actually the most critical since it is the determining factor for a company's long-term operating cash flow or price/cost margins. Of course, that assumes that the up-front fixed (e.g., startup) costs of building a particular network system are not so much higher than other competing systems that the project would never get off the ground. But this is not likely when alternative system costs are compared on a per-subscriber basis for a large-scale urban market. In that case, the high up-front fixed costs are spread over so many demand units that the average fixed cost represents a very small portion of the average total cost (the sum of average fixed and variable costs).

The goal of economic analysis is to identify and design the wireless access system that achieves the lowest investment in network facilities for a given demand level (assuming the level of service quality is a competitive one). This usually means that, for a given market area, a network design is selected that provides area coverage for the least amount of network facilities. The network is engineered in accordance with technical network parameters (RF spectrum bandwidth, radio carrier channel size, user channel size, co-channel interference factors, frequency reuse patterns, etc.) corresponding to a particular technology (e.g., TDMA/CDMA) and network architecture (e.g., macrocell/microcell). Depending on the market area (e.g., city) to be studied, a geographic terrain and climate are assumed (e.g., flat, hilly, rainy, dry), along with assumed levels and distributions of man-made RF interference factors (e.g., traffic patterns and loads, buildings). A subscriber density must also be assumed (e.g., subscribers per square kilometer and calls or call attempts per hour).

Based on the size of the radio coverage area, the network startup or initial construction phase includes investments in the core network hardware and software represented by the MSC and the associated trunk network connecting to the initial number of base stations deployed. BSTs are placed to prevent unacceptable signal fading and signal propagation associated with geographic topology (e.g., lakes, rivers, hills, valleys, trees) and other physical RF barriers (e.g., buildings, tunnels, bridges). Any number of problems are associated with the lack of line of sight for the RF signals between the base stations and handsets, and considerable engineering discretion is used in solving them in any specific instance. For example, when a large building or other structure blocks a given radio transmission path, the problem may be handled by the placement of an extra radio antenna on top of a building or along a section of street to go around or even under the building by transferring the signal to underground wireline facilities.[34]

Once the network system operating parameters and assumptions have been developed for any given market area and the network has been engineered, the vendor equipment can be sized and priced to estimate the initial or first cost for building the network. First cost is also called the *engineered, furnished, and installed* (EF&I) system cost and represents the total cost of "turning up" a network system. By assuming an initial market penetration rate, the relative cost per subscriber for different wireless access systems of similar service capability and service quality can be determined.

In the case of satellite networks, the EF&I costs of satellite development and launch dominate the first costs of the system (or transponder lease costs), followed by Earth station siting and construction costs. By their very nature, the initial capacities of satellite systems are huge. During the build-out phase for satellite systems, the per-subscriber system costs fall even more rapidly than those experienced by land-based systems because average costs for satellite networks are more sensitive to the scale of operations. In most metropolitan land-based wireless access systems, the per-subscriber system costs level out relatively early compared to satellite systems (e.g., 50,000 versus 1 million subscribers). This makes it imperative for satellite operators to sign up as many subscribers as possible through advance marketing programs. This is the opposite of the situation for most land-based systems, which often are more concerned with keeping up with demand early in market rollouts. In both land-based and satellite-based wireless access systems, once system build-out has been reached, the variable capital cost of adding individual subscriber connections is quite low.

A further evaluation of the EF&I costs of different wireless access systems can be made by holding constant the total available RF spectrum and the size of

---

34. For a discussion of common modeling assumptions regarding natural and man-made RF interference factors, see W. Lee, Mobile Communications, and Steele, Digital Mobile Communications.

the service coverage area (using the same assumed levels of terrain and man-made interference factors) and then systematically varying the subscriber density. That will reveal how different systems (e.g., CDMA/TDMA, macro-cell/microcell, ESMR) perform for dense urban applications versus less dense suburban and rural applications.

The analysis can become considerably more complex by combining different wireless access technologies in all or certain portions of the radio coverage area (e.g., wireless multimode systems using both CT and cellular technology). Furthermore, due to advances in digital signal coding and compression techniques, directional antenna placement, and sophisticated variable powering of handset-to-base-station signal strength to account for near/far conditions, the capacities of most wireless access systems are constantly being improved, usually resulting in reduced per-subscriber system costs. The different combinations of the various methods available to simultaneously increase system capacity and lower unit costs make it hard to distinguish definitively which type of wireless access system can achieve the highest capacity and lowest cost per unit of available RF spectrum. Different types of wireless access systems have different methods of channel access and utilization, different power levels, frequency reuse patterns, and co-channel interference factors, all of which affect the overall economics of system construction.

In another stage of the cost analysis, by systematically increasing the available spectrum per coverage area, there is the possibility for increased channel spacing and less concern about controlling co-channel interference, which adds to system costs. It is useful to examine the trend in cost per demand unit for increments in available spectrum, including an examination of the resultant per-subscriber costs for increasing levels of subscriber density and penetration with and without the possibility for increasing the available spectrum.

The entire study process would yield an evaluation of the relative cost and efficiency of spectrum use at various levels of system utilization. While such an approach in the abstract would clearly be preferred before the FCC decided on its spectrum allocation and licensing scheme, it cannot happen that way in practice because the performance characteristics of the technology itself are so fluid. It is simply not possible to wait for the "right" wireless access method to come along before licensing spectrum since no one really knows what the right one is. For example, several years from now, further advancements in so-called spread spectrum and broadband wireless access techniques (e.g., CDMA) may reveal that the FCC's current spectrum licensing scheme of 30-MHz blocks and 10-MHz blocks, up to a total allowed 40 MHz per market area, may not have been enough to maximize efficient bandwidth utilization.[35]

---

35. The extreme example being the advent of the digital broadband radio, which Gilder likens to an electronic eye, capable of scanning a wide range of radio frequencies in real time to ensure

4.10 ECONOMICS OF WIRELESS ACCESS

The engineering and capital budgeting analysis for prospective wireless access systems involves considerable effort and numerous assumptions about some very young technologies, all in the presence of uncertain future demand. The competitive environment and the FCC's continuing spectrum auctions have raised the stakes considerably for would-be wireless access network providers to decide now which technology to select for a market rollout. Consequently, detailed and specific engineering and financial analyses being performed in the industry are being held close to the vest. However, based on publicly available data (including those from investment houses in their efforts to calculate prospective market penetration rates and net cash flows to establish valuation benchmarks for the investor community), indications are that the state of the art in engineering economics and financial modeling of network systems is not very far along.

There are several reasons for that. First, there is the "cart before the horse" problem. The FCC set spectrum allocations and licensing schemes before the technology of digital wireless access has progressed to the point of it being known how much spectrum should be allocated to narrowband and broadband wireless access services. The fact that the technology is so fluid, coupled with the deadline for spectrum auction bids, puts a tremendous amount of pressure on industry players to commit now to a given wireless access technology and network architecture so that financial modeling can precede the spectrum auction awards.

Consequently, prospective wireless access system operators have had to contract with one or another equipment manufacturer to obtain bid prices for the new (and, in some cases, untested) technology in advance of the development of production equipment. That has led most major players to set their stakes in the ground based on one preferred technology and/or equipment vendor, rendering moot the issue of analyzing the costs of alternative systems.

While it is still possible to pursue financial analysis to evaluate the relative costs of different network configurations within a chosen technology, it occurs in a much more limited context than a full evaluation across technologies. Given the FCC's announced spectrum policy, coupled with the fact that a technology choice must be made relatively quickly, the industry's network models and financial analyses are being conducted in a rather unsystematic fashion.

In the economic and financial phase of the analysis, the network engineering design is now ready for application to a dynamic capital budgeting plan in a business case setting. Once the static cost of initial construction is combined with an analysis of the incremental costs of the system build-out over time, a

maximum efficient utilization of available spectrum. See G. Gilder, "Telecosm: Auctioning the Airways" (*Forbes* ASAP, April 11, 1994).

dynamic picture of the stream of expenditures associated with a given wireless access system is sufficiently developed so an informed decision can be made about committing investment dollars to the construction program.

Initial system costs for wireless access network construction for land-based systems are dominated by investment in siting and constructing the network nodes, especially MSCs and BSCs, related hardware and software, and the trunk network required to aggregate and "backhaul" subscriber usage to the BSC and MSC. After initial system construction, the cost drivers associated with system growth during build-out are the addition of transceivers (e.g., BTSs) and trunking facilities to expand system coverage and capacity incrementally.

Once build-out has occurred and the system has matured, operating and marketing expense factors dominate. Usage-based interconnection charges paid to the PSTN operator will likely be a significant cost driver during both the growth and the maturation phase. Bypassing the local telco network (for example by interconnecting to a competitive access provider or long distance carrier) may be a way for a wireless carrier to avoid paying the high rates for PSTN access on the originating end of a call, but it is not so easy on the terminating end of a call where there is no way of knowing where the calls are going to terminate *ex-ante* on the PSTN.

The expenditure streams associated with the three primary phases of wireless access development (startup, build-out, and maturation) can be estimated according to the time path of forecasted demand. The demand forecast is based on pricing assumptions. Because it is so difficult to forecast market penetration rates over time and total demand levels at any future point—especially in what is arguably going to be a highly contentious market due to the number of participants—sensitivity analysis to account for forecasting error is crucial. Sensitivity analysis involves randomly changing the initial demand assumptions over a range of possible values to be able to judge the potential for forecasting error to affect prospective cash flows.

Returning to the dynamics of system costs, it is interesting to note that when initial construction and build-out of AMPS cellular systems began in 1984 the per-subscriber costs were very high at first, at $2,000–$3,000, and fell rapidly thereafter, leveling at about $700–$1,000 per subscriber, with very little marketing expenses. After only 10 years of AMPS cellular systems being in existence, competition for customers has become fierce, with the marketing expense per new subscriber now almost equal to the total amount of current capital costs per subscriber, about $700 (making the total cost of a new subscriber about $1,400). Thus, even before the AMPS market has matured (it is growing rapidly), the nature of the business has already been transformed from one of simply keeping up with demand to one of actually vying for demand.

AMPS subscribership is still rapidly expanding (51% last year). But system capacities, many of which have been increased through the use of FDMA/TDMA techniques and the partitioning of cells into sectors, are generally able to handle the rising demand with little additional capital cost. That has created some very high cash operating margins from the current base of cellular subscribers. The cellular experience buoys the financial outlook for future wireless access systems, which are actively seeking investment dollars to build new networks.

Since new digital wireless access networks have the same fundamental cost structure as AMPS-D or GSM digital cellular systems (see Figure 4.1), the per-subscriber costs of new ESMR, macrocellular and microcellular systems are expected to track along a similar time path as system construction and build-out occur, although at a different level, depending on the specific features and costs of different types of wireless access systems.

4.11 CRITIQUE OF THE APPROACH

To date, the financial modeling of wireless access systems has focused almost entirely on static calculations of per-subscriber capital costs of the stand-alone wireless network. There would appear to be at least two areas of network and financial modeling that could use substantial improvement: (1) the common assumption that all subscribers (and their associated network costs) are alike; and (2) the lack of consideration of shared trunking alternatives, including wireline network interconnection. These issues need to be addressed for a full evaluation of the prospects of wireless alternatives.

Regarding the first point about static calculations of per-subscriber average costs, there needs to be more emphasis on dynamic process models based on the pattern and level of network usage, not on an "average" subscriber. A model based on usage would better describe the underlying network engineering relationships between network components and how they vary with growth in usage. There is at least one such model, but it has not yet been applied to actual data in the United States [6].

In other words, the network model should be able to answer the following basic question: As peak network usage grows, what is the incremental cost of handling that growth for each major network component (such as BSC, BTS, and trunking)? In contrast, current models focus on a different, but related, question: As subscribers are added to the network system, what is the average cost per subscriber? The answer to the second question may be useful, but it is much less instructive than the answer to the first one.

The efficiency of a wireless access system to handle demand growth is best measured by incremental capacity costs caused by network usage, not the average cost per subscriber. Once a wireless access network system is built, the primary cost drivers are the additional network facilities required whenever system capacity is strained by additional usage. For any given cell site, certain system components will be exhausted due to capacity constraints, causing the placement of additional antennas, transceivers, and associated trunking facilities. When cell sites themselves are exhausted, cell coverage areas are reduced to expand frequency reuse, causing new cell sites to be placed. It is expensive to equip entirely new cell sites. This explains the dynamic cost structure of wireless access systems.

Existing network and financial models are static and tend to focus on spectrum and network capital costs per subscriber or per population (in the industry jargon, "per pop") for discreet levels of market penetration. Thus, the focus is on primarily fixed and sunk costs of system startup. In reality, ongoing network cost drivers, which are important for determining operating cash flows, are based on two primary considerations not usually reflected in existing cost models. For the incremental cost of expanding area coverage and the incremental cost of usage, the per-subscriber and per-minute costs of the latter are quite different and distinct from those of the former. It is the time and spatial distribution of the frequency of call attempts and the calls themselves during busy periods that cause costs to be incurred. For example, the MSC is a computer that controls network usage, assigns frequencies, adjusts power levels, and controls call hand-off. In the case of calls from or to roaming units (meaning away from the home base station area), more work is involved to complete calls because the MSC must interact with a network database and an intelligent network system, which may or may not be located at the MSC site. The remote transceiver sites similarly must transmit calls between the handsets and the BSC, using subscriber radio channels and trunking facilities.

All the major components of wireless access systems have an operating capacity that is sensitive only to peak period usage; it is the exhaust of the available capacity that defines the trigger point for incurring additional network investments necessary to relieve that exhaust. Thus, it would be useful to view the cost of the total network and its major components as varying with usage levels. Contrast that to the common approach of current network models that assume an average usage level (in industry jargon, erlangs per subscriber) and then assume that as subscribers are added, network usage increases exactly in proportion to the existing base of subscribers. In addition, the assumed amount of usage per subscriber is a small fraction of that used in standard wireline models of the telcos and is usually based on what is known about mobile cellular subscriber usage.

This is somewhat unrealistic. What is known from the mobile cellular experience is that early subscribers tend to be heavy users of the service because they value it more and are willing to pay high prices and can afford higher total phone bills. Later subscribers joining the system during system build-out value the service less, are willing to pay less, use it less, and tend to roam less. That all network costs are usage sensitive and that different users have different usage patterns cannot be reflected in the type of broad averages assumed in current studies. A richer analysis would build costs from the bottom up by taking usage and roaming costs and assigning them to types of users. User demographics (e.g., high use/low use, roaming/not roaming, moving fast/moving slow) naturally varies from one market area to another or even within market areas by BTS location. Models based on actual usage characteristics would be better able to reflect the effects on system capacity and costs from adding subscribers and/or calls. Hence, to the extent that there is a difference between usage and subscription rates, the former should be tied to the demand forecast, which drives the economic cost model in a business case.

Furthermore, the use of an average historical usage rate per *mobile* system subscriber would not be expected to be representative of the actual usage one would eventually expect from an average wireless *access* system subscriber. Wireless access will be cheaper to use and more versatile than mobile access; therefore, per-subscriber usage will be higher. Because wireless access is suitable for all modes of portability, it is more useful and convenient in both portable and stationary situations compared to cellular mobile service. Eventually wireless access is going to become a substitute for fixed wireline telephone service. This would call for assumptions of higher usage levels than those being assumed in current cellular models, but somewhat lower than monthly network usage levels associated with flat-rate local telephone service. The reason is that wireless access systems will offer more features and similar quality but lower prices and more convenience than mobile cellular systems.

In fact, it is entirely possible, if not probable, that eventually wireless usage levels per subscriber would actually grow to levels higher than that associated with current local telephone service. The reason is that the added convenience of communicating with anyone, anywhere, anytime, would increase the overall propensity to communicate. It is well known that telephone usage begets more usage: How many times do you play telephone tag or need to follow up on a call? That is some time away, however, if wireless access network operators plan to charge for usage and do not offer flat-rate options, as local telephone companies do. Consumers like flat-rate options for local phone service and have experienced many decades of satisfaction with it. Flat-rate wireless pricing may already be getting started; the first digital PCN operator (Mercury—UK) has a zero usage price in off-peak periods.[36]

To summarize the point, the focus of current-cost models on per-subscriber capital costs requires a host of somewhat unnecessary assumptions. Fundamentally, the primary cost drivers of a wireless access system are based on usage. Changing the modeling approach to capture and reflect the costs of increasing capacity incrementally on the network system would yield a much more realistic operating scenario for capital budgeting and business case analy-

---

36. Such pricing has its problems. Over two years of advertising its zero off-peak rates for usage, demand soared and Mercury had to install additional network facilities to handle the (non-revenue-generating) traffic.

sis. In this costing approach, a clearer picture of the cash flow from wireless system operations develops. Increasing demand for wireless access and usage translates into an increase in certain portions of the engineered capacity of the system (e.g., advancing the placement of BTSs, expanding capacity of traffic aggregation and trunk and backhaul facilities) and increases revenues incrementally as well.

Another area for improvement in wireless access system models is to model explicitly the cost of PSTN network interconnection and shared trunking arrangements. The cost of PSTN interconnection could be incurred per minute or per interconnecting trunk and should be included in any financial analysis since it will be, in most cases, an unavoidable incremental cost of usage growth, whether for call originations or terminations. That raises an important strategic issue for wireless network modeling. If a wireless access system operator must incur interconnection costs to the PSTN, why not plan to interconnect in the most convenient and cost-minimizing way? Very little explicit modeling of local PSTN joint service arrangements has occurred to date, but it could be an important source of cost savings to new network operators.

A primary driver of incremental cost for wireless access systems involves aggregating and trunking traffic among remote radio nodes (BTSs/BSCs) and between those nodes and the central nodes (MSCs). There are also the network control functions, which may require trunking to and from a centralized database. Instead of the standard assumption of a stand-alone wireless access network system, including trunking facilities, why not consider as a strategic alternative the sharing of network facilities owned by incumbent wireline carrier networks, like telcos and cablecos? Interconnecting to and leasing capacity on the ubiquitous intelligent networks employed by PSTN operators or other *competitive access providers* (CAPs) have the potential to reduce substantially investment costs in stand-alone facilities of the wireless access network.

4.12 PUBLIC POLICY FOR WIRELESS NETWORKS IN THE NII

A number of public policy implications flow from the preceding discussion and analysis in key areas: NII market structure and spectrum allocation, network compatibility standards, interconnection and access pricing, common carriage, and universal service.

4.13 MARKET STRUCTURE AND SPECTRUM ALLOCATION

The Clinton administration's stated objective for the NII is to have a competitive market as the vehicle to drive investment in the telecommunications sector. The FCC has certainly followed suit by allocating RF spectrum to foster at least three major players in the market for so-called broadband PCS wireless access services. This is in addition to new and expanded allocations to true wireless broadband service providers such as wireless cable and satellite systems.

Whether intended or not, the FCC's spectrum allocations of up to 40 MHz for individual licensees of PCS services effectively preclude them from the two-way broadband services market. If wireless is to someday serve the mass market for multimedia or video telephony, it will have to come from wireless cable and satellite service providers or some combination of these and other land-based systems, perhaps coupled with inhome wireless systems using unlicensed spectrum (e.g., infrared). As wireless technology progresses and as the government can be convinced to let go of more of the fallow frequency spectrum, the role of wireless access may be expanded considerably over that already planned with PCS networks.

The FCC can facilitate the process by extending its newly found flexible-use policies beyond the relatively small amount of PCS spectrum to a much wider range of spectrum encompassing existing licensed bands, starting with those broadcast frequencies that appear to have greatest potential for two-way service in a digital environment (e.g., wireless cable) and those that are under-utilized (e.g., UHF TV). Revisiting the reasonableness of old licenses and the old spectrum endowments could not only bring more money into the government coffers, it would also expand competition and investment in the NII. In adopting its flexible-use rules for PCS and allocating unlicensed spectrum at no cost to new service providers, the FCC has begun to move down the right path. Let us hope that it will continue the journey.

4.14 NETWORK COMPATIBILITY STANDARDS, INTERCONNECTION, AND ACCESS PRICING

Critical to the success of the NII and the role of local wireless access services in the NII is its ability to offer convenient nationwide calling capability. Wireless access systems could someday provide the ability to call anyone, anywhere, anytime. Similar to what has already occurred for narrowband ISDN standards, national and international coordination of network compatibility is crucial to the success of a technology and a public infrastructure. Industry players must agree on rules for governing both the wireless network interface and the user network interface to the PSTN. The government's role is to establish a fair process to see to it that the industry sets a reasonable standard in a reasonable period of time. It is the voluntary nature of standards setting and the compliance process that will minimize the risk of adopting an inferior standard or having no standard at all.

Pricing for network interconnection and access to the PSTN must be nondiscriminatory and competitively neutral. During the transition to full competition in all aspects of the PSTN, regulations regarding cost-based, nondiscriminatory tariffs for PSTN interconnection are essential to ensuring a level playing field for entrants and incumbents alike. If such rules are developed and enforced, there is no reason to restrict in any way competition between incumbents and entrants. The FCC's licensing of wireless PCS and broadcast spectrum allocations are biased against incumbent operators so that direct competition for local telephone service and television will develop. This should be a temporary measure until nondiscriminatory pricing rules for PSTN access and interconnection are adopted. Otherwise, legitimate economies of scope from technological integration of network operators in the NII may be unduly delayed or foregone altogether, to the ultimate detriment of consumers.

The cost of new wireless technology is driven primarily by the portability demands of the calling party and secondarily by the requirements of locating the called party wherever that party is. That means the success and the cost of achieving portability critically depend on network interconnection. Even when the called party is not on the move, wireless network interconnection to the PSTN is critical to successful call completion.

Since new wireless access systems are predominately competitive local operations providing services to the public for random call originations, it will be difficult to successfully avoid paying for call terminations on the PSTN because it simply cannot be known where the calls are going to end up. Bypassing the local PSTN operators for call terminations to avoid paying network access charges has always been problematic, even for major national long distance companies. It will be a long time before the various competitive wireless access companies will be able to successfully piece together national bypass arrangements on both the originating and the terminating portions of calls. This situation would require that most Americans use wireless access and that there is close service coordination among what are ostensibly competing local companies. While some national wireless consortiums with national spectrum licenses will claim to be able to provide seamless national service, there will invariably be a need for local interconnection for some (probably most) calls.

Depending on future regulatory rules concerning pricing for interconnection, PSTN access charges potentially are substantial. The imperative of the administration's NII policy—that wireless or other private networks interconnect or are otherwise compatible with one another and the PSTN—is well founded. The cost and price of that interconnection within the context of the NII have yet to be directly addressed. If the government truly wants to solve the interconnection problem for new wireless access operators, it will require some creative plans to gradually reduce the PSTN interconnection tariffs. A system of cost-based rates for PSTN interconnection will substantially improve the financial prospects of new competitive wireless access networks and, at the same time, level the playing field between incumbent local telcos and new entrants. The transition to nondiscriminatory cost-based PSTN interconnection charges will not be easy because it involves reforming the current system of cross-subsidies to basic local exchange services, but the process must begin soon to eliminate artificial barriers to entry to new technologies like digital wireless access.

The most obvious economic solution to achieving both a competitive market for local telephone service and low-cost interconnection would simply be for the government to quit regulating local market entry and, at the same time, deregulate rates. This would start a chain reaction in the market that would begin to solve both the problem of how to increase local telephone competition and lower PSTN interconnection costs. Basic local phone rates would rise to at least a cost-compensatory level (perhaps capped by regulators at that point), thereby attracting more local market entry, which in turn would stimulate bypass and competition for local interconnection, thereby keeping its cost down as well. At that point, the main issue remaining for the government to achieve the vision of the NII is how to protect universal and affordable access to the new competitive infrastructure.

4.15 COMMON CARRIAGE AND UNIVERSAL SERVICE

The goals set for the NII hinge on principles of common carriage and universal service. Normally, the FCC forbears from regulating private radio networks, instead treating them as private contract carriers. However, common carriage is implied for new wireless access network operators because of the FCC's rapid network build-out requirement for area coverage in accordance with the terms of the license to use the spectrum. What remains more problematic from a policy perspective is the lack of a related universal service requirement. In other words, even if new wireless access networks provide the area coverage required as a condition of their license, there is still no obligation to provide service to everyone or to provide it at regulated prices. Indeed, the FCC's own new flexible-use policies provide new wireless system operators the freedom to use their system capacity for services targeted to only businesses or other lucrative niche markets within the coverage area, thereby totally ignoring the mass market of residential subscribers. In such situations, a sort of red-lining could occur due to private market incentives to discriminate in the name of profit opportunity rather than any conscious avoidance of serving certain neighborhoods.

Universal nondiscriminatory access to the PSTN is part and parcel of the tradition of regulated common carriers in the United States. On the other hand, private contract carriers like cable television companies and wireless systems have neither the obligation nor the inclination to provide service in very thin rural and remote locales. The available cost data indicate that the financial health of both wired and wireless access systems is strongly and directly related to subscriber density. That is not true, however, for satellite systems, which depend more on total system demand without particular regard to where the demand is coming from. Thus, satellite systems of the future may be well suited to provide universal coverage in rural and remote areas because they do not feature the very high subscriber connection costs that land-based network systems do. Within the context of the NII, it remains a matter of public policy as to whether the level of service via two-way digital satellite systems for rural and remote areas is acceptable and comparable to the level of service provided by land-based urban systems.

In light of this and the fact that the NII policy generally prefers private market solutions to public assistance programs, perhaps the FCC should consider a rural area policy that provides certain benefits to those network operators willing to serve remote and rural subscribers that otherwise would not be able to obtain access to the NII without a government subsidy.

In the case of telephone companies serving rural areas, the FCC typically relaxes rules restricting PSTN operators to allow them to provide wireless services within their monopoly local service areas by granting them waivers to use spectrum normally reserved for competitive entrants or to use spectrum normally reserved for other uses but that lie fallow in rural areas.

If the current state of cellular mobile service in rural areas is any indication, the FCC may need to do more. For example, extending spectrum rights to regional licensees serving metropolitan areas would encourage them to extend their coverage area, perhaps in conjunction with the rural PSTN operator using toll connect trunks back to the urban center. Beyond allocating more spectrum to rural radio services, the FCC could tailor its system powering restrictions to meet the needs of rural operators. Radio system interference is less likely in rural areas than in dense urban areas. An increase in the allowed power levels of rural radio systems would increase the coverage area per antenna site, thereby improving the financial viability of rural wireless systems.

Barring success with such policies, the government, as a last resort, may choose to subsidize PSTN network upgrades in rural areas under a related NII initiative.

4.16 THE POLITICS OF THE NII

The important message for public policy is that, until the service requirements of the universal NII have been specified, the question as to which is preferred, wireline or wireless access service, cannot be answered. If, as many believe, the NII involves only socially efficient access to narrowband digital voice and data services, then digital wireless technology is preferred for dedicated subscriber connections to the wireline intercity PSTN. Notwithstanding the fact that wireless access costs are lower, the real bonus for the consuming public from this scenario is portability.

If, however, access to broadband service, especially bandwidth-on-demand type access service, must be added to the narrowband service mix for the NII, then wireline access technology is likely to be the winner in the race for preeminence in the future NII.

An interesting irony flows out of this conclusion: Acting in their own business interests, wireless access network providers of all types, narrowband and broadband (e.g., wireless cable and satellite services) would not want to back a definition of service for the NII that included broadband capability. If they did, the long-term winner in the race to be the infrastructure network provider is likely to be wireline access.

By promoting a narrowband access infrastructure, narrowband wireless network operators would be the least-cost alternative, and digital wireless broadcast networks would also be the least-cost alternative for the traditional (huge) niche market for one-way video service.

Thus, if the social cost of infrastructure is the issue for the NII, and if policymakers envision bandwidth on demand as a long-term infrastructure imperative, integrated two-way broadband services are best provided by wireline operators (e.g., cablecos and telcos). In that scenario, even though the role of wireless access services in the NII is not a dominant one, the indisputable convenience aspects of portability coupled with the affordability of new wireless technology will ensure that the mass market will still be served by the interconnected adjunct networks of wireless access operators.

That conclusion leads to another interesting twist for the public policy stance of the wireless industry regarding the NII. By voluntarily opting out of the government NII juggernaut, wireless network system operators may actually be selecting the right path. After all, the NII concept implies government interference in such critical areas of universal service and so-called carrier of last resort obligations, common carrier regulations for pricing, standards, and network interconnection, none of which applies to private contract carriers, which is what many new wireless carriers are planning to be. Since wireless technology has inherent cost and market advantages (e.g., portability, convenience) over its wireline counterpart, its importance in future consumer markets is virtually ensured, and relatively little may be gained by the wireless industry becoming one of the tools of the federal government's regulatory competition policy in the NII. New digital wireless carriers also run the risk of encountering burdensome state regulation if they are similarly used by state governments as a tool to bring competition to the market for local telephone service.

The bottom line for wireless technology, whether preferred by policy makers for the NII or not, is that it will be around and will develop and thrive in the mass market. Considering this inescapable conclusion, and considering that the private sector tends to be distrustful of government involvement in an otherwise competitive business, wireless network operators of all stripes might consider it a blessing that they are not tagged as the vehicle for driving onto the public information superhighway.

References

[1] FCC Report No. DC-2586, "FCC Seeks Comment Regarding Allocation of [200 Mhz] Spectrum Below 5 Ghz Transferred from Federal Government Use," April 20, 1994, Notice of Inquiry (FCC 94–97).

[2] Uddenfeldt, J., "The Evolution of Digital Cellular Into Personal Communications," *Proc. Sixth World Telecommunication Forum*, ITU, Geneva, October 1991.

[3] Lee, W., *Mobile Communications Design Fundamentals*, New York: Wiley, 1993, p. 317.

[4] Reed, D., "Putting It All Together: The Cost of Personal Communications Services," FCC OPP Working Paper No. 28, November 1992 and CTIA statistics (1994).

[5] "Wireless Architectural Alternatives: Current Economic Valuations Versus Broadband Options, The Gilder Conjectures," draft, Telecommunications Policy Research Conference, Solomons, MD, October 1994.

[6] "Costing Digital Cellular Networks," draft, INDETEC Inc., Del Mar, CA, 1994.

[7] CTIA statistics (1994).

[8] McGarty, T., "A Precis on PCS Economics and Access Fees," draft, NPC SC Seminar, MIT Lincoln Laboratory, Lexington, MA, May 18, 1994.

[9] Dondl, P., "Standardization of the Satellite Component of the UMTS," *IEEE Personal Communications*, October 1995, pp. 68–74.

[10] "The Trouble With Cellular," *Fortune*, November 13, 1995, p.186.

[11] Semmoto, S., "PHP—A Better Solution?" *Annual Review of Communications*, International Engineering Consortium, 1993–1994, Vol. 47.

[12] *Business Week*, September 12, 1994.

[13] Gerard, Klauer, Mattison & Co., "Wireless Cable Industry Report," New York, 1995, pp. 10–11.

[14] Nordberg Capital, "Third Generation Television," New York, October 21, 1993, p. 13.

[15] Gerard, Klauer, Mattison & Co., "Industry Report," p. 8.

[16] Vivian, W., and A. Kreig, "Wireless Cable: Today's Low Cost, High Capacity Pathway Worldwide to Residences, Schools, and Businesses," Wireless Cable Association International, Washington, D.C., draft, July, 1995, p. 11.

[17] Staple, G., and R. Frieden, "The New Space Race," *Infrastructure Finance*, June/July 1994, p. 38.

[18] "Globalstar Offering Gets Cool Reception As Investors Balk at Mobile-Phone Plans," *Wall Street Journal,* October 2, 1995, p. A3

[19] Pelton, J., "Are Broadband Satellite Communications at Least a Part of the Answer?" Int'l. Engineering Consortium, Annual Review of Communications, 1993–94.

Comparison of Wireless and Wireline Network Systems

5

5.1 SYSTEM COSTS AND FUNCTIONALITY

The costs and functionality of wireless access systems can be compared to their wireline counterparts to assess their prospective roles in the future information infrastructure. Numerous studies detail the costs and capabilities of digital wireline access systems using fiber optic, coaxial, and copper cable.[1]

FTTH systems are the Cadillacs of wireline access systems because of the cost performance and virtually limitless bandwidth offered by an all–fiber optic system. As discussed in Chapter 3, FTTC refers to a wide variety of network systems. Wireline FTTC systems employ fiber optic cable in portions of the shared trunk network, connected to copper and/or coaxial cable to complete the connection between the subscriber's premises and the network node or switch. While there are many different FTTC systems employing a wide range of novel network architectures and proprietary features, all must conform to a generic interface to the PSTN.

5.2 SUMMARY OF WIRELINE NETWORK SYSTEM COSTS

A survey of wireline system costs on a per-subscriber basis is presented in Table 5.1. The costs shown are estimates of initial network construction costs. These *installed first costs* (IFC) costs are the costs required to procure the net-

---

1. See B. Egan, "Economics of Wireless Communications Systems in the National Information Infrastructure" (U.S. Congress Office of Technology Assessment, draft, November 1994); D. Reed, *Residential Fiber Optic Networks: An Engineering and Economic Analysis* (Norwood, MA: Artech House, 1992); L. Johnson, *Toward Competition in Cable Television* (American Enterprise Institute, 1994); and Hatfield Associates, "The Cost of Basic Universal Service," Boulder, CO, July 1994.

work system components, install the system, and make it operational. The cost estimates presented in Table 5.1 are the long-run *average incremental cost* (AIC) per subscriber, which, assuming that system capacity is fully utilized, is defined as the total project cost (or IFC) divided by the number of subscribers. In reality, due to the high upfront fixed costs of installing network systems, the average system costs per subscriber will fall as more and more subscribers come onto the system until the engineered system capacity is exhausted. Long-run AICs reflect the steady state of average costs per subscriber and are also a good estimate of the incremental cost per subscriber of system capacity growth. Chapter 6 provides a detailed discussion of state-of-the-art costing methods.

Table 5.1
Wireline System Costs per Subscriber

Type of Wireline Service	AIC per Subscriber
Base case current cost	
POTS: new telephone network access line	$1,000
POCS: new cable network access line	$700
N-ISDN	
N-ISDN telco access line upgrade	$100–200
N-ISDN upgrade, including digital switch placement	$300–$500
Mediumband digital service	
ADSL	$500–700
Fiber optic network access line upgrades	
Telco FTTH for POTS only	$3,000+
Future (1998–2000)	$1,000+
Telco FTTH (two-way broadband)	$5,000+
Future (1998–2000)	$2,000+
Cable network FTTH (N-ISDN + two-way broadband)	$1,500+
Future (1998–2000)	$1,000+
Telco FTTC for POTS only	$750
Telco FTTC for POTS + POCS	$1,350
Cable hybrid fiber/coaxial network (POCS only)	$50–100
Cable hybrid fiber/coaxial network (POTS + POCS)	$200–300

The costs for narrowband and broadband digital networks in Table 5.1 represent the per-subscriber cost of upgrading subscriber access lines for telephone and cable networks. CPE costs are not included. The costs presented are based on many industry sources and generally represent the consensus view. For purposes of comparison, Table 5.1 begins with benchmark estimates of current AICs for existing telephone (POTS) and cable television (POCS) networks using traditional analog technology. Note that these benchmark costs are *total* incremental costs per subscriber, not the incremental costs associated with a network upgrade.

As indicated in Table 5.1, telco access line upgrade costs for broadband FTTH or FTTC systems are much higher than those for narrowband systems (i.e., N-ISDN).

The cost estimates in Table 5.1 also include the mediumband technology ADSL. ADSL is a modem-based technology that uses sophisticated DSP techniques to increase the telecommunications capability of standard two-wire copper telephone lines. ADSL provides a two-way channel for narrowband digital telephony integrated with one-way mediumband service to support VDT and VOD services (technically a 1.5-Mbps downstream channel for single-channel VCR-quality video service). Applications of second-generation ADSL offer increased bandwidths up to about 640 Kbps upstream and from 4–19 Mbps downstream, enough for several digitally compressed video channels. ADSL subscriber connections will be limited to a distance of about 12,000 ft. Third-generation systems will offer even more bandwidth, but they will be limited to subscriber connections of very short distance.

Table 5.2 provides the AICs for wireless network systems. As in the case of the wireless access system cost estimates provided in Chapter 4 (except for satellite systems), the wireless system costs presented in Table 5.2 do not include the cost of CPE or the additional costs (in the case of all non-PSTN network operators whether wireline or wireless) represented by payments to local telcos for interconnecting to the PSTN to achieve ubiquitous service capability. The current (substantial) prices charged by incumbent local telephone companies to other carriers for PSTN interconnection (so-called access charges) can easily dominate the ongoing costs of operating a new digital network system. Thus, per-subscriber capital investment costs notwithstanding, the level of PSTN access charges can drastically alter the financial prospects (if any) for new digital network operators. Data generally are not available for the current or forecasted costs that must be borne by wireless operators for PSTN interconnection.

Table 5.2 summarizes the per-subscriber wireless access system costs presented in Chapter 4. In a comparison of the relative costs and effectiveness of wireline (Table 5.1) and wireless (Table 5.2) access systems, certain major inherent differences must be taken into account. First, on the wireless side, there is the unambiguous advantage of portability, which simply is not possible with the wireline alternative.

Table 5.2
Wireless System Average Incremental Cost per Subscriber

Wireless System	Capital Costs per Subscriber
Current AMPS	$700–1000
PCN/PCS	
Macrocell environment	
AMPS-D (TDMA)	$300–500
CDMA	$350 (for urban system with over 50,000 subscribers)
Microcell environment	
TDMA	$500
CDMA	$500
Wireless cable	
MMDS (television only)	$350–$450 (50% CPE and 50% installation)
	$525 ($380 reusable if subscriber discontinues service)
LMDS (television only)	$40 (cost per urban home passed)
	$110 (cost per suburban home passed)
	$700 (CPE cost per subscriber)
Two-way MMDS, LMDS	Not available or experimental
Satellite	
DBS (television only)	$300–$800 (includes CPE)

On the wireline side, there is the inherent advantage that the technology is potentially capable of providing a fully integrated interactive broadband system. As stated in Chapter 4, it is not reasonable to assume that wireless access will be able to serve as an integrated broadband system capable of bandwidth on demand applications for everything from voice to video telephony. That is not to say that it is not possible, because indeed it is. It is only to say that the spectrum allocations and licensing schemes of the FCC do not allow for it in the context of known wireless access systems.

Digital wireless technology may still become a formidable competitor of integrated broadband wired networks for providing full-service broadband capability to the mass market. Now that the FCC has granted LMDS system operators the ability to provide narrowband two-way radio telephone service in competition with the traditional wired POTS service of incumbent telcos, the fortunes of wireless access systems as full-service infrastructure networks may be changing for the better. With 1 GHz of usable spectrum, LMDS system operators have enough potential radio spectrum to provide a full range of digital narrowband and broadband services to the mass market on the same network system. However, line-of-sight issues remain a unique problem for serving everyone within the radio coverage area of the LMDS transceiver.

With its large slice of bandwidth, LMDS is unique. Other cellular operators with licenses to a slice of broadband RF spectrum, like CMRS providers, are limited to 40 MHz of broadband spectrum. This paltry amount is not nearly enough to provide broadband multimedia service to a mass market.

5.3 NARROWBAND NII ACCESS

An apples-to-apples comparison of wireline and wireless access systems would have to eliminate infrastructure options that require either broadband services alone or integrated network systems for broadband and narrowband services. That leaves two relevant options for infrastructure wireless access systems: (1) narrowband digital data and POTS; or (2) a combination of one-way distributive video and digital data and POTS.

Limiting technological options in that way sheds some light on the wireless versus wireline debate. Based on the cost data from Tables 5.1 and 5.2, wireless access for narrowband data and POTS is clearly preferable to wireline access. That is not surprising considering the obvious differences in system construction costs (e.g., the high cost of laying a physical cable circuit versus placement of an antenna at the subscriber location). Even if the IFCs of wired network access systems were the same as those for wireless systems, the long-term cost advantage would still lie with the wireless alternatives, because in a wireless environment much of the network system costs are fungible in the sense that they are available for reuse if a subscriber chooses to terminate service. Furthermore, a major portion of wireless access system costs are not committed until a subscriber requests service installation.

5.4 BROADBAND NII ACCESS

What about access for one-way broadband services or for the combination of one-way broadband, POTS, and digital data services? The results are somewhat mixed, but they tend to favor the wireline alternative.

In the case of stand-alone video systems, both hybrid fiber coax systems and their wireless counterparts (e.g., MMDS, LMDS, DBS) are fairly closely matched in terms of total cost per subscriber. Again, however, the wireless operator has a lower sunk network investment, and a more variable capital cost structure. As would be expected, the per-subscriber costs for the physical dis-tribution network are somewhat higher for the wireline alternative, while CPE costs are somewhat higher for the wireless alternative. Those costs, however, do not account for potential declines in future wireless CPE costs (or network distribution costs for that matter). Because cableco and telco networks are ubiq-uitous, R&D and manufacturing efforts have concentrated on new digital equip-ment and devices for use in conjunction with wired networks. That will change as digital wireless access networks are deployed more ubiquitously.

Because the future market for digital interactive multimedia is not particu-larly concerned with one-way video service, the relative costs of wired and wireless video distribution systems will not be discussed further, except to point out that digital satellite systems are a cost-effective alternative to wired cable systems for broadcast video service. Were it not for wired cable's huge head start in the market, satellite video systems would likely be dominating the mass market for entertainment video service. Terrestrial wireless cable systems (MMDS, LMDS) also enjoy a cost advantage over wired cable systems and are even cost competitive with satellite systems.

The more important comparison for the future multimedia environment and the new information infrastructure is between wired and wireless techno-logical alternatives for the combination of two-way narrowband digital service and one-way broadband video. Now the wireline alternative seems to have the cost edge. The cost of upgrading a wireline broadband video distribution net-work (e.g., CATV) to provide narrowband two-way service is somewhat lower than the cost of adding a two-way wireless capability to a wireless video net-work (satellite or wireless cable), even without incurring the costs of physically integrating the narrowband and broadband service on the same radio access link. The reason is that, in light of planned spectrum allocations, there do not appear to be many system cost efficiencies from integrating PCS and broadcast video services on the same wireless network access system. On the other hand, there do appear to be cost efficiencies from such integration on the wireline side.[2]

It is important to note that this conclusion presumes that the broadband wireline operators themselves (e.g., cablecos) do not have to incur the cost of a narrowband switching capability. In other words, the wireline access company is just that—access. The switching capability would be provided by intercon-nection to the PSTN, thus avoiding the fixed capital investments associated with the switching function. A cableco's PSTN interconnection arrangement, necessary to obtain a switching function and ubiquitous call completions is not likely to come free, but that is also true for wireless cable and satellite systems that want to provide telephone services. Competitive telecommunications com-panies simply will not be viable as infrastructure network providers without

2. This is Reed's conclusion in "Economics of Wireless Communications Systems."

achieving the capability of ubiquitous call terminations. That invariably requires PSTN access and the costs that go along with it.

Interestingly, the conclusion that wired video systems have an advantage over their wireless counterparts when being upgraded to provide narrowband two-way services does not necessarily hold true for the situation where a wireline POTS network is upgraded to handle one-way video services. Upgrading a narrowband telephone network to include integrated wireline video service may not be cost effective relative to a nonintegrated approach whereby two-way narrowband digital service continues to be offered on a separate network and broadband video on a wireless one.[3] It is possible that the reason is that, until a lower cost, more mature technology comes along to integrate video with the telephone network, large incumbent local telephone companies are investing heavily in nonintegrated wireless alternatives such as wireless cable and satellite networks.

Thus, based on the available data, it is safe to conclude that, in the future, wireless access systems (i.e., wireless cable and satellite) will be the preferred technological choice (i.e., the most cost-efficient method of providing *dedicated* subscriber connections to the PSTN) for either stand-alone digital video network systems or for digital narrowband services. If that is so, why are not they being used as such? The answer is simply that the technology is too new. Over time, that will change, and wireless alternatives will begin to displace their wired counterparts.

This conclusion does not necessarily hold for network facilities that are shared among a number of subscribers and among a number of narrowband and broadband services. In those cases, it is a close call between wireless and wireline technologies. However, once sharing of network facilities reaches a very high level in a multinode network, like the PSTN trunk network, the calculus dramatically shifts in favor of wireline (i.e., fiber optic) trunk connections.

The importance of this discussion is that, until the service requirements of the universal information infrastructure of the future have been specified (or are otherwise discovered via the market mechanism), the question as to which system is preferred—wireline or wireless access service—cannot be answered.

If, as many believe, the NII only contemplates socially efficient access to narrowband voice and data services, then wireless technology is probably preferred for dedicated subscriber connections to the wireline intercity PSTN. Notwithstanding the fact that wireless access costs are lower, the real bonus for the consuming public from this scenario is portability.

If, however, broadband service, especially bandwidth on demand, is added to the narrowband mix for the NII, then wireline access technology is the

3. Johnson, in *Toward Competition*, and Reed, in *Residential Fiber Optic Networks*, find that the sum of the costs of stand-alone telco and cable systems is actually slightly less than an integrated system.

winner. Interestingly, the Telecommunications Act of 1996 specifically calls for a universally available and affordable switched broadband network infrastructure. It remains to be seen if the government sticks to the letter of the law when it discovers the price tag.

An interesting irony flows out of this discussion: Acting in their own business interests, wireless access network providers of all types, narrowband and broadband (e.g., wireless cable and satellite services), would not want to back a definition of service for the NII that included broadband or interactive multimedia capability. If they did, the winner in the race to be the infrastructure network provider would be wireline access.

By promoting a narrowband infrastructure, narrowband wireless access providers are clearly the least-cost alternative. Broadband wireless access alternatives would also become the least-cost alternative for their traditional (huge) niche market—distributive video service.

Thus, if service cost is the issue for the NII, and if policymakers envision bandwidth on demand as a long-term infrastructure imperative, integrated two-way broadband services are best provided by wireline operators (e.g., cablecos and telcos). In this scenario, though the role of wireless access services in the NII is not dominant, the indisputable convenience of portability coupled with the affordability of new wireless technology will ensure that the mass market will be served by the interconnected adjunct networks of wireless access operators.

That conclusion leads to another interesting twist for the public policy stance of the wireless industry regarding the NII. The bottom line for wireless technology, whether or not it is preferred by policymakers for the NII, is that it will be present and popular in the mass market. Considering that inescapable conclusion and the private sector's general distrust of government involvement in a an otherwise competitive business, wireless network operators of all stripes might consider it a blessing that they are not tagged as the vehicle for driving onto the public information superhighway.

5.5 CAPITAL RECOVERY AND FINANCING PROSPECTS

Given the cost data in Tables 5.1 and 5.2, it is useful for purposes of illustration to point out what is implied for the demand side of the capital budgeting equation. As a rule of thumb, for every $1,000.00 ($1) of per-subscriber network access line upgrade costs, fully $14.00 ($.014) per month of additional revenues per household served would be required to allow for full capital recovery of the original investment costs over a 10-year discounted payback period at a 12% *rate of return* (ROR). This hypothetical situation includes the rather heroic assumption that new revenues would begin flowing immediately on completion of the network construction, which is why the cost and implied capital recovery estimates represent a best case for cash flow analysis.

Table 5.3 provides a rough cost summary and estimates of the associated construction timelines for deployment of mass market broadband network upgrades for cablecos and telcos, along with how much new sales revenues per month each of them would require from every household passed by the broadband network. (Note that passing a home does not necessarily mean that the home has subscribed.) The numbers are cause for alarm if one is planning to go it alone in the face of stiff competition from both integrated and nonintegrated infrastructure network alternatives.[4]

The second and third columns in Table 5.3 ("The Next Generation") provide a range of likely costs for upgrading basic cableco and telco analog networks to provide one-way broadband services in the case of telcos and two-way narrowband telephone services in the case of cablecos. This basically puts cablecos and telcos in a position to compete with one another on a more or less equal footing for integrated service to households. According to Table 5.3, cablecos have a tremendous cost advantage in the near term. However, when the costs of network upgrades for integrated service offerings are considered, it is important to keep in mind that there is little, if any, positive cash flow opportunity from providing traditional local telephone services. In the case of long distance service, the costs of interconnection to the PSTN are also substantial. Thus, as expected, we do not observe cablecos scrambling into this market (despite grandiose announcements to the contrary that appear from time to time in the trade press).

The fourth and fifth columns in Table 5.3 ("and Beyond") present the costs and implied capital recovery requirements for second-generation cableco and telco network upgrades to provide two-way broadband service capability. Notice that the higher end of the cost range for cableco network upgrades is nearer to the lower end of the range for telcos. That makes the ultimate choice between the "passive" nonswitched network architecture preferred by cablecos and the "active" switched architecture preferred by telcos a tougher call for cablecos' long-term capital budgeting strategy.

Based on those data, it is clear that, except for narrowband ISDN and local cable network two-way interactive services, it is costly indeed for any of these companies to go it alone in building the types of integrated multimedia networks for the mass market that are contemplated in the popular press and that are the objective of national infrastructure policy.[5]

---

4. These data previously appeared in the *New York Times*, February 21, 1993, p. B1.

5. One policy and strategic planning conclusion for infrastructure investment is that perhaps the relatively low-technology, but very low cost, network solution is N-ISDN. This technology is capable of supporting most digital information and transaction services (including slow-scan video telephony) that are being contemplated for the residential mass market. Even though N-ISDN is not broadband, it is likely to be much easier for network operators to generate positive cash flows.

Table 5.3
Estimated Upgrade Costs for Cable and Telephone Networks

Parameter	The Next Generation...		...and Beyond	
	Cable (Fiber Optics and Coaxial Cables)	*Telephone (Fiber Optics and Coaxial Cables)*	*Cable (Two-Way Fiber and Coaxial Cables)*	*Telephone (Entirely Fiber Optic Network*
Cost to install*	$50–300	$1,500	$1,000–1,500	$1,500–5,000
Monthly revenue†	$1.40	$10–$20	$14–$17	$20–$35
Time frame	3–10 years	5–10 years	10–20 years	10–30 years
Services	Telephone, data, cable TV	Telephone, data, cable TV	Telephone, data, cable TV, two-way video, high-resolution TV	Telephone, data, cable TV, two-way video, high-resolution TV
Overall cost	$5 billion–$30 billion	$75 billion–$150 billion	$100 billion–$150 billion	$150 billion–$500 billion

* Per subscriber.
† Extra monthly revenues per subscriber needed to justify the investment.

Thus, in a sense, the race is on, at least on paper. But then again, in a sense, it is not. Who wants to go first to wire up America with broadband?

Based on the cost data in Table 5.3, even under the heroic assumptions of quick mass market deployment, the additional per-household monthly revenues required to pay for the original investment is staggering, considering the base of per-household revenues spent on telecommunication services today. The average household in the United States spends about $45 per month on telephone services and about $25 per month on cable television services. Advertisers pay $25 per month per household to support over-the-air broadcasting, or so-called free TV, and another $7 per month for broadcast radio.

In total, not counting what an average household spends on electronic devices, about $100 is up for grabs in a competitive marketplace. This amount is not growing much at all, nor is household disposable income. In fact, over the last decade, the percentage of household income spent on telecommunications services has been flat, at about 2%. The percentage of household income spent on cable TV service has also been flat in recent years now that the huge growth rates have begun to reach a market saturation point. Per-household broadcast media revenue has been flat or slowly declining.[6] However, there are other po-

---

6. In recognition of the dire straits broadcasters were in in the late 1980s and early 1990s, the government has been relaxing rules, thereby allowing broadcasters to own more media proper-

tential revenue streams involving video media like movies, video tape, and video game rentals and sales, which could add another $20 billion in potential revenues. Revenues from information and transaction services, like home shopping, home banking, and other advertising services also exist, but there are no solid data on the market potential for such new services. It is reasonable to assume, however, that they are potentially substantial. Witness the rapid growth of direct mail advertising, which is now estimated at $20 billion annually and is continuing to grow.

Overall, the current demand and revenue data from the telecommunications sector indicate that a competitive service provider of two-way residential broadband network services face an uphill battle. New revenue growth is always going to be subject to the ability of households to afford to pay for fancy new services and the terminal devices that support them. What is more, current revenue streams are supporting the payback for old and current capital investments and may not be immediately available to fund new construction budgets if alternative investments are more attractive.[7] The bottom line is that, unless an integrated broadband telecommunications network operator is allowed to freely pursue all revenue opportunities, including partnering with other service providers to save on new construction costs, it is difficult to justify mass deployment of the new broadband to the home technology.

Even the telcos' own financial simulations for public broadband networks are pessimistic. Telephone company studies indicate discounted payback periods for VDT network upgrade alternatives ranging from 6–7 years for medium-band systems, with limited functionality and bandwidth, to 12–15 for more advanced broadband systems.[8] This even assumes some rather aggressive demand assumptions, on the order of 40% subscribership to a host of new services within 10 years.[9]

Researchers in the investment banking community have examined the available data, and it is apparent that they are not willing to accept the entire

---

ties and consolidate their operations. In addition, the FCC has been allowing them into some new markets. The prime example of all this activity is the lifting of the restrictions that barred broadcasters from sharing in profits from the syndicated programming business.

7. Not the least of which might be investing funds in foreign markets, where infrastructure investments may have a relatively bigger payoff.

8. See Chapter 3.

9. Even when Ameritech put forth its most favorable estimates of prospective cash flows for its various proposed VDT projects in a recent public disclosure to the FCC, it was obvious that there was little to be optimistic about. Ameritech projected 7- to 9-year discounted payback periods for rather primitive systems, even assuming a 40% demand penetration rate over 10 years and including business market revenues. This chapter specifically does not address any business markets, because it is assumed that business customers generally will enjoy competitive service alternatives and are, therefore, not a public policy issue or concern for the information infrastructure.

risk for capitalizing new broadband infrastructure ventures. They will only consider such high price tag projects when the borrower provides the lion's share of financing. Even then, the coupon rate for external bonds is very high and will potentially be coupled with a demand for an ownership stake (e.g., stock warrants). That is why the only large-scale projects are primarily financed by internal sources of funds from the deep-pocketed incumbents like telcos and cablecos.

In press announcements, the major industry players have "committed" (on paper) to major network infrastructure investments. To date, the RBOCs alone have stated their intentions to spend about $60 billion in broadband PSTN investments. The major long distance companies, including AT&T and MCI, have announced similar amounts. The cablecos, following suit, have announced many billions of dollars for digital broadband infrastructure investments. Lately, the real strategies of the long distance carriers have emerged. Rather than invest billions building a local digital network infrastructure, the long distance companies have vigorously lobbied state and federal regulators to force the incumbent telcos to lease capacity and resell their local network connections at discounted rates. No one has really committed to spending the amounts of money that building a nationwide network infrastructure would require, and, with so many announcements, it is likely that there is a lot of market signaling going on. Most likely, some of the major "commitments" are really just a repellent to scare future rivals enough that they do not ultimately take the investment plunge, lest there prove to be a first mover advantage after all.

Most of the financing for new PCS ventures has required the backing of deep-pocketed incumbents as well. Smaller wireless infrastructure projects, including wireless cable and digital satellite systems, have been having trouble finding external financing. Only recently have the RBOCs shown significant financial interest in wireless cable investments. They usually require a significant equity stake, or they purchase an existing system outright (or the license where a system is not yet built).

The increasing competition being allowed by regulators in traditionally monopolistic markets is largely responsible for the riskiness of new broadband infrastructure investments. Recent attempts by large industry players to broaden the base of external investors in new infrastructure projects like global digital satellite systems do not bode well for financing infrastructure investments, even such relatively small ones such as digital satellite systems. Two of the leading contenders in the race to deploy satellite systems, Globalstar and Iridium, have both failed recently to attract investor interest in recent bond offerings, even at fairly high coupon rates [1].

Thus, even though there is a clear technological trend toward industry convergence, based on the twin facts that broadband infrastructure investment projects involve extremely expensive up-front costs and the industry is becoming increasingly competitive, it is not likely that private enterprise will be willing to take the financial plunge anytime soon.

Reference

[1]   "Globalstar Offering Gets Cool Reception as Investors Balk at Mobile-Phone Plans," *Wall Street Journal,* October 2, 1995, p. A3.

The Economics of Broadband Networks

6

6.1 TECHNOLOGY, REGULATION, AND INDUSTRY STRUCTURE

Just as the technology of future communication networks will be fundamentally different from today's, so too will the economics. Today's local telecommunication networks are fragmented and compartmentalized, offering specific services in a monopoly market setting; those of tomorrow will be integrated, offering a wide range of interactive multimedia services. The primary reason for the difference is that significant advances have been made in DSP. In essence, fundamental networking is being transformed from service-specific networks and functions to commodity functions, and network control is moving from centralization to decentralization.

Historically, the ability for people to enjoy most telecommunications services was inherent to the network itself, which was uniquely designed to deliver one or a few specific services. In the future of integrated multimedia networks, the features and functions of most network services will be determined locally by the capabilities of network terminal devices found in the household.

As the rather homogeneous technology and process of digital computing cause the convergence of digital telecommunications and television, economies of scale and scope will expand. Perhaps, most important, regulation is finally beginning to release its grip in favor of market forces so that network infrastructure development can follow a more natural, consumer-friendly (i.e., less monopolistic) course.

These factors have significant economic implications for the future of the industry structure and for pricing and costing practices. Although the fundamental changes taking place on the supply side of the multimedia marketplace are becoming much clearer (see Chapters 3–5), it is much less clear what significant changes will occur on the demand side of the market. In the future, customer control and use of networks will be limited only by the capabilities of application software, network peripheral devices, and CPE. This is by far the trickiest part of the business to forecast and was the subject of much of the discussion in Chapter 2.

Earlier chapters covered broadband and multimedia network technology and trends and presented broad-gauge average cost estimates for many types of network systems. This chapter discusses the underlying economics of those systems, including key economic concepts that are useful for insight into the industry's evolving market structure. Particular emphasis is on costing and pricing issues and the economics of regulation, with an eye toward providing the reader with the basic economic tools for evaluating industry structure and performance.

To date, the economic concepts and tools available to government policy-makers and regulators either have not been used, because they have not been well understood, or have been misused by those wishing to gain a regulatory advantage in the marketplace. This chapter attempts to provide a concise exposition of proper economic concepts and tools and recommends regulatory frameworks within which those tools can be applied to promote both competition and infrastructure network investments. A discussion of the political economy of the regulatory process itself is reserved for Chapter 9 which covers telecommunications public policy.

6.2 ECONOMICS AND INDUSTRY STRUCTURE

To evaluate the type and number of companies providing public telecommunications network systems in a competitive environment, it is important to have a good understanding of the underlying cost structure of the business. Because a lot of confusion attends the definitions of the terms commonly used to describe different types of costs, a glossary of terms related to economic costs is provided in Appendix A.

The telecommunications network business, like other businesses whose production processes rely on large-scale capital-intensive and long-lived investments, exhibits significant economies of scale and scope. The existence of economies of scope is key to the integrated network business, because it means that as more and more types of services are delivered over a common shared network facility, the average costs of all the services together falls. That may create a significant competitive advantage for a multimedia network operator.

Figure 6.1 illustrates the concept of economies of scope. The firm in this example has a total cost of $10 for producing both products A and B. The *total incremental cost* (TIC) of each of the two services is $3. Therefore, the remaining shared cost (e.g., the network) is $4.

Should either product in Figure 6.1 be discontinued, the total stand-alone cost to produce the other product would be $7. That is because, when one

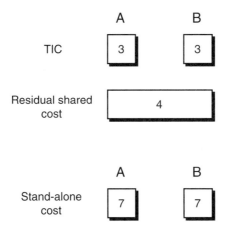

Figure 6.1 Economies of scope.

service is dropped, the shared network cost of $4 shifts from being a shared cost to being a direct cost of the remaining service. Now, only one service is causing and bearing the cost. Significant economies of scope obviously give the firm with the most services on the same network a competitive cost advantage. This would argue for a relatively small number of firms in the particular industry segment.

Another key economic determinant of industry structure in a network business is the size of the network itself. Network businesses, unlike most other types of businesses, can exhibit strong positive externalities. That means that as subscribers are added to a network, the value of the network increases overall for any existing subscriber because it can now reach more people. That explains why a network, like the Internet, once it reaches a critical mass of subscribers, suddenly becomes the only network in town to which anyone wants to subscribe. It also provides some explanation as to why, when America Online purchased an existing large-scale public network that was part of the Internet backbone, it was able, almost overnight, to dominate the field of online service companies, surpassing long-time rivals CompuServe and Prodigy. (Aggressive marketing was the other reason).

It is common sense that a subscriber, especially a new online service subscriber, would want to be a member of the larger network, because, all other things being equal, the larger network offers more value. Most formal economic simulation models of industry structure in network businesses show that the long-term equilibrium condition is a single network. Furthermore, as more subscribers are added to a growing network operation, supply-side impacts of cost economies of scale and scope serve to reinforce the demand and value aspects of the larger network system.

6.2.1 Regulation and Industry Structure

Regulation has always had the power to derail the market forces of supply and demand when it comes to technology adoption. The most simple case in point is when the government withholds RF spectrum from private market use or when it restricts the use of spectrum to certain types of services, regardless of whether those services are the ones the market would naturally demand. While largely effective at creating and sustaining telecommunication and media monopolies, regulatory mandates represent a terribly blunt instrument for competition policy and often lead to huge losses of economic efficiency, productivity, and consumer welfare. Beginning with the Telecommunications Act of 1996, this situation will gradually change to allow market forces to work. Breaking down regulatory barriers to market entry will ensure that the marketplace will sort out the winners and losers among multimedia technologies and network operators.

After the government's largely failed policy of creating "a little" competition with its adoption of a duopoly market structure for cellular telephone service, federal lawmakers and regulators are now calling for at least three or more competing local network alternatives. Current cellular service license allocations call for up to five digital service providers in a single service area, not counting new wireless cable services. Interestingly, if left to natural market forces of supply and demand, we may never see that many. In their efforts to "let a thousand flowers bloom" in the local telecommunications market, regulators could overdo it and again put themselves in the position of thwarting natural market forces. It is safe to bet that, for the next few years at least, the government will not allow cable and telephone monopolies to reemerge. That could change if multiple providers of local broadband network infrastructures do not emerge, or if all but one is forced by the market to go out of business.

6.2.2 Technology and Industry Structure

The underlying technology of multimedia network infrastructures is the same as that for the computing business, namely, DSP and microchips. That is why telecommunication networks exhibit substantial technological economies of scale, albeit less than those that the computing business exhibits. Recall that network upgrade costs in the local telecommunications business is dominated by the cost of provisioning network connections that are usually dedicated to individual subscribers. While the speed of subscriber lines can increase with advances in optoelectronics and DSP, it is still a fairly labor- and capital-intensive process to install and maintain the access lines themselves. In addition, while the huge on-going advances in memory chip capacity translate more or less directly into faster and better computing, that is not necessarily so for local telecommunications network infrastructures. You cannot just put a new chip into the old phone device or onto the phone line to make it work faster or perform new functions. Telephone and cable television networks are so standardized that such increases in speed and capacity can only be implemented wholesale throughout the entire network. That will gradually change as new network architectures and designs allow for more flexibility in the way service is provisioned to effectively meet the specific needs of individual subscribers and subscriber groups.

Explaining all the reasons why local telecommunication networks have heretofore not enjoyed the radical gains from technological cost economies that the computing business has is beyond the scope of this discussion. Suffice it to say that things will be changing for the better. Much faster network speeds, including those of individual subscriber connections, are on the horizon. That will cause unit costs to fall rapidly, in much the same way as modem costs have fallen in the past.

The primary implied result of increasing the speed of local telephone connections and the speed and memory capacity of consumer telecommunications devices distributed throughout the network system would be a much more decentralized network and industry structure compared to that which exists today. On the other hand, the very same technological economies of scale inherent in the speed and capacity processing of memory chips, coupled with the significant cost economies inherent to a single central node for network databases, routing functions, programming, and information content, would argue for a continued centralized approach to providing so-called intelligent network services.

The overall implication for the future is that, depending on the amount and the type of a particular service or application individual subscribers want, both locally distributed and centralized network systems will emerge. In fact, both will thrive over the long term. This situation is similar to the one often debated about the long-term survival of a localized software system for desktop computing (e.g., Microsoft), versus the newer phenomenon of network-based computing (e.g., Sun's remote computing software, Java). Again, due to the heterogeneous nature of consumer demand, both types of software systems will coexist. Still, the question remains as to which type of system is likely to dominate the mass market for residential demand applications. The answer is probably the networked computing solution. Beyond the issue of remote versus desktop computing applications, it would be expected that network servers for database and programming functions likely will continue to be highly centralized for some time to come, at least for most residential applications (e.g., sources for entertainment video).

Network centralization and network economies of scale, no matter how great, will not be the death knell for CD technology or other truly portable local telecommunications technologies. Just as video tape rentals were supposed to displace movies and cable pay channels, and cable television was going to displace over-the-air broadcasting, both the fixed and portable technologies of multimedia will peacefully coexist. In fact, they will often be complementary to each another (see Chapter 2).

Multimedia technology trends indicate that, in certain segments of the industry (e.g., infrastructure networks), the market structure will be highly concentrated with only a few service providers' networks). In other segments (e.g., programming, information services, and publishers) the market structure will be highly diversified, with many service providers. While individual subscribers may choose to purchase the bulk or even the entirety of their multimedia services from a single integrated network supplier, there will still be many others who will continue to use nonintegrated alternatives or some combination of integrated and nonintegrated alternatives, depending on price and service quality considerations. Even in the presence of tremendous technological economies of scale and scope in large-scale multimedia delivery systems, the underlying structure of consumer demand will not permit a monopoly industry structure to develop. It is a given that individual consumers will always prefer individualized mixes of services. In the information age, as tempting as the vision of the ubiquitous information superhighway may seem, one network cannot be all things to all people.

6.3 NETWORK COST STRUCTURES

To understand the pricing structures in the telecommunications network business, it is important to have a good understanding of the underlying cost structure of the network itself. That is not true of most other types of unregulated business, in which pricing practices are based largely on the (ever changing) perceived value that consumers have for the product or service.

The physical cost structure for wired and wireless digital network systems were presented in the stylized illustrations provided in Chapters 3 and 4. By far the most dominant cost component for any of these network systems was the subscriber connections, which have to be provisioned to each individual household. An extreme example is found in today's cable television systems, in which less than 10% of the total system cost is represented by centralized head-end facilities. Comparatively, in today's telephone company networks, about 20% of the total cost is for centralized switching equipment. As network connections are upgraded to provide broadband and interactive multimedia services, that percentage is expected to decrease substantially. Such a decrease is primarily because of the relatively expensive electronic equipment required for upgrading dedicated subscriber connections located in the outer reaches of the network system.

The costs of subscriber terminal devices and the associated household wiring and cables also dramatically affect the per-subscriber costs of obtaining interactive multimedia network services. However, as discussed in Chapter 2, all indications are that those devices will be competitively supplied and will be required for any individual subscriber to hook up to any type of multimedia network system (e.g., telcos, cablecos, satellite, cellular, and wireless cable networks). Therefore, the cost of subscriber premises equipment, while being a significant factor in determining the absolute level of per-subscriber system costs, will not be a determining factor in gauging the relative success of one type of network system over another. An important qualification for this conclusion is that some network termination equipment located at the subscriber's premises can be reused (e.g., satellite dishes and antennas in wireless systems), resulting in a potentially substantial market advantage in a competitive environment where subscriber disconnection/reconnection (called "churn") can be costly. Because very little is known about the production costs for the myriad of devices that consumers ultimately will be using (except that they will represent a substantial portion of the total cost of service), such costs generally are excluded from the discussion of network costs.

Some basic cost characteristics are exhibited by all infrastructure network systems. It is convenient for the purposes of analysis to view network systems as broken up into at least two separate pieces: individual subscriber connections (access lines) and the shared trunk line network featuring all-digital high-speed network routing and switching nodes. Customers may purchase or lease access line facilities to the core network gateway or host node in any quantity they desire to meet their perceived service requirements. Customers will also be able to purchase their own hardware and software, which can dynamically allocate the carrier channel bandwidth inherent to their network connection to satisfy whatever requirements they may have for voice, data, and video services.

Once customers obtain an access line or lines, the network operator will be obliged to serve the customer by making available sufficient core network capacity to provide an acceptable level of reliability and service. For the network operator, the engineering problem will be determining how to effectively allocate the engineered capacity of the network gateway or host node and the shared network trunk lines. That is also true of network peripheral devices and remote nodes. One of the most vexing problems for managing the network and for costing and pricing purposes is determining how much of the cost of network capacity expansions should be allocated or assigned to subscriber demand units, including current and future network usage or lines.

6.3.1 Network Capacity Costs

Once the network infrastructure is built and subscribers have purchased access lines according to their individual demands for service, nearly all the additional network system costs are caused by capacity expansion. Generally speaking, the *incremental cost* (IC) of network usage is zero or very close to zero. In other words, the network facility is in place and paid for whether or not it is actually being used. The real cost issue arises when demand for network usage grows and network capacity nears exhaustion.

In situations where the network operator is selling service on a private network facility dedicated to single subscribers, it is easy to know how to allocate the costs of network capacity expansion. It would simply be charged to that user who caused the costs to be incurred. Public networks do not work that way and therein lies the most difficult costing and pricing problem of all: What is the best way to calculate the appropriate cost and price when you cannot determine who is responsible for incurring the costs of capacity additions?

Furthermore, with such long planning and network construction horizons, public network operators do not install capacity with any particular customers in mind. Rather, capacity is installed in anticipation of serving demands for network usage as it is delivered to the network randomly and anonymously from the public at large. Also, when capacity expansion occurs, it involves the installation of very large amounts, because the requisite network equipment is available for purchase only in large quantities, or "lumps." Thus, for a variety of reasons, it is impossible to properly assign economic costs to those subscribers who actually may cause network capacity to be exhausted. This problem is bad, but it pales in comparison to the costing and pricing problems of future multimedia network operators. Today, the usage of any particular network infrastructure is primarily for one or a handful of services (i.e., telephone calls and television). At least in that scenario, it is easy to know what services are causing the capacity costs to be incurred. Interactive multimedia networks will be used for wide range of services, which compounds immensely the cost allocation problem.

An effective way to allocate shared network infrastructure costs to customers and services is based on the notion of treating the capacity of all public network infrastructure facilities and their associated costs as inventory costs. Even though capacity expansion of telecommunication network facilities requires the installation of rather large lumps of capacity, it is reasonable to treat all the spare capacity in the network as a fixed cost of inventory. If a network operator is efficient at running the business, or if regulators insist that a certain amount of spare capacity exist for quality-of-service reasons, they will strive to ensure that the network never really runs out of capacity by having a consistent margin of spare capacity (e.g., 30%) throughout the network system.

The spare capacity of 30% that can be expected to be used in the future will always exist (unless the business is dying, in which case there is no capacity issue anyway) and is therefore properly treated as an on-going fixed cost of inventory. In other words, every additional minute of network usage is treated as taking out of inventory a minute of spare capacity, which, over time, will be replaced by adding more spare capacity. However, all additional minutes would have costs allocated to them according to the ICs of network capacity additions, not the costs associated with the embedded network investment, even if it is the old network facilities that are being used.

Many other businesses operate this way. Take, for example, petroleum companies that run gas stations. The cost allocated (and the price charged) for additional gallons of gas will not be based on the cost of the oil stored in the underground tanks at the station but on the cost of replenishing that supply, because the gas taken out today must be replaced in inventory by "new" gas. An analogous situation is one in which a telephone company charges for network usage based on the ICs of new fiber optic transmission systems even though the usage served is actually carried on the old copper trunk line.

6.3.2 Application of the Capacity Costing Method

Capital costs associated with network construction derive from the initial one-time investment made in network facilities. The economic, or opportunity, costs may be expressed as recurring annual costs, because the money that has been invested in these assets could have been invested elsewhere and earning a return. The components of annual capital costs include capital repayment (or depreciation), return on capital, and income taxes.

The primary reason for incurring capital costs is the advancement or deferral of network facilities construction caused by changes in demand. The effect on total costs caused by demand changes can be accurately measured through the use of capacity cost calculations.

Capital investments come in indivisible lumps of capacity, such as a module of memory or an entire mainframe computer processor. The purchase of a lump of capacity begins to create excess capacity. Over time, the firm grows into the excess capacity by serving higher levels of demand. When the existing capacity is about to be exhausted, another lump of capacity is acquired, and the growth cycle repeats. At any given time in the growth cycle, some amount of excess capacity exists. Several questions arise when the treatment of excess capacity is considered.

- How should the cost be spread to units of service?
- Who should pay for the excess capacity?
- How should the cost of excess capacity be recovered?
- How should the cost of excess capacity be incorporated in the cost of products?

Capacity costing is a method of identifying the cost of excess capacity caused by the units of capacity in service based on long-run cost causation. The capacity cost calculation provides a good approximation of long-run marginal cost for both the long-term and short-term changes in output, provided that investment components will be exhausted through increased use during the term considered. Capacity cost theory identifies capital costs caused by services based on the capacity of plant utilized by each service.[1]

Capacity cost practices rely on the divisibility of time to make the costs of lumpy investments appear divisible. The underpinning of capacity cost is that the long-term capital cost of a decision to expand or contract the volume of a service is the change in cost due to: (1) advancing or deferring the timing of the next growth in capacity; and (2) any effect on the size of the capacity. Only capital costs are addressed for now; other costs, such as operating expenses (e.g., maintenance and repair costs), will be assumed to be zero, but they could be added to the capital costs to arrive at total annual costs.

While the example herein applies to a telephone network, it could also apply to any other type of network. In fact, the general approach applies to any resource, divisible or indivisible, variable or fixed.

Capacity cost techniques are developed based on the capacity growth cycle, which depends on both the size and the timing of network capacity additions. Capital investments are available in indivisible lumps of capacity, such as the getting-started costs of installing and bringing on line a new telephone switch, a feature-enabling software package, or a switch line unit capable of terminating a specified number of lines. A new lump of capacity is usually purchased when the existing installed capacity nears exhaustion (reaches its optimum or objective fill). For example, an increase in the number of terminating lines on an existing switch line unit will exhaust the capacity of the line unit and necessitate the purchase and installation of a new line unit (new capacity) in preparation for future increases in demand. The firm will grow into the spare capacity over time. The capacity growth cycle for a switch line unit is illustrated next.

In Figure 6.2, assume the line unit has the absolute capacity to terminate 512 lines. Further assume 12 terminations on the line unit are required for administrative spare capacity, leaving a usable capacity of 500 line terminations per line unit. The first line unit must be purchased before any demand can be served. At the beginning of year zero, before any demand has been served, 500 units of spare capacity are in inventory. During the first year, 250 line terminations are demanded by customers. The remaining 250 units are in inventory at the beginning of year 1. During the second year, demand increases to 500 units and exhausts the existing *usable* capacity of the line unit. The purchase of a

---

1. The term *capacity cost* as used here is analogous to the term *average incremental cost* (AIC) used by B. Mitchell in *Incremental Costs of Telephone Access and Local Use* (Santa Monica, CA: Rand Corp., 1990). See also R. Park, *Incremental Costs and Efficient Prices With Lumpy Capacity: The Single Product Case* and *Incremental Costs and Efficient Prices With Lumpy Capacity: The Two Product Case* (Santa Monica, CA: Rand Corp., 1989). Beware of other uses of the term *capacity cost*. For example, in some applications, the term refers to the cost of capacity divided by utilization. That is *not* how the term is used here.

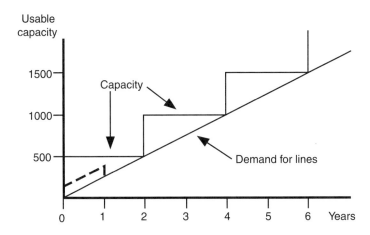

Figure 6.2 Capacity growth cycle: hypothetical switch line unit.

new line unit at the end of year 2 increases the inventory of plant to 1,000 line terminations, 500 of which are now available for growth. Since demand is projected to grow by 250 lines per year, a new line unit will be placed every two years, repeating the cycle.

Two conditions, both of which are satisfied in this example, must be present for the capacity cost approach to result in a nonzero value. First, forecasted demand for the network component must exhaust the existing capacity prior to planned replacement for technological or other non-use-related reasons. If that criterion is not met, there may be no future cost impacts (advancements or deferments in timing) caused by using an increased or decreased amount of existing capacity. Second, the changes in demand must be of sufficient duration to affect future component placements. For example, in Figure 6.2, there is no investment impact of serving 100 units of additional demand during the first year under a one-year contract that will not be renewed, because serving the additional demand will not affect the size or timing of the next addition to capacity. Accordingly, the capacity investment of this short-term increment to output is zero.

The long-term cost effects of decisions to expand or contract the usage of network components are the changes in cost caused by changes in the size or timing of future capacity. For example, if a customer appears at the beginning of the first year and signs a 10-year contract to terminate 250 lines on the switch, serving the additional customer will cause the demand curve to shift from the old level to a new, higher level.[2] The cost impact of serving the additional cus-

---

2. The output commitment in this example is of an intermediate term. However, as we will see shortly, whenever the output commitment spans the time between planned component placements, the resulting cost is well approximated by assuming long-term (permanent) output changes. This provides an opportunity for simplifications and standardization.

tomer is measured by identifying the changes in the timing of future capacity caused by the new customer. Figure 6.3 illustrates the future capacity growth caused by the new customer.

Serving the new customer at the beginning of year 1 has no immediate effect on the firm's investment because sufficient spare capacity exists in inventory to serve the increased demand. However, Figure 6.3 shows that the demand from the new customer causes the first switch line unit to exhaust after one year instead of two years, as projected by the original demand curve. The new customer has demanded one year's worth of growth (250 lines) and, therefore, moves the projected future placements of new capacity forward in time by one year. Thus, the incremental capital cost of serving the new customer is the cost caused by advancing the timing of the forecasted future capacity placements. This advancement cost is calculated using the general definition of the long-run unit IC: the present value of the differential cost divided by the present value of the differential output flow. This number, it turns out, will be closely approximated by a simple calculation: the annual cost of a switch line unit divided by its usable capacity.

Although more rigorous proofs are left to more advanced treatments of the subject [1], the result is easy to understand. The new demand causes the firm to carry an additional switch line unit 50% of the time (one of every two years, as depicted by the shaded areas in Figure 6.3). The cost impact of carrying one additional switch line unit 50% of the time is nearly the same as the cost im-

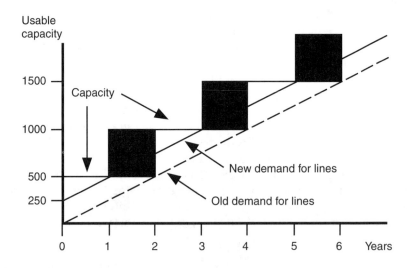

Figure 6.3 Future capacity growth caused by the new customer.

pact of carrying 50% of a switch line unit all of the time.[3] More generally, the long-run unit IC of long-term demand for one socket on the switch line unit is computed by dividing the cost of the line unit by its usable capacity and converting the incremental investment to an annual cost. For example, if the installed cost of a switch line unit is $50,000 and the usable capacity is 500 units, the capacity investment is $100 per unit ($50,000/500).

In summary, the capacity costing technique is a simple and accurate method for calculating the long-term costs of network capacity additions and is a useful guide for setting the minimum long-term level of price that will result in total recovery of the original investment costs over the life of the investment.

6.4 PRICING NETWORK CAPACITY

Since network costs are most sensitive to peak period demand, higher peak period prices are one way to reflect that increased cost. Peak-load pricing is the preferred theoretical option because the costs of capacity expansion are caused only by peak-period capacity requirements. The problem here is that network systems with wide area coverage have multiple nodes and links, which always exhibit noncoincident peak loads. In other words, while one subscriber may indeed cause a peak condition on a certain day in a certain node or transmission trunk line in the network system, other users may be causing it in the next instance; it simply is not possible to always know who is causing what in terms of network capacity. Even if it were possible to know, no subscriber wants to be on a network system where usage charges are like a reverse lottery, where you know the price of usage almost all the time but run the risk of placing that one call at the wrong time and getting zapped with a huge bill. Even if the network could somehow alert the next user that they had better wait to call or they'll get zapped with a full-capacity charge, it would still be perceived as inferior to a competing network operator that was willing to guarantee no such capacity charge in return for a subscribers willingness to pay a little more for all usage along the way.

Time-of-day rates, common in the long distance business, are not very efficient in solving the overall peak-usage problem because while some switching locations are busy during peak-rate periods, others are not. But all locations would face the same (high) peak-hour prices. Where there is idle network capacity, such a pricing scheme would needlessly (inefficiently) repress usage. A

---

3. The exact calculated cost using the general definition of unit IC will depend on the duration of the demand and how full the unit is when the new demand appears. If there is an equal probability of being at each level of capacity utilization when the new demand appears, as the duration of output approaches infinity, the long-run unit IC approaches the simple capacity cost.

more sophisticated pricing scheme would be required based on very localized peaks and busy conditions in the network.

These types of noncoincident peak problems are bad enough to cope with through sophisticated peak-load pricing schemes, but they would be overwhelming in the fast packet switched networks of the future (or even today, considering the statistically shared Internet). On the Internet and future fast packet and so-called connectionless networks, each individual packet of information traveling through the network has its own routing information. That means that different packets originating from the same location may arrive at different times over different routes at their common destination. Any one packet could experience a potential busy condition or congestion delay condition at any point along its path.

The problem of noncoincident peaks is also going to be aggravated in future multimedia networks that provide for more direct customer control. Large surges and shifts in customer usage over time and locations are more likely due to smart customer-controlled bandwidth and routing processes. One pricing solution for this important supply problem is some form of dynamic (potentially real time) peak-capacity charging system.

The only type of peak pricing scheme that may be truly capacity cost based is one that allows for real-time information to the message originator. Some experts have already been suggesting such an approach for efficient pricing of capacity on the Internet [2]. That reminds me of a plan devised by some of us working on this problem at AT&T Bell Labs back in the late 1970s. At that time we considered a convenient (i.e., not too annoying) way to tell telephone callers in real time that they would have to pay a higher charge for calling during network busy periods. One proposal was to have a low-powered light on the telephone (like the message waiting indicators commonly found on hotel room phones). The brightness of the light's glow at any moment (governed by the network itself) would be directly related to the charge for the next call. In that way, the phone network operator could discover the relative value (i.e., willingness to pay) that the caller placed on each call and charge accordingly. Subscriber value feedback is an important factor for implementing an efficient pricing structure, but the consumer annoyance associated with that feedback is also important.

Pricing mechanisms that account for value feedback have been proposed for the Internet. Value feedback is particularly useful in this environment. Recall that in a packet switched environment, like the Internet, all the infrastructure network routing nodes and transmission links are statistically shared in real time by numerous subscribers and services. There is no way for the network to know high-valued versus low-valued usage; therefore, all messages, once they have been originated by the user, are treated on a first-come-first-served basis. That means that certain high-valued usage, for which the originator would have been willing to pay, could be blocked or delayed at any point in the network by lower-valued usage.

The circuit switched phone network has a service advantage over packet networks like the Internet in this regard because, once a call has been connected, the end-to-end telephone network circuit path is dedicated to that single call until someone hangs up the phone. While this type of network system is considerably less efficient than a shared network system such as the Internet (when no one talks, the circuit is idle but still tied up), it does mitigate the problem associated with statistically shared networks that can and sometimes do cut off one or another message, in whole or in part, due to contention between simultaneous transmissions. Anyone who has experienced the frustration of interrupted or lost e-mail messages and file transfers on the Internet can relate to this problem.

In peak load pricing schemes, neither the circuit switched telephone system nor the packet switched network systems are prepared to handle problems associated with continuous transmissions. Once a message has been initiated on the network system, it is difficult, if not impossible, to continue to charge according to changing real-time network traffic load conditions. After a transmission has begun (e.g., sending a video), it is difficult and, from a user perspective, unacceptable to signal the message originator that someone else might also be trying to send a potentially more important message on a congested network facility and that the first message originator would therefore have to pay more to complete the entire transmission.

In a nonhierarchical, statistically shared network system like the Internet, it would appear that some form of price averaging would be absolutely necessary to be technically practical and acceptable to consumers. Nevertheless, if there is one thing that packet networks are designed to do—and do at almost no cost—it is to keep instant and accurate records of how many packets are processed and where they are coming from and going to. That being the case, it is only logical that, to the extent practicable, these data should be used for deaveraging usage prices to better reflect the relative costs associated with a subscriber's (or subscriber group's) usage.

Forced blocking of calls during busy periods is an unacceptable solution to most congestion problems, even though the small amount of call blocking that occurs on the phone network is much preferred to the situation on the Internet, which, when the routing nodes get overloaded, simply "drops" packets which are never recovered and about which the user is never notified. Business customers with high-valued usage will opt out of the public network system and transfer to a competitive private network system to obtain more reliability. That is fine as far as it goes, but it may needlessly sacrifice the public network system's underlying efficiency and economies of scale and scope. Supplying the demands of the large business community with dedicated or nonswitched bypass networks facilities would defeat the primary rationale for public network integration and would also raise societal issues as to the equity of having one high-quality network system for the rich corporations and individuals and one public network system for the public at large.

A brief digression may be useful at this point to contrast today's capacity constrained networks with a hypothetical super network with so much capacity that it would never reach exhaustion. In theory, as communications technology continues to advance, such a super network infrastructure could be constructed to serve all possible demand and still be nonblocking. In other words, switching and transmission would be so fast that congestion would not be possible (or not perceptible). That, of course, would eliminate the peak-capacity cost problem. In such a scenario, the economic implications for pricing would be completely changed, and pricing strategies would be driven solely by software applications and demand-side considerations. Being the first to serve a customer's demand in a timely manner would lead to a sales contract based on the value of the service to the customer. That would likely result in some sort of two-part pricing scheme for network access and usage, even though usage cost at the margin would be effectively zero. Presumably, there would still be a cost associated with customer access to the super network and various administrative and operations functions and activities. Notice that expense, rather than capital-related costs (depending on how one classifies network software costs), would dominate the marginal costs of providing customer service.

The implications are clear for those wishing to compete in such an environment. In this abstract world, capital (or software) does not physically wear out; it simply becomes technologically obsolete. Accordingly, a network operator recovers costs and makes a profit through customer price discrimination or value-of-service pricing. That may even result in an environment featuring totally flat-rate pricing, in which there would never be a usage charge and the network operator would get all its money in the form of monthly subscription charges. Actually, this scheme is not so hypothetical. Many computer network providers and online information service providers already sell service on a flat-rate basis, but it tends to be a premium priced service. Most computer system pricing schemes are related to usage.

This digression is purely hypothetical, but it is useful to see the changing pricing and cost-recovery implications of significant advances in communications technology. For now and the foreseeable future, network capacity cost will remain important. Beyond that, however, the emphasis will still be on demand-based over cost-based pricing. It is no coincidence that where regulation does not exist in the Internet community, many network and information service providers offer flat-rate service options to residential customers, while telephone companies continue to charge for long distance on a usage basis. Both services use the PSTN. Until regulations on telephone company price structures are eliminated so that the market can sort out the ultimate pricing structures, it is difficult to speculate as to what mix of flat-rate and usage-sensitive prices will emerge.

Futuristic scenarios aside, for the foreseeable future, the main cost and capacity problem for managing a future digital network infrastructure will continue to be congestion of the network trunks and routing and switching nodes. It is inefficient and probably prohibitively expensive for a network provider to construct network switching capacity to handle all demand contingencies.

6.5 PRACTICAL PRICE STRUCTURES

What types of pricing schemes are likely for future network infrastructures? The starting point for pricing reform is the plethora of pricing structures that currently exist in regulated telephone and cable companies today. In the case of telephone companies, the tariff rates are based on regulatory price discrimination in which the exact same services are offered under different tariff rates to different subscriber groups. This regulated system of tariffs is fraught with hidden inefficiencies and cross-subsidies.

Based on the previous capacity cost discussion, it is clear that today's broadly averaged regulated telecommunication pricing structures do not comport well with the cost structures of future digital networks. Interestingly, as complex as network peak-load pricing problems appear to be, the larger problem here is that telephone company tariffs are completely out of sync with underlying network costs. The key to effective pricing reform, therefore, is in fact tariff-rate integration and simplification. In particular, in the future integrated broadband network environment, cost-based rate structures may not include any time-of-day or distance-sensitive rate elements the way regulated tariffs do today. The artificial distinction between residential and business tariffs for what are essentially the exact same services also should be eliminated. The same goes for many business services that are effectively the same but employ different tariff rates targeted to certain types of customers.

Peak capacity rates, however, need to be added to tariff rate structures. Volume discounts, in the form of continuous tapers or declining block prices may be employed to achieve an efficient tariff rate structure based on the number and capacity of a subscriber's network connections and level of usage. Some, if not most, subscribers have a serious aversion to the uncertainty and complexity of a continuously variable rate structure. One way to simplify a continuous tapering in a rate structure and still achieve a high degree of pricing efficiency is to offer service in packages—a set of flat-rate options that would at least allow customers to choose a threshold level of service that meets their needs and a corresponding single flat-rate monthly charge. The flat-rate monthly charge would be based on the number and capacity requirements of individual subscriber connections and a specified level of usage (including zero). Such a set of flat-rate charges makes sense for pricing subscriber connections, but it may not have any cost basis for network usage because there is still no rate element that reflects the cost of peak-period usage.

While costs may be a useful guide for reforming regulated tariff rates, in an increasingly competitive market place, the prices for telecommunication services on the public network should ultimately be determined by both market supply and demand conditions.

In the future, customers will request subscription to one or another integrated digital network providers. The final services that customers will demand are voice, data, video, voice messaging, control, and so on—whatever services are generally inherent to the joint capabilities of network connection and subscribers' terminal equipment. It will then be the network operator's job to provision a level of network capacity for access, network usage, and control sufficient to handle the customers' anticipated demands. Assuming that price regulation for local service still exists, these are items for which the local service provider must have a tariff rate structure.

6.5.1 Access

In the future integrated digital network environment, access service is probably best viewed as the purchase of ports on a host computer. There will be high-capacity ("big") and low-capacity ("little") ports, and the monthly flat-rate prices (customer charges) for each will reflect that capacity differential. The service provider may also offer certain enhanced customer access service capabilities through remote nodes. Remote nodes, located on the customer side of the host digital network processor, may be used to provide certain features and functions. The use of remote nodes may provide customers with so-called smart access. This access is distinguished from usage of the intelligent core signaling network (discussed later). The features and functions that smart access provides will likely be in the category of enhanced services derived from software applications of the remote processor. Prices to recover the software and processing costs of smart access will likely be unregulated and determined by the market. Pricing of ports for access bandwidth, however, will require a tariff.

A cost-based pricing structure for both big and little access ports would include a nonrecurring installation charge and a recurring monthly flat-rate customer charge. Due to provisioning efficiencies, the charges may exhibit volume discounts in the form of a tapered rate schedule or declining block prices. Network access prices should be based on the amount of potential bandwidth capacity available to the end user. Of course, the per-equivalent-channel charge for big ports will be less than for little ports, but both may be offered as uniform (nontapered) rates. In any event, even with volume discounts, the most complex type of rate structure for access services would be a two-part monthly tariff. Volume discounts may be based on provisioning costs of installed capacity for a given technology. This could be tricky, however, since the subscriber's use of sophisticated electronics and signal compression can vary the effective bandwidth capacity of the same physical access facility.

6.5.2 Usage

Usage of the core signaling network is another matter. The cost structure presented earlier would argue for a novel rate structure to be cost based and therefore efficient. The digital network provides customers with circuit and channel switching service capabilities as well as packet data service and database lookup type services. Many other enhanced services may be obtained from the intelligent network; however, only basic network functions are likely to require a tariff. Purchasing network usage is best viewed as purchasing computer processing time, since all network services are provided to subscribers using public network host nodes and the signaling and control portions of the network. This may become problematic, however, when interconnected competing networks are required to share subscriber usage and routing data. Competing network operators do not always like to share their subscriber usage information.

A cost-based rate structure for usage may only require a uniform usage charge based on capacity costing methods. One possibility is a uniform rate per unit of bandwidth in terms of bits per second of processing required. That is basically the same as charging for computer processing cycles, except that each bit processed is not charged. Rather, they would take the form of a flat monthly usage charge for transmissions requiring at least a certain customer predetermined bandwidth (e.g., less than 144 Kbps, 1.54 Mbps, or 90 Mbps).

Another possibility is a uniform usage-sensitive charge for bit processing. That requires measurement of usage at the host switch. If a reasonable translation between computer processing cycles and a minute of network usage for voice or broadband service is available, then the tariff usage rates may be in the form of uniform per-minute (hour, second) charges by type of service (including database lookup). This uniform rate structure is similar to usage charges on some of today's computer networks.

There is still an important capacity cost issue. The network peak cost problem requires a pricing solution that dynamically allocates peak network capacity for high-valued usage. Off-peak usage should be priced according to market conditions, but it should not be priced below short-run operating (avoidable) costs. There are several possible pricing solutions that may dynamically and efficiently allocate network capacity. Two likely forms are customer capacity charges and interruptibility rates. Customer charges are simply monthly flat-rate charges to guarantee a certain amount of network bandwidth and throughput so that no blocking of usage is possible.

It is important to note that this charge is based on units of bandwidth capacity of the host processor switch, which is made available to a customer on demand, not on how much is actually ever used. This charge is also not likely to be based on available access capacity, since the transmission rate may be under the individual customer's control. The network operator will not have control over the customers' configurations or use of access services, but does know how processing at the host digital switch occurs. In the future, customers' control of bit rates will derive from their selection of terminals, electronics, and optical lasers. The rates for actual usage would be the same uniform charges as before.

The problem of sharing data and reimbursing the relevant costs of competing interconnecting carriers remains a thorny issue, since all network providers may not want (or even be able) to engineer for the guaranteed end-to-end throughput to which the subscriber's originating service provider has committed.

In the case of customer interruptibility rates, customers with heavy usage requirements but who do not require immediate guaranteed completion (e.g., some data services or large manufacturing firms) would receive a discount from the usual uniform usage charge with the understanding that their service will be blocked whenever the network host processor is busy. Thus, the public network becomes efficient at screening high-valued from low-valued customer usage at the margin by using a fairly simple rate structure.

The guaranteed capacity rates described here would probably be transparent to most small business and residential customers, who generally would not be willing to pay a monthly fee for guaranteed network capacity and would simply expect to get blocked or delayed occasionally the way they do today. This does not necessarily imply that messages will not get through, because one network access service or feature can be used for delayed call completion, for example, via a call store and forward service.

As the telecommunication industry evolves, regulatory costing and pricing practices must move in a direction that will allow for the realization of the benefits of an efficient public network for all customers. Fully distributed cost allocation practices are inappropriate to accomplish that goal, so IC practices should be adopted and costing and pricing flexibility phased in as rigid regulated structures are eliminated. Integration and flexibility are the key to the network of the future, and the same is true for costing and pricing if such a public network is to be developed to its fullest extent. If society in general is to benefit from a technically sophisticated integrated public telecommunications network, regulators must allow market-based pricing and marginal costing standards for new and enhanced services. Such a standard will lead to competitive rates and encourage network use by a maximum number of telephone subscribers. If artificial costs continue to be loaded onto new and enhanced services, they will become available more slowly, and access to them will likely be too costly for many who otherwise benefit from their use.

The prescription for a cost-based rate structure for the future integrated network infrastructure suggests that much simpler rate structures than those that exist today can be very efficient. Fairly simple nondiscriminatory rate structures not only are possible but are desirable for customer understanding.

6.6 NETWORK INFRASTRUCTURES AS PUBLIC GOODS

The public-good aspects of the future network infrastructure are often referred to in terms of the information superhighway.[4] In some respects the analogy is a useful one, and in some respects it is not. First, while it may be true that the interstate highway system might not have ever been built if it were not for the government to promote and fund its development, that is not so for the vast intercity network of trunk lines and switches of the phone network. It is also not true for the "on ramps" of subscriber access lines for either cable television or telephone service. On the other hand, it is true that both highways and telephone network infrastructures have social value (i.e., positive externalities) beyond that which strictly private interests would have.

At first blush, the highway analogy also seems somewhat inappropriate because prices are charged for using the telecommunications network infrastructure (at least for long distance calls), and there are no prices for using public highways. But technically, the roadways are not really free to use since gasoline prices include implicit payments for roads. The analogy does seem appropriate from a usage perspective—once the public infrastructures are built, the vast majority of usage on the network (highway) is free (local phone calls usually have no charge). Since anyone with a phone or a car may freely use the public network or the highways, it is difficult to say that they are not public goods based on the fact that it is possible to exclude the public from using them. Phone prices could be raised to the point that exclusion could be forced, but the same is true for gas prices.

One aspect of the information superhighway that could be directly analogous to public roads generally is the use of radio frequency spectrum (until the recent auctions that granted temporary private property rights to certain portions of that spectrum). Either may be used at will by the public as long as people have a license and as long as they do not interfere or collide with each other.

Like public roadways and many other public goods, there are possibilities for congestion to occur. That would not be the case for a so-called pure public good (e.g., air), where no one's consumption can affect anyone else's. This is

---

4. For a discussion of universal broadband networks as public goods, see B. Egan, *Information Superhighways: The Economics of Advanced Public Communication Networks* (Norwood, MA: Artech House, 1991), ch. 3.

true for both wired and wireless telephone networks, except that the phone company acts as the traffic cop for the wired network system by engineering the network to make sure that no one interferes with anyone else. Even in private computer networks, whether wired or wireless, interference and collision avoidance is the name of the game.

Roadways and the transportation infrastructure have long been recognized for their considerable contribution to the public welfare in terms of economic productivity and growth in jobs and income. But the same can be said about a ubiquitous, highly efficient telecommunications network infrastructure. In the same manner in which good roadway systems lower distribution costs for goods and services, the infrastructures for telecommunications networks are critical for reducing transactions costs (e.g., costs of finding and transmitting information) and exhibit substantial spillover effects to stimulate long-term job and income growth in most sectors of the economy.

Pure public goods, like pure competition, are rare to observe; therefore, each is a matter of degree. On that basis, then, the highway analogy is useful in many respects for understanding the public value of an advanced telecommunications network infrastructure. Those theoretical purists who claim that the public telephone network system cannot be considered a public good because it does not meet the strict economist's criteria for public goods are not adding any constructive criticism to the debate about the true costs and benefits of infrastructure investment.

Everyone has a personal opinion as to the social value of an advanced telecommunications network infrastructure. Personally, I do not take the information superhighway analogy, or any other argument for that matter, as evidence that the government should get involved in owning or financing any part of the future telecommunications infrastructure. I do take such arguments as compelling reasons for the government to quit overregulating the sector generally and, more specifically, to stop restricting the profits and scope of operations of infrastructure network providers, so that private investment will be stimulated. Only then could any objective evaluation be made to determine if government intervention is required to further stimulate infrastructure development.

It is difficult to produce an objective quantitative measure of the net macroeconomic benefits of telecommunications infrastructure investments. Suffice it to say that the overwhelming evidence from studies that have been performed to date indicate that the public payoff from private sector investments in information technology, in general, and telecommunications, in particular, is huge.

6.6.1   Public Benefits of Infrastructure Investments

Enhancements to the telecommunications infrastructure are claimed to carry with them many benefits. This section provides a selective overview of the types of benefits that can be expected from applications of advanced information technologies, focusing on those that are particularly well suited for use on the public network infrastructure.

Many government officials have been calling for applications of telecommunications technology to enhance the effectiveness of other basic service infrastructures, most notably, public administration, health care, education, transportation, and the network of researchers in universities, private industry, and government agencies. In public administration, many potential direct and indirect benefits are anticipated from advanced telecommunications infrastructures. Some are tangible and quantifiable, like mechanization and modernization of government information and communication systems; others, like a more informed and participatory electorate, increased convenience and efficiency in individual interaction with government agencies, and improvements in the criminal justice system are difficult to assess but are still socially valuable [3].

In medicine, important advantages are seen in making the knowledge of skilled specialists available to patients those specialists would otherwise be unable to treat. For example, advanced imaging technology combined with high-bandwidth transmission could make it possible for specialist experts in rare diseases and difficult surgical procedures who are based in metropolitan areas to participate electronically in the diagnosis of diseases and even to guide local doctors performing operations in remote areas. However, some researchers[5] see the greatest health benefits from telecommunications coming from applications designed to improve the efficiency of health care by improving public access to information and thereby increasing consumers' participation in the treatment and prevention of illness.[6]

Anticipated contributions to the productivity of the U.S. research and educational systems are also motivating forces behind many government initiatives. For example, President Clinton has set a goal that all classrooms in America will have access to the Internet by the year 2000. Some state governments have set even more aggressive goals. Substantial public moneys have already been committed to the support of telecommunications networks to further higher education and scientific research. The people best suited for working together on any given research project frequently are scattered among a variety

---

5. A. Melmed and F. Fisher, in "Towards a National Information Infrastructure: Implications for Selected Social Sectors and Education" (Center for Educational Technology and Economic Productivity, New York University, December 1991), review several examples of intangible benefits to public services administration through network technology adoption.

6. There have already been uses of remote imaging technology over high-speed fiber optic network links, and the practice is expanding. In "Towards a National Information Infrastructure," Melmed and Fisher list several categories of health care costs, totaling hundreds of billions of dollars, that are potentially "avoidable" using advanced telecommunication networks and digital databases.

of sites throughout the country or the world. Advanced telecommunications networks could help overcome time and space constraints on collaborative endeavors by facilitating file transfer and database sharing among researchers in different locations.[7] As in health care, educators foresee benefits in making scarce and specialized talent available to wider audiences. For example, a lecture on quantum mechanics delivered by a physicist at Princeton might be viewed by graduate students at Berkeley, Stanford, and any other university that wanted to pick up the transmission. Telecommunications technology is already being used to offer college degree programs up to the master's level to residents in rural (or even urban) communities.

While current services allow for only limited interaction between students and faculty, greater bandwidth would permit real-time interaction.[8] Similar services have been developed for elementary and secondary schools, where the need for substantial improvement in educational services is generally acknowledged.[9]

In the case of the transportation and postal infrastructures, advanced telecommunications also offers the hope of easing the load on already overburdened systems, which is already happening to some degree. Many messages that formerly would have traveled by U.S. mail or by private delivery services such as Federal Express and UPS are now sent by fax and e-mail. Millions of people telecommute, and many projections indicate that this number could increase to the point where it represents a substantial portion of the work force. Telecommuting produces direct savings in terms of miles traveled and time spent traveling [4]. Additional benefits are fossil fuel savings, reduced pollution and traffic congestion, and reduced associated health care costs (e.g., respiratory and stress problems). Loss of the benefits of face-to-face contact with coworkers is a constraint on telecommuting, as is the inability of telecommuters to simultaneously send and receive messages for multiple applications over the (bandwidth limited) public network. Those drawbacks could be reduced substantially, however, if the public network were upgraded to make telephonic video interaction and simultaneous, multichannel communication possible.

---

7. Wider application of networking technologies in and among research organizations may also lessen the tradeoff between specialization and interdisciplinary collaboration. See T. Allen and O. Hauptman, "The Substitution of Communication Technologies for Organizational Structure in Research and Development," in J. Fulk and C. Steinfeld (eds.), *Organizations and Communication Technology*, Newbury Park, CA: Sage Publications, 1990, pp. 265–274.

8. Telecommunications-delivered educational services also can be used to increase the effectiveness of home education for grades K–12, which, due to parental dissatisfaction with the traditional system, has been estimated to be growing at 30% annually.

9. See A. Melmed and F. Fisher, "Towards a National Information Infrastructure," for a partial survey of this literature.

In addition to home health care, learning, and working, many observers foresee significant benefits from applying telecommunications to other every-day activities such as shopping,[10] banking and bill paying, news and information consumption, and a whole host of leisure activities. The anticipated benefits will be realized in the form of time savings, convenience, and product and service enhancements, analogous to the benefits and time savings realized from current networks of automatic teller machines.[11] Many consumer activities take place outside the marketplace. While it is empirically, and sometimes conceptually, difficult to measure the value of nonmarket activities, estimates range as high as one-third to one-half the value of market activities [5].[12] Therefore, the payoff from applications of information technology to these activities is potentially huge.

Of the prophesized benefits from enhancements to the telecommunications infrastructure, hoped-for improvements in productivity and international competitiveness for American business probably have received the most attention. These benefits have also been the subject of the most serious attempts at documentation and measurement. There are not many published studies of the macroeconomic productivity effects of telecommunication infrastructure investments, but there have been a number of studies on the broader category of information technology [6].

---

10. Teleshopping is already possible over Prodigy and smaller dedicated shopping networks in some areas. However, consumer acceptance is low, due in part to the lack of ease and convenience of these types of electronic transactions on the relatively low bandwidth public network.

11. For individual households, technology adoption improves the standard of living through time savings by doing things "better." Two primary sources of welfare gains from time savings include the increase in leisure time, the time spent in an alternative productive activity previously foregone, and the quality of life generally.

12. Often, measurement of the value of many personal activities is relatively straightforward because they are available for purchase from third parties in the marketplace, such as child care and housecleaning. Others are hard to evaluate because they are problematical for third-party performance. Some examples are: personal care activities like getting ready for work (and getting back to "normal" after work), leisure and recreation activities like enjoying a movie or reading a book, thinking, and learning. Interestingly, but perhaps not so surprisingly, personal care activities take up a substantial portion of an individual's time, in fact, double that of paid work time according to a University of Maryland study. Even personal care time, which at first glance does not seem to have any potential relationship to electronic technology adoption in a household, may he indirectly related. No doubt, some portion of that time might he saved by telecommuting, shopping, or learning from home. See B. Egan, "The Case for Residential Broadband Telecommunication Networks" (Columbia Institute for Tele-Information, Research Working Paper Series, Columbia Business School, 1991), sect. 4, "Demand Side Considerations," for a more detailed treatment of some of these issues.

6.7 REGULATING MARKET ENTRY AND EXIT

Besides considerations of regulatory rules promoting universal service, barriers to entry represent the most critical concern of competition policy. Barriers to market exit are also a concern, but that issue is mostly related to the government-imposed universal service and the so-called *carrier of last resort* (COLR) requirement.

Anyone who agrees that the market forces of competition are preferred to regulation would recommend that the government move forthwith to eliminate artificial regulatory barriers to market entry. In particular, that means there should be no regulations that favor or otherwise protect the incumbent monopoly provider of basic local telephone service, for example, by making it difficult for new entrants to obtain franchises or other operating certificates of "public convenience or necessity."

Two particularly unique aspects of universal service policy that are fundamentally at odds with the government's competition policy are the regulations that require incumbent firms to continue to subsidize low basic local service prices for residential subscribers and the barrier to market exit. Those policies cause substantial disincentives for new firms to enter the market. It is obvious why the incumbents' below-cost prices for basic service would discourage entry by a competitor that must make a profit to survive. Another typical provision of universal service policy is that, on market entry, a firm must commit to provide service to all households on demand throughout a specific geographic area at broadly averaged (usually low) monthly rates without regard to the cost or profitability of serving any particular customer. Assuming that this type of barrier-to-exit rule will be enforceable over the long term, it raises the risk of competitive entry considerably. Probably the best conclusion that one could hope for is for the barrier to exit to be strictly limited to service provided to low-income or otherwise unprofitable subscribers and that subsidies be paid to any carrier that meets this truly unique obligation, lest it become a serious competitive disadvantage vis à vis competing private network providers, which are not subject to the same rules.

Other important economic barriers to entry include: (1) the incumbent's sunk network investment, historically paid for from monopoly ratepayer funds, which could create a formidable and potentially prohibitive financial barrier for new entrants to overcome; and (2) the incumbent's not allowing new entrants to interconnect to its PSTN facilities, thereby preventing competitors from competing on service quality because its subscribers cannot achieve ubiquitous calling capability.

The first of these two barriers is a genuine issue for competition policy, since it is a benefit of the monopoly era bestowed on the incumbent by cost-based regulation. Practically speaking, however, in a world where profitable market opportunities are being pursued by an almost limitless amount of capital, sunk-cost arguments would probably not constitute valid arguments for continued regulation of the incumbent if there were no other regulatory barriers to entry. For example, assuming the LECs were profitable in any given market area, it is ludicrous to think that other corporate giants with deep pockets, like AT&T, would not feel free to enter the market and pursue the same profit opportunities. The second barrier listed is a genuine concern, too, except that it is generally illegal to deny interconnection to such public network facilities. The courts would view such a denial as a violation of principles governing private access to so-called essential facilities of public utilities.

Thus, the only barriers to entry with which regulators should be concerned are artificial (those created in accordance with or facilitated by the regulatory process itself). Entry barriers that derive solely from the operating efficiencies and business acumen of firms, be they large incumbents or small entrants, are not the province of regulatory policies. It is important that regulatory policies clearly recognize the distinction between entry barriers that are too high to clear and competitive rivals who just cannot jump.

For example, an incumbent's reputation for being a high-quality, reliable service provider constitutes a barrier to entry for certain competitive entrants who may have a relatively poor track record for timely restoration of service and reliability. No one would seriously propose that regulators do anything to eradicate this barrier to entry. The following excerpt from Schumpeter captures the spirit of this principle [7]:

> The first thing to go is the traditional conception of the modus operandi of competition. Economists are at long last emerging from the stage in which price competition was all they saw. As soon as quality competition and sales effort are admitted into the sacred precincts of theory, the price variable is ousted from its dominant position. However, it is still competition within a rigid pattern of invariant conditions, methods of production and forms of industrial organization in particular, that practically monopolizes attention. But in capitalist reality as distinguished from its textbook picture, it is not that kind of competition that counts but the competition from the new commodity, the new technology, the new source of supply, the new type of organization (the largest scale unit of control for instance)—competition which commands a decisive cost or quality advantage and which strikes not at the margins of the profits and the outputs of the existing firms but at their foundations and their very lives.

The Schumpeterian view of competition is that of an evolutionary process. The perennial gale of "creative destruction" pits firm against firm and product against product in a constant struggle for market dominance. Market dominance is rewarded with high profits, not penalized with competitive handicapping. In that context, the following point cannot be overemphasized: The level playing field refers to an equality of opportunity to compete—not an equality of marketplace outcomes. A regulatory preoccupation with market share mistakenly focuses on the latter.

6.8 ASYMMETRIC REGULATION AND CONSUMER WELFARE

The regulatory remedy to barriers to entry is to regulate the incumbent to guard against its potential for keeping out competitors. Because the same regulations would not apply to entrants, the regulations are said to be asymmetric.

Generally, the purpose of economic regulation is to augment competitive market forces by intervening in markets to emulate those forces when the market itself fails to discipline firms for exercising monopoly power. But that does not always call for regulatory intervention. The rule of thumb is that regulation should be implemented as long as it can be shown that the total costs of implementing it are exceeded by the total benefits received from it.

Of course there is no such thing as perfect competition, so the best a regulator can really hope for is to intervene to create effective or "workable" competition. Dr. Alfred Kahn describes effective competition in this way [8]:

> Effective competition and economic efficiency alike require that lower-cost firms be encouraged, because of their own lower costs, to reduce their prices to take business away from other higher-cost competitors.

The most effective way to promote economic efficiency in competitive markets is to allow competitive market forces to entice and discipline the behavior of firms. Effective competition requires that all firms compete on the merits of their respective efficiencies. By contrast, imposing artificial restrictions on one competitor but not another (asymmetric rules and responsibilities) can mask the relative efficiencies of firms and thereby allow inefficient firms to displace efficient firms. While it may give the appearance of competition, this form of rivalry does not constitute effective competition. A level competitive playing field should be encouraged and maintained, not by handicapping the efficient players, but by allowing fair and equitable competition to sort the efficient firms from the inefficient firms.

In some cases, certain *public service obligations* (PSOs) are regulatorily imposed on firms as an instrument to affect certain social policies that may not otherwise be addressed in a competitive marketplace. Examples of this are the *universal service obligations* (USOs) and COLR obligations (discussed at length in Chapter 7). In any event, the pursuit of these social policies should be largely transparent to the competitive process. The primary objective of regulation in a competitive environment should be to foster an equal opportunity to compete among market providers without preordaining marketplace outcomes. In other words, regulators should endorse the principle of competitive parity. Competition is a means—frequently an effective means—by which to enhance social welfare and economic efficiency in telecommunications markets. Competition is not, however, an end to itself. Maximizing economic welfare is not synonymous with maximizing the absolute number of competitors in telecommunications markets.

In the telecommunications industry, asymmetric regulation has primarily taken the following forms:

- Pricing constraints necessary to support various social policies;
- Geographically averaged rate structures that do not reflect corresponding cost differences;
- COLR obligations that require the incumbent firm to stand by with the service capacity in place to serve consumers on demand, either where other competitors have chosen not to provide service, or in the event of a failure on a rival's network;
- Information disclosure requirements that force the incumbent firm to reveal to competitors, in advance, plans for new service offerings and associated prices and strategies.

Aside from the obvious efficiency effects of handicapping one firm over another, today's problems represent the results of a long history of asymmetric regulation. Unfortunately, federal regulators never adequately addressed the fundamental question: Can there be competition—real competition—when the parties to it do not enjoy the same freedoms, bear the same responsibilities, or endure the same constraints? As a result, there has been a needlessly prolonged and socially costly policy of asymmetric regulation in the interstate long distance market in the name of protecting the cross-subsidies on which much public policy was founded.

Similarly, in a postcompetitive environment, all firms should share (in a competitively neutral manner) in the burden of the costs of continuing what were formerly PSOs of the monopoly incumbent. Providing universal telephone service and ensuring the presence of COLRs are important public policies. However, the pursuit of these policies is not costless. If incumbent suppliers bear all the costs, they are disadvantaged relative to competitors. Consequently, competitors may be able to survive in the marketplace even though they are not the least-cost provider of service.

Alternatively, incumbents may fail to survive even as they meet or beat the efficiencies of their rivals. To ensure that only the most able suppliers survive in the marketplace, the cost of achieving social goals should be borne symmetrically by all market participants, perhaps through a revenue surcharge, or, preferably, through a value-added levy on all telecommunications firms, or, even better yet, funded from general tax revenues. The overriding consideration is that administration of these social obligations not be carried out in a manner that distorts the competitive process. In other words, administration of these policies should be largely transparent to the competitive process.

The true social costs of asymmetric regulation are difficult to measure because they involve market transactions that do not occur but would have otherwise. In other words, the cost of asymmetric regulation is the value of the benefits of competition that are foregone. Nonetheless, the fact that the FCC continued to regulate the long distance market for more than 20 years after competition first emerged, is direct evidence of the failure of this policy. Mark Fowler, a past chairman of the FCC, acknowledged the failure of asymmetric regulatory policies in an article he cowrote after leaving office [9]:

> It can be argued, for instance, that some of the commission's regulatory actions in the interexchange market that were designed to promote competition during transition, such as highly discounted access pricing for OCC's [other common carriers] and restrictions on competitive pricing responses by AT&T, in fact have encouraged entry by uneconomic providers and uneconomic construction of excess capacity. If this is true, the gradualist approach to deregulation of interexchange markets will have resulted in substantial, unnecessary costs for society that never would have been incurred in a truly competitive marketplace. Moreover, this approach will have directly increased consumer costs by requiring regulated firms to charge higher prices to protect competitors during the transition.

Asymmetric regulation may lead to at least four types of social cost. First, asymmetric regulation fosters technical inefficiency or productive inefficiency, because it can preclude the least-cost provider from being the least-price provider. That may occur, for instance, because the incumbent firm may have regulatorily imposed on it PSOs not borne by its competitors. Hence, asymmetric regulation derails the competitive process and thereby harms consumers.

Second, the regulated incumbent firm's PSOs tend to inflate its costs relative to those of its competitors because it is required to deploy capital ubiquitously without regard to profitability. On high-cost, low-density routes, the incumbent firm is frequently the exclusive provider of service. Yet, because regulation requires that costs be averaged for rate-making purposes, the incumbent firm's obligation makes it easier for a relatively high cost entrant to compete with the incumbent firm in the high-volume, low-cost market segments. From the entrant's perspective, saddling the incumbent firm with a PSO is a means to raise its rival's costs. Moreover, the competitive entrant can default to the incumbent in the event of a network failure because of the latter's ubiquitous deployment of capital and common carrier obligations. From the consumers' perspective, there may be no risk in using the services of a lower-priced, low-reliability provider because they can always turn to the regulated common carrier in case of need and pay nothing for the priviledge.

Third, certain information disclosure requirements (e.g., changes in tariffs and service offerings) placed on regulated incumbents often constitute a market advantage to new entrants. That leads to subtle inefficiencies that are difficult to measure because they involve welfare losses associated with foregone innovation. For example, the incumbent firm may fail to invest in innovation because the information disclosure requirement precludes it from capturing the returns from innovation. When competitive entrants are granted advance knowledge of the incumbent firm's product plans and strategies, they can wrest first-mover advantage from the incumbent firm by delaying it's product introductions or making their own offerings first. In this fashion, asymmetric regulation fosters imitation and stifles innovation. This argument goes beyond the standard critique that the competitive entrant may meet with market success even when it does not have a better mousetrap. The problem is that the rate at which mousetrap innovation occurs is artificially retarded due to asymmetric regulation. This entails dynamic efficiency losses resulting from a suboptimal level of investment in innovation.

Fourth, asymmetric regulation provides new entrants with a nonmarket means to compete with the incumbent firm. That constitutes an extraction of profits from the incumbent and a transfer of profits to the entrant (sometimes referred to as "rent seeking" or "regulatory predation"). The entrant may have an artificial competitive advantage in the regulatory arena relative to the incumbent, not because of its lower cost, but because the incumbent must bear the burden of proof with regulators and customers while the entrant bears the burden of proof only with customers.

Hence, asymmetric regulation gives rise to an inferior breed of competitor—more adept at imitation than innovation and more prone to battle in the hearing room than in the marketplace. For those reasons, it is critical that regulators endorse a policy of symmetric regulation and competitive parity to achieve effective competition.

The first principle of regulatory reform for curing the ill effects of asymmetric policies involves a simple and straightforward proposition: Regulations should enable the development of competition in the industry without mandating it directly or promoting it artificially. The practice of asymmetric regulation generally is inconsistent with this principle. Therefore, in a situation where it is not known what the outcome of a given asymmetric policy will be, regulators should err on the side of not adopting (or otherwise withdrawing) the policy.

The second principle is that regulatory policies should facilitate production by low-cost providers. To that end, regulatory policies should be nondistortionary and competitively neutral.

The third principle (which is actually a corollary of the second) is that burdens formerly placed on the incumbent firm (e.g., cross-subsidization, rate averaging, PSOs) should become industrywide or economywide burdens under competitive market conditions. Continuing to treat these obligations as the exclusive responsibility of the incumbent firm violates the second principle of competitive neutrality.

The fourth principle is that regulatory policies that promote static (allocative and productive) efficiency may not be perfectly consistent with policies that promote dynamic efficiency. Here it should be observed that the proverbial level playing field is inherently a static concept.

The fifth principle stresses the importance of limiting incentives for undesirable arbitrage of regulatory rules. To the extent possible, the regulatory process should be immune to strategic manipulation by the incumbent and competitive entrants alike.

Sixth, regulatory rules, including incentive regulation plans, should explicitly include provisions for their own sunset once movement toward a competitive market has commenced. The overhang of excessive regulation imposes direct costs on society and raises the risk of incurring the indirect costs of potentially harmful marketplace intervention. The preservation of interests established under the regulatory rules tend to be perpetuated by the very availability of such rules.

Incumbent carriers should be given the ability to respond to competition. In particular, incumbents should be afforded pricing flexibility regardless of market share, with appropriate safeguards against predation. Such safeguards generally require that prices exceed incremental production costs. The incumbent LEC should be free to set prices at will within predetermined rate bands or be afforded complete pricing flexibility, with IC serving as the price floor. The overriding objective is to ensure that the least-cost provider is not precluded from being the least-price provider.

6.8.1　Consumer and Competitive Safeguards

One of the most popular set of asymmetric rules to protect competitors are those that regulate the incumbent's prices, forcing them to remain at an artificially high level so that competitive entrants can attract the incumbent's customers and make a profit merely by undercutting the incumbent's prices. That is the practice the FCC followed, much to the delight of competitors, in the continuing regulation of AT&T in the interexchange market. This type of regulation is absolutely the wrong policy. It can inadvertently put competitors in the driver's seat for determining when, if ever, regulation of the incumbent should be relaxed. Many economists have noted that entrants in the toll market are happy to soak up the rain of profits dripping from the monopoly price umbrella of the incumbent. Note, however, that when the 1996 Telecommunications Act was still pending before Congress, and the shoe was on the other foot, AT&T took the low road and lobbied heavily to restrict the RBOCs. At that time, *The Economist* magazine, in describing the pending bill, stated it thusly [10]:

> ...the biggest question—which companies will win and which will lose—still divides the industry's analysts.
>
> No wonder. A few months ago the future seemed to belong to the Bells. The test of the presence of competition in the Bells' home market seemed vague enough to allow most of them into the long-distance market within a year or so, long before their rivals had time to make much of a mark. A single phrase in the final version of the bill may change that. It adds a demand for a "facilities-based" competitor to the checklist. That may mean that giant long-distance companies such as AT&T and MCI can move into local markets as much as they want, reselling capacity on the Bells' lines, but that the Bells cannot reciprocate. Only when the long-distance competitor begins to build its own networks ("facilities") will the Bells pass the competition test. That phrase could keep the Bells out of long-distance markets for many years.
>
> This may be exactly the opportunity that AT&T's chairman, Robert Allen, had in mind.

Now that the legislation has passed you can bank on the prediction that AT&T will vigorously contest every application of the Bell companies to enter the long distance business. Now that there is a competitive checklist that requires facilities-based local exchange competition before Bell companies are allowed into the long distance business, AT&T will aggressively lobby federal and state regulators for large discounts on LEC local service offerings, leading to limited incentives for new entrants to build local service facilities; that will cause long delays in the LECs' ability to meet the competitive checklist. Whether or not it is the *interexchange carriers* (IXCs), CAPs, or others who ultimately play this strategy, conditioning streamlined regulation on market share gives rise to poor incentive properties.

In the postdivestiture period, not only did this regulatory practice force AT&T to run the entire race in the outside lane, it also forced AT&T to slow down whenever it got too far ahead. Indeed, this asymmetrical regulatory structure forced AT&T to adopt a strategy of "optimal mediocrity"—a careful balancing of the benefits from innovative new products and services with the costs of additional regulation should it fare too well in the marketplace. It would be difficult to envision a greater efficiency loss than that of diverting the talents and resources of AT&T and its employees in this manner.

In fact, market share is not a reliable indicator of market power. This point is made succinctly in a recent paper by Schankerman [11]:

The market share of a firm is an endogenous variable and is determined by the same fundamental factors that govern market power. Market share does not cause market power any more than market power causes market share. The fact that market power and market share both reflect the underlying efficiency levels of all firms in the industry cannot be overemphasized. A policy which conditioned regulatory streamlining on the incumbent's market share would have the effect of penalizing efficiency and commercial success, and would represent major retrogression from the recent provision of efficiency incentives under price caps.

Reliance on market share as an indicator of market power is particularly troublesome in regulated markets wherein (1) prices may be maintained below efficient levels; and (2) entry and exit restrictions are in place. Landes and Posner have also recognized this point [12]:

> The causality between market share and market power is reversed. Instead of a large market share leading to a high price, a low market price leads to a large market share; and it would be improper to infer market power simply from observing a large market share.

To summarize, market share is a necessary but not a sufficient condition for exercising market power.

6.8.2 The Role of Costs in Determining Economically Efficient Prices

Promoting economic efficiency is actually a key element in furthering the public interest. To attain economic efficiency, two conditions must be satisfied: (1) it must produce each service using the lowest possible value of resources; and (2) the value that consumers place on each service must be at least as great as the cost incurred by the firm in the production of the service. Of particular importance in promoting fair competition and economic efficiency among firms is the IC criterion, which requires that each cost incrementally imposed on a firm be compensated by associated revenues. This condition is necessary for economic efficiency. Economic efficiency should be promoted within firms as well as among firms, whether regulated or not.

Economic theory establishes that rates should equal or exceed marginal cost. Due to economies of scale and scope, however, rates equal to marginal cost are not financially viable. Thus, this requirement is usually satisfied in practice by showing that each rate exceeds its average volume-sensitive cost. When a single rate is selected for a service, setting that rate to exceed the service's AIC normally more than satisfies this criterion since AIC includes both volume-sensitive and volume-insensitive costs. Unless there is an explicit public policy to the contrary, the regulated rates for each service should be set so that the revenue from the service in total covers the IC of the service in total. This provides safeguards against both cross-subsidization and predatory pricing. In order for the firm to remain financially viable, rates must (on average) exceed this minimum requirement to recover the costs shared by multiple services. Just how much contribution toward shared costs is obtained from each service will depend on competitors' prices and the relative value of the services to customers. Open market entry and exit generally will constrain rates and earnings from being too high, and the resulting competition will promote economic efficiency. Intervention into setting rates should be confined to governing any portions of the business determined to be naturally monopolistic (if any) or to accomplish specific public policy objectives.

Although this approach to regulated rate setting results in flexible pricing for most services, any attempt to set rates based on more rigid cost formulas such as fully allocated costs is not compatible with today's increasingly competitive environment.

6.9 NETWORK INTERCONNECTION AND NETWORK COMPONENT UNBUNDLING

When the sale of certain services necessarily requires unique capabilities, features or functions of some underlying input, the final service is often a bundle that contains "essential" input in a single service offering with a single price. If only one firm has substantial market power over the underlying unique input(s) or essential facility, it may be desirable for regulators to require that the firm unbundle the essential facility and offer it for sale to others wishing to compete in the final service market. Also, to stimulate competition in final services and to prevent the monopolist from pricing the underlying essential facilities so high relative to the price of the service using those facilities that it forces an equally efficient "dependent" competitor to lose money on its competing service, regulators may regulate the price of the essential facility so that the competitor will not be squeezed out of the market. This form of anticompetitive behavior is prevented by competition and the profit motive in unregulated sectors of the economy. In regulated markets, where no substitute for essential facilities exists, an appropriate "imputation rule" will protect against it. The imputation rule is a pricing rule that safeguards competitors against price squeezes by the owner or controller of the underlying essential facility.

For a facility (or, more correctly, for any resource or capability) to be essential, the competitor must be dependent on the facility owner, meaning that the competitor is unable to compete in the final market without acquiring the essential resource. This dependence may be due to law, regulation (e.g., the legal exercise of patent rights), the inability to economically substitute for or circumvent the facility, or technical limitations that tie one competitor to another. When a resource provided by the incumbent monopoly service provider is required by its competitors in order to provide a particular service to end users, the incumbent is said to control (or have substantial market power over) an essential facility. That may occur when the incumbent provides both competitive and noncompetitive services, and its competitors must purchase a noncompetitive service element to provide the competitive service in question.

Technically, once markets are opened to competitive entry, it may be possible for any competitor to acquire or build a functional substitute for any portion of the incumbent LEC's PSTN (e.g., switch, trunk transmission line, or loop), and there will always be disagreements on what exactly constitutes an essential facility. Obviously, what is or is not essential is in the eye of the beholder. From the perspective of a potential LEC competitor in any particular situation, the "essential" classification will contain different facilities at different times. Any competitor needing access to underlying PSTN facilities to compete with the LEC will claim that it is essential.

There are two important economic rules regarding essential facilities. The first is that the facility must be made available to potential competitors for purchase through unbundling of the essential facility or component. The second rule is the imputation rule with respect to pricing services using the essential facility. These rules should, however, be strictly confined to essential facilities or their equivalent.

Unbundling is a concern when a prospective competitor wants to enter the market, with the intention of supplying only a few specialized services, but cannot technically or economically do so without relying on at least some of the LEC's infrastructure. An example would be the ability to terminate calls to the LEC's customers. Also, a problem may arise when a current competitor of the LEC arranges a contract with a large customer to provide the customer with only a few specialized services, once again, where one of the services requires, for economical or technical reasons, the LEC's facilities. If unbundling is not required, the supply of the specialized services by the competitor can be prevented by the LEC. If the LEC is not willing to supply the specialized services alone but instead requires the competitor to purchase other services (which may not be necessary to the competitor) in addition to the specialized services, then the competitor may be prevented from serving the market most efficiently. Bundling services (which require the use of an essential facility) with other services may retard effective competition and may encourage inefficiencies.

In general, the competitor requesting an unbundled component should be required to provide evidence that the component is necessary to provide a competitive telecommunications service and that it is feasible to make the component available. In practice, a competitor should have to establish the following:

- The component is a monopoly component available only from the incumbent LEC.
- The component is required for a competitor to compete in the incumbent's existing markets.
- The component is technically feasible to unbundle.
- There is no reasonable technical or economic alternative to obtain or circumvent the component.

It is important to note that the capabilities of competitors change with time, so a service or particular use of a facility that today is deemed to be essential may become nonessential in the future as technology advances and the capabilities of the competitor change.

6.9.1 Regulating Prices of Essential Facilities

For purposes of imputation, a facility is essential if a firm has no economically or technically reasonable alternative to employing its competitor's facility to assemble a retail product or service to compete with that competitor.

A component is not essential if the competitor can provide the component itself, obtain it from other sources, or substitute alternative technologies or practices for it at comparable cost or efficiency. In such a circumstance, the component is considered to be nonessential or optional. "Essentiality" is not determined by management or strategic choices of competing firms. For example, just because Ford and General Motors cars both require steering wheels does not mean Ford must sell its steering wheels to GM. GM has its own capability to make or otherwise acquire steering wheels; thus, neither the steering wheels nor the capabilities to make them are essential facilities.

The proper price floor for a competitive service requiring the use of an essential resource can be determined by using the general rule of imputation: When a firm uses an essential facility in its own competitive retail service, the retail price of the service should equal or exceed the IC of the retail service plus any *lost-contribution margin* (LCM) that could have been obtained from selling the essential facility to its competitors

A nontelecommunications example can help explain how imputation rules work. Assume that a company has patented a particular chemical compound indispensable to the production of a new medicine. Many firms can supply the medicine in the competitive marketplace, but to do so, all firms need access to the newly developed chemical compound. Assume further that the firm developing the chemical compound is in the business of producing and selling medicine in competition with the other firms. The prices and ICs of the essential-component provider (the provider of the chemical compound) and one of its competitors are shown in Table 6.1.

Table 6.1
Imputation Rule: Scenario 1

Parameter	Essential-Component Provider		Competitor A
	Medicine	Compound	Medicine
Price	$11	$8	$11
AIC (compound)	(4)	(4)	(8)
AIC (competitive components)	(3)	0	(3)
Contribution to shared and common costs	$4	$4	(0)
LCM	($4)	0	0
Price less price floor	(0)	$4	(0)

In this example, the essential-component provider and competitor A both sell medicine in the retail marketplace. They both charge a market price of $11. The essential-component provider sells the patented compound to its competitor for $8. It costs $4 to produce the patented compound,[13] whether the essential-component provider supplies it as a part of its own retail medicine or provides it to a competitor (that is, there are no economies or diseconomies of vertical integration). Other components can be provided competitively by either the essential-component provider or its competitor at a cost of $3.

First, notice that the essential-component provider satisfies the IC + LCM rule. In this example, the $4 contribution from the sale of the patented compound is lost when the essential-component provider sells medicine. That is, for the essential-component provider to realize a unit sale of medicine, the competitor loses the sale, and therefore the essential-component provider loses the contribution that would have been made by selling the compound to the competitor. By adding the lost contribution ($4) to the IC of the medicine ($7), the price floor becomes $11. Since the medicine is selling for $11, the retail price of the medicine satisfies the price floor. Neither producer can force the other out of business without violating its respective price floor (the price floor of the competitor is its IC for reasons discussed previously).

Table 6.2 continues the example. Consider what would occur if competitor B sells the compound (not the medicine) in a market not related to the medicine market (i.e., in a market in which the patent holder does not compete).

---

13. The difference between the $8 price and the $4 cost can be considered to be a contribution toward the research and development costs leading to the discovery and production of the compound.

Table 6.2
Imputation Rule: Scenario 2

Parameter	Essential-Component Provider		Competitor B
	Medicine	Compound	Compound
Price	$11	$8	$10
AIC (compound)	(4)	(4)	(8)
AIC (competitive components)	(3)	0	0
Contribution to shared and common costs	$4	$4	$2

In this example, the producer of the compound has no contribution loss when it sells the compound to an independent firm. This is because a sale by the independent firm does not compete with the medicine market. Thus, the traditional constraint enforced by competition, that price equal or exceed IC, is the only constraint that must be satisfied by either party. This constraint is adequately enforced by competition and needs no oversight from regulators.[14]

The producer of the compound and the medicine could not engage in a price squeeze if the medicine contains no essential compounds. Any attempt to raise the price of the compound above the market levels would simply encourage an equally efficient or more efficient provider of the compound to fill the market need. The concern for a price squeeze arises only when one party has an essential compound that is a component of the products that the party and its competitors offer in the marketplace.

Imputation rules are not necessary when the LEC's PSTN facilities are not essential for a competitor to provide its competing service(s). That means that an LEC should not be required to follow imputation standards when it employs facilities that may have been essential for some purposes but are not essential for the immediate application. For example, in the example in Table 6.1, there may come a time when the patent on the chemical compound expires. At the time of expiration, other competitors may have unrestricted access to producing the chemical compound rather than purchasing it from the original compound inventor. When that occurs, competitors could, with equal efficiency, produce the compound or, at their discretion, continue to buy the compound from the original compound provider. Regardless of whether the competitors choose one or another source of supply for the compound, the compound

---

14. This does not imply that imputation substitutes for the body of antitrust law and its underlying economic principles; it simply states that there is no separate need for price regulation regarding price squeezes.

would no longer be considered an essential component. Thus, a component may be essential under some circumstances and not under others; the conditions change over time and across circumstances.

Figures 6.4 and 6.5 are useful illustrations of the imputation rule for local telephone companies as it has been applied in the past to interconnecting long distance companies and long distance (toll service) resellers. Figure 6.4 shows that the pricing floor for the LEC's own long distance service must be no less than the sum of its IC of providing the service plus the profit it would have made by selling interconnection service to the long distance company, which would in turn have provided the service to the end user.

Figure 6.5 depicts the situation in which a long distance reseller must purchase some wholesale PSTN service from the LEC (e.g., private line) to provide a retail long distance service in competition with the LEC. From Figure 6.5, a potential price squeeze is prevented when the LEC's own retail service price equals or exceeds the sum of the price of the wholesale (resold) service and the difference between the LEC's cost for its retail service and its cost of its wholesale service.

No special regulatory constraints need apply to services or components of services that are subject to competition on par with that generally found in unregulated markets. In addition, competitive services should be priced so that revenues from competitive services cover the IC of competitive services. This condition provides assurance that there is no opportunity to cross-subsidize competitive services through revenues received from essential facilities.

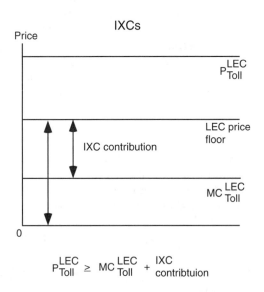

Figure 6.4 Imputation test: toll service.

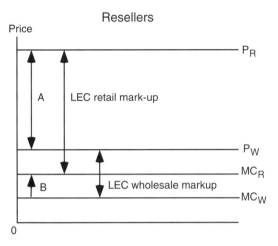

Figure 6.5 below:

Rule: A must be $\geq$ B or

$$P_R \geq P_W + (MC_R - MC_W)$$

P_R: LEC retail price
P_W: LEC wholesale price to resellers
MC_R: LEC marginal cost of retail
MC_W: LEC marginal cost of wholesale

Figure 6.5 Imputation test: resellers.

6.10 PRICING SUBSIDIZED SERVICES

Even though imputation rules are often used to prevent cross-subsidies, many such cross-subsidies currently exist in regulated rate structures. It is difficult to continue to subsidize services in a competitive environment without government support and intervention. Most often, that support is the provision of public services such as the transportation network (including highways, roads, and bridges). Such public services are generally owned and operated by the government rather than private industry and are financed by general tax funds rather than cross-subsidies within the privately owned and operated economy. It is rare that such subsidies are maintained within the private economy. When private subsidies do exist (such as with agricultural activities), the funding is again normally provided out of general tax funds. Even more rare is the requirement that an individual company subsidize its services or portions of its services. When such circumstances do arise (and this is usually the case when averaged rates are required of a public utility), the company having such a requirement is a protected monopoly. The cross-subsidy is maintained by protecting the source of the subsidy payments from competitive entry and erosion. In the current telecommunications environment, a rather novel attempt is underway to allow the maximum amount of open competitive entry while maintaining public policies that require a subsidy. For that reason, there is a strong movement among regulators and legislators to identify external sources of funding for subsidized services rather than continuing to rely on internal cross-subsidies. Many jurisdictions around the nation are investigating the means of measuring the need for subsidies and are identifying the appropriate mechanisms for funding those subsidies.

Previous attempts to fund subsidies within telecommunications from external sources are failing because the source of the subsidy payments is too narrowly targeted and therefore can be bypassed. For example, access charges on IXCs are, effectively, taxes on long distance carriers designed to fund subsidies among the services of local exchange carriers. When such narrowly targeted taxes are used, there are often substitute services that fall outside the taxed jurisdiction (we typically call this *bypass*). Inefficient and uneconomic activity is encouraged as long distance carriers attempt to find ways to reduce costs by circumventing the taxes. This problem is largely avoided in the economy at large by levying taxes over a very broad base, such as sales taxes on all vital sales, taxes on all income, or property taxes. These issues must be addressed in finding sources for subsidies within the telecommunications market. Competition and maintenance of current subsidies are fundamentally incompatible with one another and unsustainable. Either restrictions on the degree of competition or an elimination of cross-subsidies is necessary. If subsidies are not obtained from sources protected from competition or from broadly based tax-like arrangements, they are not sustainable. In addition, the rules governing the unbundling and pricing of subsidized services must be carefully and consistently constructed if a sustainable market in which competitors sell services obtained from one another is to emerge.

The best (i.e., most efficient and sustainable) approach to maintaining subsidies in the face of competition is to align rates of services sold to competitors so that the rates charged meet or exceed ICs for each item sold. Of course, if this rule were not further modified by the imputation rule (which raises the price floor above IC), competitors could be placed in a price squeeze in those areas where costs are higher than the retail price of the incumbent LEC (presumably regulation causes some prices charged by the LEC to end users remain subsidized while the prices charged to the competition are not subsidized). If, however, the competitor were to receive the same subsidy as the LEC for pricing service below cost, a price squeeze also could be avoided. The result would be competition on the merits of the relative efficiencies of each competitor and the maintenance of the mandated subsidy. Of course, as before, the subsidy should come from external sources rather than internal cross-subsidies.

Imputation rules would also apply to subsidized services sold to competitors. A modification of the chemical compound and medicine example can illustrate the imputation rule in this case (see Table 6.3). In this example, the competitors consist of competitor A, who sells the compound as is in competition with the essential-component provider, and competitor B, who uses the compound to manufacture medicine in competition with the essential-component provider. In both cases, it is assumed that the market prices for the medicine and the compound are $6 and $4, respectively. The prices of the products are the same regardless of which competitor supplies the product.

Table 6.3
Imputation Rule: Scenario 3

Parameter	Essential-Compound Provider		Competitor A	Competitor B
	Medicine	Compound	Compound	Medicine
Price	$6	$4	$4	$6
Subsidy	$1	0	0	$1
AIC (compound)	(4)	(4)	(4)	(4)
AIC (competitive components)	(3)	0	0	(3)
Contribution to shared and common costs	0	0	0	0
LCM	0	0	0	0
Price less price floor	0	0	0	0

In this example, a subsidy amount is calculated based on the difference between the desired medicine price and its cost (for the moment, the cost is based on that of the essential-component provider). It is also assumed that it is the medicine that is to be subsidized, not the compound as it may be utilized in unrelated applications. The compound is sold to both competitors at cost (an assumption to be dropped later).

Note that the source of the subsidy in this example need not be external to the firms. Internal (cross-subsidy) sources of funds may be used to supply the $1 subsidy and still satisfy competitive parity so long as both firms have equal opportunities to supply such funds.

The application of the imputation rule suffices to prevent a price squeeze and establish competitive parity among competitors. Of course, in this simple example, the lost contribution margin is zero since the compound is priced at cost.

Building on Table 6.3, consider the situation in which it is less costly to supply competitors in a wholesale manner than to engage in a vertically integrated production process (there is a diseconomy of vertical integration), as outlined in Table 6.4. A positive contribution from selling the compound results. All numerical values remain the same as before, except that the cost of the compound is $1 lower when the compound is sold to a competitor than when it is vertically integrated in the production of medicine.

Table 6.4
Imputation Rule: Scenario 4

Parameter	Essential-Compound Provider		Competitor A	Competitor B
	Medicine	Compound	Compound	Medicine
Price	$6	$4	$4	$6
Subsidy	$1	0	0	$1
AIC (compound)	(4)	(3)	(4)	(4)
AIC (competitive components)	(3)	0	0	(3)
Contribution to shared and common costs	0	1	0	0
LCM	(1)			
Price less price floor	(1)			0

Because the compound provides a $1 contribution, the LCM is imputed in establishing the price floor for the medicine as sold by the essential-component provider. The essential-component provider is, in this example, squeezed out of the medicine market by competitor B. That is, competitor B can at least break even in the medicine market while the essential-component provider would need to price below the imputed price floor in order to compete. That price squeeze could be avoided if the price of the medicine were increased to $7, in which case the essential-component provider would just meet its price floor. Of course, the competitor of medicine (competitor B) could increase its price to $7, thereby showing a positive contribution of $1. Should it choose to do so, competitor B could use this competitive advantage to force the essential-component provider out of the medicine market.

If the medicine producer is a profit-maximizing firm (as opposed to a firm with regulated earnings), it would not need to be forced out of the medicine market in that it would find the compound market more profitable (medicine provides a contribution to shared costs of zero, while the compound provides a positive contribution). Thus, competitor B's advantage is based entirely on the relative efficiency of producing medicine in a vertically disintegrated rather than integrated manner. Competitor B's competitive advantage is precisely equal to the difference in resource cost associated with each means of production. In that case, a total of $6 of society's resources are required to produce the medicine by transferring the compound to competitor B, while $7 worth of society's resources are required if the compound is produced in a vertically integrated manner. Finally, note that there is no need for an imputation rule to apply to competitor A.

Pricing an essential component above IC does not prevent competition as long as the imputation rule is followed. The imputation rule, by definition, includes in the retail price of the essential-component provider's service the same contribution as charged to its competitor supplying the same service. In general, all multiproduct firms need contribution toward non-ICs and must price above ICs to remain financially viable. Such pricing practices are entirely consistent with competition and economic efficiency.

If the telecommunications business is to become competitive, cross-subsidies will need to be eliminated or allowed to be driven out by the competitive process. Either all subsidies need to be eliminated or they need to be neutrally funded from sources external to the individual service suppliers. Only then can effective competition flourish without price squeezes. This must be achieved with respect to cross-subsides within and among services, classes of service, customer segments, and geographical areas.

In addition, the regulatory structure will need to change to accommodate the rapid changes in the telecommunications industry. Regulation by categories of service will need to be restructured to deal with essential facilities, unbundled from services. Imputation rules and other price constraints on essential facilities will need to be limited. Also, there must be recognition of the fact that a facility's "essentiality" is not a constant and depends on the markets served, not the facility itself (recall that the chemical compound in the examples given is essential only in the market for medicine).

Finally, the regulatory process must be removed from the competitive arsenal of rival firms; competitive contests are best played out in the streets of the marketplace, not before panels of even the most informed judiciary bodies. Obviously, there will be a continuing need for a complaint process and sanctions for harmful misbehavior. However, such processes should not substitute for competition for the customers' favor by improving efficiency and profitability.

That does not mean that end users will pay rates as high as cost in all cases. What is not sustainable with the opening of local exchange markets to competition is cross-subsidies. That is not to say that rates cannot be subsidized in other ways. For example, the current flow of subsidies could be made more sustainable if all current sources of subsidy funds were declared to be "taxes," from which the recipients of subsidy funds could be maintained. In general, all carriers (not just the local exchange carrier) could contribute to a fund or funds from which end user rates would then be subsidized. For example, it would be a great improvement to economic efficiency if rates were restructured to be higher in high-cost geographical areas and lower in low-cost geographical areas, with subsidies being given to those providing service to end users in the high-cost geographical areas to maintain affordable rates. By allowing the subsidy to be bid down as the cost of serving the high-cost areas falls, efficiency is further improved. In particular, essential facilities or services (to the extent they exist) should be subject to unbundling upon a bona fide need for obtaining access to the essential component in order to compete with the local exchange carrier.

As the essential facilities are unbundled, prices should be set no lower than the IC of providing the essential facilities (some recovery of shared and common costs would, of course, be appropriate). This, of course, could create a price squeeze in that the purchasers of the essential facilities would pay more than the subsidized services provided by the local exchange company using these essential facilities. In essence, the competitors to the local exchange company would be at a competitive disadvantage by having to purchase essential facilities at or above cost while a local exchange company subsidized its services employing those same essential facilities. This price squeeze problem can be solved in one of two ways. First, subsidized services using essential facilities could be restructured to eliminate the subsidy or need for a subsidy. The second method is more challenging.

In general, the following three characteristics cannot be sustained concurrently: (1) cross-subsidization of one service or group of customers by others; (2) open competition and unrestricted competitive entry; and (3) mandatory unbundling of subsidized essential facilities offered to competitors without restrictions on use or competition.

One of those three characteristics must give. In today's environment, it seems most appropriate to sacrifice the requirement of cross-subsidization and convert the need for subsidies to accomplish public policy purposes to external subsidies such as a universal service fund.

In other words, an affordable rate structure can be sustained by deaveraging rates to align them to cost (and restructuring rates to eliminate cross-product subsidies) and reinstating an acceptable rate structure through the use of an external subsidy fund. This external subsidy fund need not include new general taxes but would include funding mechanisms that acquire funds from some of the same sources providing the internal cross-subsidies today. The difference would be that those sources would be extended to all carriers in a competitively neutral and inescapable fashion.

In many of its market areas, an LEC's residential basic exchange rates are priced below IC. Hence, a price squeeze is created by regulated rates that are below cost. In relation to the example in Table 6.4, the LEC is pricing in a manner similar to the essential-component provider, in which the compound is priced at or above IC and the medicine is priced below IC. Thus, careful structuring of subsidy funds is required to achieve effective competition.

To eliminate the price squeeze, residential rates should be adjusted so that they are above IC. However, subscribers of residential service may not tolerate what may, in some cases, be a large, one-time increase. Instead, LECs could increase residential rates as much as is currently practical, with the understanding that further rate increases will follow. Eventually, residential service should be priced above IC. In the interim, the structure of subsidy mechanisms should maintain competitive neutrality on the one hand and be sustainable for the duration of the transition on the other.

6.11 PRICING NETWORK INTERCONNECTION

Regulators should adopt an interconnection, pricing, and compensation system for local network competition that, at a minimum, meets the following two economic criteria. First, the governing interconnection arrangement should produce a result as close as possible to that which would prevail under competitive conditions. Second, to the extent that explicit public policy objectives are not compatible with a competitive environment, the method of attaining such public policy objectives should minimize the resulting loss of economic efficiency.

Thus, network interconnection pricing arrangements should be comparable to those that would be observed in competitive industries. For example, an interconnection arrangement is essentially a relationship in which one carrier distributes a service to another carrier that, in turn, resells the service to a buyer further down the distribution chain, perhaps adding value in the process. One can judge whether a proposed interconnection arrangement is compatible with such arrangements naturally occurring in markets unencumbered by regulation.

Reasonably precise measurements can be performed regarding certain constraints that represent necessary conditions for economic efficiency. For example, a widely accepted necessary condition for economic efficiency is the IC test for cross-subsidization. This test requires that each service generate sufficient revenue to compensate for the IC of providing the service. In competitive markets, there is no need to monitor such a condition because self-interested firms generally have every incentive to avoid such subsidies; the profits of a competitive firm would be lower were it to attempt to engage in cross-subsidization. Where either franchised monopolies or natural monopolies exist, public policymakers require that this necessary condition for economic efficiency be met and that the prices be set accordingly. Thus, all interconnection prices should be set to avoid subsidizing interconnection; prices should equal or exceed IC. Beyond that general rule for a pricing floor, the ceiling on interconnection prices is generally based on the total cost the interconnecting firm must incur if it is to circumvent the LEC's PSTN altogether (i.e., the cost of bypass).

While the reason for the lower limit on prices is well known (to protect against inappropriate subsidies), the reason for the upper limit is not so well known. The upper limit on prices exists because prices in excess of the ceiling are viewed to be both unfair and inefficient. The price is unfair because an interconnecting carrier is forced to pay a cost in excess of that which it would voluntarily incur to circumvent or bypass the competing carrier. The price is inefficient to the extent that the bypass alternative consumes a lower value of resources than the supplier of interconnection. Presumably, such a circumstance would occur if: (1) a carrier were declared to be a legal monopoly; (2) bypass were disallowed; or (3) one carrier was otherwise forced to use the facilities of another.

As long as interconnection rates are going to be strictly regulated, proxies for competitive rates need to be developed and implemented by regulators. All interconnecting carriers will rely on the PSTN facilities of carriers having the USOs and COLR obligations to handle traffic in areas that are not financially or strategically appealing to them. The accommodating carriers will continue to need universal service and COLR funding from all network users who are not end users unless and until end user charges pay for the USO and COLR obligation costs. While that does not argue for identical rates to all carriers, it does suggest that all carriers should equitably contribute to the USO and COLR obligation costs if such contributions are a component of interconnection prices.

IXCs, CAPs, and other carriers will probably expect to have access to all tariffs that are available to their competitors. An LEC should not be able to restrict access to tariffs based on the identity of the interconnecting party. Indeed, any differences in interconnection prices offered to similarly situated competitors could distort the competitive process, so all tariffs should be available to similarly situated competing interconnecting carriers (the important exception being access to explicitly subsidized services).

All competing interconnecting carriers should be allowed to select from the same array of tariffs. Those tariffs should recover ICs, contribute toward the fixed costs of the LEC's PSOs, including the universal service and COLR responsibilities. The only exception to that rule (equal charges to potentially competing interconnecting carriers) should be based on unavoidable differences in the cost of serving the interconnecting firms.

Each interconnecting carrier should pay the other for exchanging traffic and for any other services within the limits of competitive price ceilings and floors, as described earlier. Additionally, the carrier with universal service and COLR responsibilities should be compensated for the cost of those responsibilities, including all subsidies, and a fair contribution to the shared and common costs of the ubiquitous network.

In summary, the following principles will form the foundation for an economically sound interconnection arrangement.

- Any party demanding or requesting interconnection with a second party should pay at least the IC such interconnection imposes on that second party. Interconnection prices may, however, exceed IC (e.g., to fund USOs or COLR obligations).
- The price of interconnection should be no higher than the cost of bypass. That is, the price of interconnection should not exceed the cost of a party supplying to itself the facilities requested from another. Ordinarily, this criterion would be self-policing, but it may need to be enforced or monitored if self-supply or bypass is disallowed by legal or regulatory mandate.
- The rules governing reciprocal compensation should be completely symmetrical so that neither party is artificially handicapped in a competitive contest.[15] That is, each party should win or lose contested business on the merits of its own relative efficiencies rather than relying on rules that are not competitively neutral.
- Interconnection price arrangements between carriers should be sustainable without relying on artificial and inefficient use or end user restrictions. In general, the only appropriate use or user restrictions would be those consistent with the outcome of a competitive market, including those based on unavoidable differences in cost.
- In general, both retail rates and carrier access and interconnection charges need to be restructured and rebalanced as competition increases so that rates are sustainable in a competitive environment and so that public policy objectives are properly funded in a manner compatible with competition.
- Interconnection charges for new exchange carriers should be lower than the LEC's interconnection charges, so that the LEC may continue to receive funding for the USOs and COLR obligations. This should be true even if interconnection rates are deaveraged to reflect the different costs across geographical areas. Only the carrier incurring the cost of USOs and COLR obligations should receive contributions specifically designated to fund those costs.

Some industry players have recommended that a simple approach to carrier interconnection is to essentially set a zero price for all traffic exchanged between interconnected networks. While it is simple and seemingly easy to ad-

---

15. This does *not* imply that rates should be identical, only that the rules for developing those rates should be symmetric.

minister, such an arrangement is contrary to competitive outcomes and economic efficiency. The incentives in this arrangement are not to become the most efficient provider of service but to maximize the opportunity to bill (and keep) revenues. For example, the incumbent LEC acquired both high and low geographical concentrations of revenue by building a large network (pursuant to its USOs and COLR obligations) and was able to maintain affordable rural rates through statewide average tariffs or limited tariff differentials between urban and rural areas. By competing only in the densest and most lucrative market areas, a new entrant might be able to bill, say, 50% of the LEC's revenue while making only 10% of the LEC's investment (and incurring 10% or less of the LEC's cost). A "bill-and-keep" arrangement takes all the contribution from the highest contributing portions of the business (those that competitors want to enter) and requires the LEC to find alternative sources of contribution to sustain its USOs and COLR obligations. In other words, the arrangement essentially erodes one of the most important sources of contribution to the USOs and COLR obligations. A bill-and-keep arrangement thus would greatly increase the need for funding the LEC's USOs and COLR obligations and would reward the new competitor in ways not possible in an unrestricted competitive environment.

In a competitive environment, the LEC could win the business where it was most efficient (and lose business where it was inefficient) through flexibly pricing to profitably meet the competition. Similarly, the new entrant would enter the areas with low revenue concentrations if it could more efficiently serve in those areas than could the LEC. In other words, each player would be attracted by profit opportunities equally in rural and urban areas. Entry decisions would be based on anticipated efficiency of the incumbent versus that of the competitor, rather than on where the competitor felt it would be most likely to bill and keep the most revenue (and avoid the highest-cost, lowest-revenue business).

Wholesalers in unregulated markets would not agree to an arrangement in which their retailers kept all the revenue received. Even when wholesalers supply one another's retailers (this is the situation between interconnecting retail telephone network service suppliers), they do not compensate one another by allowing the retailers to keep all the revenue received from further distribution of the goods. Rather, the wholesale and retail transactions are negotiated at arm's length. A bill-and-keep agreement generally is going to be unfair to the incumbent. The only situation in which bill-and-keep arrangements are equitable and economic is when both parties have exactly the same cost and rate structures, traffic exchanged between the two is symmetric in all respects (i.e., the exact same type of traffic and in equal amounts), and regulatory obligations are the same—a highly unlikely proposition. The risk of imbalanced compensation is too great to allow such agreements to become common in competitive markets.

In general, to avoid inadvertent price discrimination and to maintain competitive parity, all transactions among carriers should be explicit. Bill-and-keep arrangements mask the gross revenue flows among carriers by assuming the net flows are and should be zero. (A net flow is what one carrier owes the other less what is due back.)

6.12 REGULATING NETWORK MODERNIZATION

One of the most important influences on the willingness and ability of infrastructure network operators to invest in the modernization of their network facilities is the type of regulation they will face. In a monopoly environment, regulators have always faced the difficult task of how to provide incentives for the monopoly incumbent to expand and upgrade the public network infrastructure and at the same time keep prices affordable for consumers at large. In developing countries with low telephone penetration, nascent markets, and scarce capital, the regulatory prescription for stimulating network modernization can be very different from that for developed countries.

In the United States and other developed countries, the goal of universal service has largely been achieved and the focus of regulation is shifting to the introduction of competition into the market and at the same time stimulation of the regulated incumbents to modernize their public network facilities. To an economist, network modernization and deregulation are synonymous—there is no reason to regulate if the goal is increased investment. Indeed, the pursuit of competitive advantage is itself the most powerful investment incentive of all.

For a variety of political and social reasons, the government apparently does not fully believe this, and the issue of regulating network modernization remains front and center. The ways in which economic regulation may be used to stimulate infrastructure investment in a competitive market are largely the same as those discussed earlier. The discussion that follows focuses on how regulation can be further reformed to stimulate network modernization while protecting the interests of consumers and competitors.

Historically, the regulation of local telephone companies has been based on ROR. This system is best described as one that authorizes an LEC to invest in PSTN facilities and to recover that investment over time from tariff rates for network services.

In the case of cable television companies (which had been free to set their own rates from 1984 until 1992, when Congress ordered the FCC to reassert rate regulation)[16] and local telephone companies, tariff rates had to be based on

---

16. For a comprehensive look at the effects of changing cable regulation on basic service rates and investment in programming services, see T. Hazlett and M. Spitzer, *Public Policy Towards Cable Television, Vol. 1: The Economics of Rate Controls* (Cambridge, MA: MIT Press, 1996).

costs. At first blush, it certainly seems that, if the goal of policymakers is to increase infrastructure network investment, ROR regulation is certainly the way to go. What better way to get more investment than to tell the industry incumbents that they can set rates at whatever level is required to recover all their investments? This system is often characterized as "cost-plus" regulation.

But it's not that easy. While ROR regulation allows for capital recovery, it does not provide a blank check to incumbent network operators. Regulators have consistently fought hard to keep the rates for local service down and have always limited the amount of profit that incumbents could make.

In the past 10 years or so, federal and state regulators have pursued new forms of regulation, partly to accommodate the introduction of competition and partly to try to give increased incentives to regulated carriers to become more cost conscious so that local rates could remain at low levels. The new regulation is generally referred to as incentive-compatible regulation because it provides incentives for firms to cut costs. The incentive derives from the fact that, as incumbent network providers reduce their costs, all other things being equal, they are allowed to keep more of the resulting profit. "All other things being equal" means that basic service prices must not be raised beyond a certain level for many years (usually the level at which they started when the incentive plan was implemented, or even lower). By now, many such plans are in effect, and, no doubt, all states will be implementing them in one form or another. It is beyond the scope of this discussion to evaluate the various plans; Sappington and Weisman have provided the most comprehensive treatment of this subject to date [13].

6.12.1 Price Caps

Incentive-compatible regulation has come to be synonymous with so-called price cap regulation because every single plan features some form of direct price control on basic service rates. Usually, the only increase in basic service prices allowed under these plans is equal to or lower than the annual rate of inflation in the general economy.

Under "pure" price cap regulation, a firm is allowed to earn whatever profit it can, subject, of course, to competition and subscribers' willingness to pay. The only commitment that the firm makes to the regulator is that it will not raise (but it can reduce at will) the average price of basic service beyond that which some formula based on inflation and productivity might allow. In turn, the regulator commits to allowing the firm to keep whatever profits it can make as long as it does not raise tariff rates higher than the level of the cap. In addition to factors accounting for inflation and productivity, the price cap formula also includes an adjustment factor for unanticipated events, such as changes in accounting rules and tax laws, that may materially affect the financial performance of the regulated firm and that would justify a tariff rate adjustment.

The simple formula for price cap regulation is $Pt = Pt\text{-}1 \, (1 + RPI - X)$. In other words, the maximum level of tariff rates that the firm is allowed to set for the next year (Pt), will be equal to the allowed price level from the last year ($Pt\text{-}1$) adjusted upward for the rate of inflation or change in the *retail price index* (*RPI*) minus the amount of productivity that the firm is able to achieve (X). Obviously, as the value of X rises, all other things being equal, the firm's allowed price levels are reduced, and, of course, so are its profits. Thus, if a regulatory agency manages either to set the value of X too high and fix it there or reserves the right to tighten the pricing constraint by mandating an increase in the value of X over time, the price regulation mechanism is no longer pure because the profit incentive is no longer guaranteed.

By retaining a strong incentive for the regulated firm to cut costs or otherwise increase productive efficiency, pure price regulation has become synonymous with competition because the incentives are similar. The problem is that any deviation from pure price cap regulation, such as forcing the firm to share any profit increases with ratepayers, can both destroy the firm's incentives to increase productive efficiency (why increase effeciency if it has to give money back?) and retard competitive market entry.

If regulators, in their zeal to get consumer prices reduced, tighten the screws on the incumbent firm's profits by increasing the value of the X factor, the reduced profit margins will be felt not only by the incumbent but by new market entrants who relied on the incumbent's profit margins to undercut the prices of the incumbent and still make enough profit to survive. In fact, regulators in the United Kingdom may have already overdone it in this regard.

The United Kingdom was the first country to adopt price caps and has a longer history than any other country of trying to increase domestic competition for telephone service under a price cap regulatory regime for the incumbent. In the first year of price caps in the United Kingdom (1984), regulators set the value of X at 3%. It is now at 7.5%, and there are plans that could increase it to 9%. According to the formula, given the current 7.5% value of X, if annual inflation is only 2%, *British Telecom* (BT) must reduce its overall tariffs by an average of 5.5% per year.

Since 1984, average call prices have fallen by over 50%, squeezing the profits of BT and its competitors. Especially hard hit was Cable and Wireless's Mercury Communications company, which is BT's only major competitor. Just like MCI, Sprint, and other U.S. competitors, the dominant incumbent firm's regulated price levels provide some protection for entrants, who need to undercut the incumbent's prices to gain market share but still need to make a decent profit to survive. While a direct cause and effect is questionable, the result for the competitive situation in the United Kingdom has not been good. Mercury languishes in terms of both its financial performance and its market presence in the United Kingdom. In fact, BT has just announced its intention to acquire Mercury's parent, Cable and Wireless. The bottom line is that unless price caps are adopted in their pure form, they can backfire as a regulatory tool and actually serve as a barrier to competitive entry and investment in infrastructure networks.

Under incentive regulation, the incentive to cut costs is powerful. For the first time ever, regulated firms are allowed to earn profits according to whatever the market will allow. A potential problem with this plan, however, is that increased profits become more a function of cost reductions and less a function of creative market-based pricing. In recognition of that and the fact that the incumbent's unregulated competitors' pricing practices are not regulated, the FCC and state regulators are continually increasing the pricing flexibility that incumbent firms may exercise to allow them to compete.

It would seem that pure price regulation, applied only to basic monopoly services, is the solution to the most vexing problem since the beginning of regulation: How can regulation provide strong incentives for the incumbent monopolist to reduce costs and still provide incentives to innovate and invest in new products and services? The simple answer is to cap the price of monopoly basic service and deregulate everything else. Along the way, regulators are removing barriers to market entry, which further increases the pressure on the incumbent to modernize its network as all firms pursue a service-quality advantage.

Unfortunately, the powerful incentives of pure price regulation almost never exist. It turns out that regulators, in fear of losing their jobs and tarnishing their reputations, lest the public utilities that they regulate make too much money, almost always include some provisions for profit regulation in their price cap rules. So much so, that in many, if not most, cases price cap plans are really just profit regulation by another name.

6.12.2 Infrastructure Investment Incentives Under Price Caps

Almost no one would deny that an advanced telecommunications network infrastructure is an important goal of public policy. The federal government and most state governments have by now codified that perspective in legislation. However, to realize an advanced telecommunications network infrastructure, a significant amount of investment is required. It is obvious in this era of strained public budgets that the best way to achieve that goal is to provide incentives for the private sector to invest in the public infrastructure. That, too, is codified in the new law. But, for whatever reason, the new federal law is virtually silent on how regulation should be reformed to achieve the objective.

To its credit, the new law does stress that market entry should be allowed (not exactly a revelation). That provision is good as far as it goes and should result in many new market participants.

SEC. 253. REMOVAL OF BARRIERS TO ENTRY

(a) IN GENERAL—No State or local statute or regulation, or other State or local legal requirement, may prohibit or have the effect of prohibiting the ability of any entity to provide any interstate or intrastate telecommunications service.

Basic economics, however, states that market entry is not a panacea for profit incentives. In fact, basic economics suggests that increased entry will reduce profit margins. That is not all bad as long as firms, over time, are allowed to continue to pursue competitive advantage via investments in research and development, innovative production processes, and new-service introduction. Progress and growth in the computing industry have borne this out. Unfortunately, the new law does little to stimulate market activity in this area and nearly denies altogether the possibility of the RBOCs from profiting from fundamental network innovations or even partnering with equipment manufacturers to do so.

The law simply neglects to mention that the most powerful investment incentive of all is the pursuit of profits. If increased investment is the goal, the regulatory prescription should be obvious: Cap basic service prices and deregulate everything else so profits can be pursued. Instead, the new law goes out of its way to accommodate the outdated cost and profit regulations of local telephone companies that are pervasive at both the state and federal levels. For example, take the following passages contained in the new law:

SEC. 261. EFFECT ON OTHER REQUIREMENTS

(a) COMMISSION REGULATIONS—Nothing in this part shall be construed to prohibit the Commission from enforcing regulations prescribed prior to the date of enactment of the Telecommunications Act of 1996 in fulfilling the requirements of this part, to the extent that such regulations are not inconsistent with the provisions of this part.

(b) EXISTING STATE REGULATIONS—Nothing in this part shall be construed to prohibit any State commission from enforcing regulations prescribed prior to the date of enactment of the Telecommunications Act of 1996, or from prescribing regulations after such date of enactment, in fulfilling the requirements of this part, if such regulations are not inconsistent with the provisions of this part.

(c) ADDITIONAL STATE REQUIREMENTS- Nothing in this part precludes a State from imposing requirements on a telecommunications
carrier for intrastate services that are necessary to further competition in the provision of telephone exchange service or exchange access, as long as the State's requirements are not inconsistent with
this part or the Commission's regulations to implement this part.

SEC. 103. (n) APPLICABILITY OF TELECOMMUNICATIONS REGU-
LATION- Nothing in this section shall affect the authority of the Federal Communications Commission under the Communications Act
of 1934, or the authority of State commissions under State laws concerning the provision of telecommunications services, to regulate the
activities of an exempt telecommunications company.

Those and other passages are proof that the federal government was not
willing to take on the powerful interests of state regulators and their parochial
desire to continue their heavy-handed regulation of the industry.

Contrast those provisions with the single mention of price cap regulation
contained in the law:

(a) IN GENERAL—The Commission and each State commission with
regulatory jurisdiction over telecommunications services shall encourage the deployment on a reasonable and timely basis of advanced telecommunications capability to all Americans (including,
in particular, elementary and secondary schools and classrooms) by
utilizing, in a manner consistent with the public interest, convenience, and necessity, *price cap regulation* [emphasis added], regulatory forbearance, measures that promote competition in the local
telecommunications market, or other regulating methods that remove barriers to infrastructure investment.

This pitiful attempt to suggest that existing cost and profit regulations be
eliminated in favor of pure price caps will certainly not be sufficient to cause
any significant increase in private incentives to increase investments in infrastructure modernization.

References

[1] Schmid-Bielenberg, V., "Bellcore's Switching Cost Information System (SCIS) Cost Model: A
    Practical Approach to a Complex Problem," *Marginal Cost Techniques for Telephone Serv-
    ices: Symposium Proceedings*, Columbus, OH, 1991.
[2] McKnight, L., and J. Bailey, *INTERNET Economics*, Cambridge, MA: MIT Press, 1996.

[3] Egan, B., "The Case for Residential Broadband Telecommunication Networks," Columbia Institute for Tele-Information, Research Working Paper Series, Columbia Business School, 1991, Section 4. See also Melmed, A., and F. Fisher, "Towards a National Information Infrastructure: Implications for Selected Social Sectors and Education," Center for Educational Technology and Economic Productivity, New York University, December 1991.

[4] Little, A., "Can Telecommunications Help Solve America's Transportation Problems?" Draft, February 1992. This study estimates $23 billion (1988 dollars) in annual benefits to society from telecommuting. These conclusions are, however, disputed by other researchers.

[5] Juster, F., and F. Stafford, "The Allocation of Time: Empirical Findings, Behavioral Models, and Problems of Measurement," *Journal of Economic Literature*, June 1991, pp. 471–522.

[6] Nelson, J. (ed.), B. Egan, and S. Wildman, "Investing in the Telecommunications Infrastructure: Economics and Policy Considerations," *A National Information Network*, Institute for Information Studies, Aspen Institute, 1992. See also Loveman, G., "An Assessment of the Productivity Impact of Information Technologies," MIT Working Paper, September 1990, pp. 88–154.

[7] Schumpeter, J., *Capitalism, Socialism and Democracy*, Harper and Row, 1975 (first published in 1942), p. 84.

[8] Kahn, A., *The Economics of Regulation*, Vol. I, Cambridge, MA: MIT Press, 1988 (first published in 1970), p. 164.

[9] Fowler, M., A. Halprin, and J. Schlichting, "'Back to the Future': A Model for Telecommunications," *Federal Communications Law Journal*, Vol. 38, No. 2, August 1986, pp. 193–194.

[10] "Washington's Wake-Up Call," *The Economist*, January 20, 1996, pp. 61–63.

[11] Schankerman, M., "Symmetric Regulation for a Competitive Era," presented at the Conf. Telecommunications Infrastructure and the Information Economy: Interaction Between Public Policy and Corporate Strategy, sponsored by the School of Business at the University of Michigan, Ann Arbor, March 1995.

[12] Landes, W., and R. Posner, "Market Power in Antitrust Cases," *Harvard Law Review*, Vol. 94, March 1981, p. 976.

[13] Sappington, D., and D. Weisman, *Designing Incentive Regulation in the Telecommunications Industry*, Cambridge, MA: MIT Press, 1996.

Universal Service

7

7.1 THE ORIGINS OF UNIVERSAL SERVICE POLICY

A great deal of confusion and controversy surround the popular concept often referred to as universal service. Universal service, the most unambiguous goal of public policy in telecommunications, is also the most nebulous in terms of its meaning.

For many decades now, the most common notion about universal service was that affordable basic local telephone service would be available to all households wanting it. Unfortunately, the definition of "affordable" has never been agreed on, which limits its usefulness in clarifying the scope of "universal service."

Interestingly, the entire concept of universal service as it is generally understood today evolved from a very different concept in the early days of telephony. After Bell Telephone's[1] patents on basic telephony ran out, its monopoly quickly eroded until, by 1907, close to half of U.S. cities with telephone service had competing local service providers. At about the same time, Bell filed additional patents for long distance business network equipment, giving it renewed monopoly power in providing long distance connections between local companies.

Theodore Vail, a brilliant businessman who took over as chairman at Bell, sought to reestablish Bell's local monopoly by leveraging its long distance power. From those inauspicious, self-serving beginnings, the concept of universal service was born. Vail envisioned a nation of telephone customers all interconnected by a single monopoly network supplier to achieve ubiquitous

---

1. "The Bell Telephone Company was renamed the American Bell Telephone Company in 1880 and was incorporated as the American Telephone & Telegraph Company (AT&T) in 1990. We will, however, use the term 'Bell System' or simply 'Bell' to refer to the company prior to divestiture in 1984. 'AT&T' will refer to the post-divestiture company." Quoted from M. Kellogg, J. Thorne, and P. Huber, *Federal Telecommunications Law* (Boston: Little, Brown, 1992), p. 78, footnote 1.

end-to-end calling capability. In furtherance of that objective, Vail refused to interconnect with or sell equipment to non-Bell companies. That policy was effective in forcing many independent local telephone companies to either fold or sell out to Bell.

Thus, universal service polices had private sector origins based on concepts of monopoly end-to-end service provision, especially for long distance service, focusing on physical aspects of interconnection more than financial aspects like affordability. This is very different from the more recent sociopolitical version of universal service that stresses the availability of affordable basic telephone service in a competitive environment and that focuses on an individual subscriber's local phone line connection for public network access and local usage. The affordability of long distance and network interconnection are rarely mentioned by regulatory authorities, except to mandate that charges be averaged across broad categories of subscribers so as not to allow "undue" price discrimination.

Vail actually convinced the federal government that telecommunications was a natural monopoly that was critical for society's well-being and for national security. He used the slogan "One Policy, One System, Universal Service."

Current regulatory policies for universal service evolved pursuant to the Communications Act of 1934, which established the FCC and which was the government's first attempt to officially grapple with the issue. The Act's objective was to:

> [M]ake available, so far as possible, to all the people of the United States a rapid, efficient, Nation-wide, and world-wide wire and radio communication service with adequate facilities at reasonable charges.

Note the absence of any mention of "universal service" or "affordability." This first governmental effort did little to provide any specific guidance for the formulation of state and federal universal service policies. The consequence is the hodgepodge of specific policies and definitions that exist today.

The federal government's latest attempt to define universal service was codified in the Telecommunications Act of 1996. This version of universal service emphasizes the importance of access to "advanced" (not just basic) telecommunications services, but the specific wording of the Act is still vague.

7.2   UNIVERSAL SERVICE LAW

After more than 100 years of telephony and over 60 years of legislation, the Telecommunications Act of 1996 finally addressed the issue of universal service by name. An attempt to provide a definition for the concept is found in Section 254 of the Act:

SEC. 254.

(c) DEFINITION—

(1) IN GENERAL—Universal service is an evolving level of telecommunications services that the Commission shall establish periodically under this section, taking into account advances in telecommunications and information technologies and services. The Joint Board in recommending, and the Commission in establishing, the definition of the services that are supported by Federal universal service support mechanisms shall consider the extent to which such telecommunications services—(A) are essential to education, public health, or public safety; (B) have, through the operation of market choices by customers, been subscribed to by a substantial majority of residential customers; (C) are being deployed in public telecommunications networks by telecommunications carriers; and (D) are consistent with the public interest, convenience, and necessity.

This so-called definition is not very definitive. It states that universal service is an evolving concept, which makes it a moving target. This is probably just as well, considering that, in the future world of integrated digital networks and multimedia, it seems logical to expand the definition of universal service beyond basic analog telephone service. The larger problem with this "definition" is that none (or even all) of the four factors listed as considerations useful for defining universal service is sufficient to be conclusive.

Later in the section, the Act lists some principles that govern universal service policy:[2]

SEC. 254. (b) UNIVERSAL SERVICE

The Joint Board and the Commission shall base policies for the preservation and advancement of universal service on the following principles:

(1) QUALITY AND RATES—Quality services should be available at just, reasonable, and affordable rates.

---

2. Several principles are listed in the legislation. Only the primary ones are quoted here.

(2) ACCESS TO ADVANCED SERVICES—Access to advanced tele-communications and information services should be provided in all regions of the Nation.

(3) ACCESS IN RURAL AND HIGH COST AREAS—Consumers in all regions of the Nation, including low-income consumers and those in rural, insular, and high cost areas, should have access to telecommu-nications and information services, including interexchange services and advanced telecommunications and information services, that are reasonably comparable to those services provided in urban areas and that are available at rates that are reasonably comparable to rates charged for similar services in urban areas.

(4) EQUITABLE AND NONDISCRIMINATORY CONTRIBUTIONS— All providers of telecommunications services should make an equi-table and nondiscriminatory contribution to the preservation and ad-vancement of universal service.

(5) SPECIFIC AND PREDICTABLE SUPPORT MECHANISMS— There should be specific, predictable and sufficient Federal and State mechanisms to preserve and advance universal service.

The first major principle is one that had already been adopted in one form or another by most state and federal regulators. The second, third, fourth, and fifth principles, however, represent a significant departure from most existing regulatory policy statements by emphasizing advanced telecommunications and requiring funding from all telecommunications providers.
Advanced telecommunications is defined in Section 706.

(c) DEFINITION - (1) ADVANCED TELECOMMUNICATIONS

The term advanced telecommunications capability is defined, with-out regard to any transmission media or technology, as high-speed, switched, broadband telecommunications capability that enables us-ers to originate and receive high-quality voice, data, graphics, and video telecommunications using any technology.

The government apparently intends for all American households, both ur-ban and rural, to have affordable access to a digital public network infrastruc-ture capable of providing high-speed interactive multimedia services.
Although the language of the fourth and fifth provisions in Section 254 are unquestionably vague, their inclusion in the law is understandable given the

federal government's aversion to so-called unfunded mandates. Unfunded mandates refer to federal laws requiring state compliance for which no federal funding is approved. By default, such funding becomes the responsibility of the states. Even though the new law includes broad funding provisions, it is disturbing how unclear it is regarding exactly where the money will come from and how it will be collected and distributed. What is clear is that the government is not planning to finance this huge undertaking from federal coffers.

Other funding provisions found in the Act include the following.

(d) TELECOMMUNICATIONS CARRIER CONTRIBUTION—Every telecommunications carrier that provides interstate telecommunications services shall contribute, on an equitable and nondiscriminatory basis, to the specific, predictable, and sufficient mechanisms established by the Commission to preserve and advance universal service. The Commission may exempt a carrier or class of carriers from this requirement if the carrier's telecommunications activities are limited to such an extent that the level of such carrier's contribution to the preservation and advancement of universal service would be de minimis. Any other provider of interstate telecommunications may be required to contribute to the preservation and advancement of universal service if the public interest so requires.

(e) UNIVERSAL SERVICE SUPPORT—After the date on which Commission regulations implementing this section take effect, only an eligible telecommunications carrier designated under section 214(e) shall be eligible to receive specific Federal universal service support. A carrier that receives such support shall use that support only for the provision, maintenance, and upgrading of facilities and services for which the support is intended. Any such support should be explicit and sufficient to achieve the purposes of this section.

(f) STATE AUTHORITY—A State may adopt regulations not inconsistent with the Commission's rules to preserve and advance universal service. Every telecommunications carrier that provides intrastate telecommunications services shall contribute, on an equitable and nondiscriminatory basis, in a manner determined by the State to the preservation and advancement of universal service in that State. A State may adopt regulations to provide for additional definitions and standards to preserve and advance universal service within that State only to the extent that such regulations adopt additional specific, predictable, and sufficient mechanisms to support such definitions or standards that do not rely on or burden Federal universal service support mechanisms.

Other minor funding provisions are targeted at helping schools, libraries, hospitals, and small businesses, but none details the process through which universal service will be subsidized and the criteria for receiving compensation.

7.3 ECONOMICS OF UNIVERSAL SERVICE

Even though universal service has been held sacred for decades by most public policy authorities and has been aggressively pursued by telephone companies, the policy has not been clearly defined in ways that have been widely accepted. For purposes of the analysis and discussion here, the term *universal telecommunications service* (UTS) is introduced to characterize the economic concept of universal service.[3] In economics, an accurate representation of the total cost of a PSO like UTS first requires an abstract exercise comparing the cost of being obligated to provide ubiquitous telephone service (whatever that may include in terms of specific services) with the cost of a hypothetical situation in which the obligation does not exist. Because the PSO is exogenous to the firm, then presumably the difference in the firm's total cost with and without the obligation will also be exogenous, making it an objective representation of the increment to the firm's total cost caused by the government imposed PSO.

For decades, economists have failed to measure, either directly or indirectly, the societal costs and benefits of universal service policy. Yet far-reaching public and private decisions continue to rely on this nebulous policy to affect the very fabric of our daily lives. As a result of universal service policies, nearly every resident and business has affordable instantaneous telephone network access to nearly every other subscriber location. It is therefore not surprising that few public policy authorities appear willing to advocate abandoning this policy. However, the advent of competition in local telecommunications markets makes it imperative to conduct a more formal economic analysis to measure the social costs and benefits of universal service policies, because funding the obligation in a closed monopoly system has vastly different implications for competition policy than funding the obligation in an open competitive system. The new telecommunications law recognizes that fact in its universal service funding provisions by mandating that all telecommunications providers, not just the old monopoly providers, contribute funds toward covering the cost of the USO in a competitively neutral fashion.

This fundamentally alters the traditional mechanisms employed to ensure universal service and requires a careful examination of the costs of the obligation itself to determine the amount of funds that each telecommunications ser-

---

3. The economic discussion here draws heavily on R. Emmerson, "Competition and the Maintenance of Universal Service," Draft, 1995.

vice provider should have to contribute. In the old monopoly system, the cost of the obligation and, for that matter, the source of funds for paying for it, were hidden in a complex system of internal cross-subsidies from highly profitable monopoly services (e.g., long distance, the Yellow Pages) to subsidized basic residential service.

Public policy authorities need to have a clearer understanding of what universal service is, how it is funded, and what must be done to sustain it in the new competitive environment under the requirements of the new telecommunications law.

7.3.1 What Is Universal Service?

An array of issues have been classified under the banner of "universal service." It is critical to a comprehensive analysis and, in turn, to creating a comprehensive policy that all issues surrounding universal service be viewed as portions of a single portfolio rather than as problems that can be solved independently of one another. The list that follows broadly outlines the areas in which decisions are most critical if UTS is to be realized. Because UTS describes a portfolio of obligations, it is useful to discuss PSOs, which individually and collectively are a variety of specific government-imposed obligations. Without attempting to identify each and every PSO that the government might impose, what follows is a discussion of common requirements.

7.3.1.1 *Widespread Availability*

A policy deeply entrenched in U.S. telecommunications regulations is that of ubiquitous availability of basic telephone services. Under the new law, that policy will be expanded to include advanced services. That concept, however, is evolving and highly contentious and no doubt will be the subject of litigation for years to come. Therefore, it is better to stick with a discussion of basic service, both because it is easier to understand and because there is actual experience with that concept.

Franchised incumbent telephone network operators (LECs) have historically been required by regulators to extend the PSTN to pass or even reach nearly every home and business in the nation without any particular regard to the cost of doing so.

In the real world of telephone regulation, the widespread availability requirement has typically taken two forms, loosely distinguished, and not easily separable: (1) a universal service requirement to promote the overall market penetration of the PSTN or to specify a requirement that a minimum acceptable percentage of homes or population be subscribed; and (2) a COLR obligation specifying that a designated carrier must provide basic local service or be ready to provide service to any premise not served by another carrier.

What makes the COLR obligation so costly is the readiness requirement. It is very expensive to deploy network capacity sufficient to be ready to provide service on demand to any and all potential subscribers, where such demands arrive randomly. It is also not very profitable to do so considering that, on demand, service must be provided without regard to cost and that the prices paid for the service are broadly averaged tariff rates also calculated without regard to the underlying costs of providing the service.

Thus, the COLR obligation itself takes two forms: (1) The COLR must construct plant to (nearly) every premise in its prescribed service area (even if a competing non-COLR's network facilities are in place or in use); and (2) the COLR must construct plant only to those premises not capable of being served by competing facilities-based service providers. The first form may be referred to as the "hard" COLR obligation and the second the "soft" COLR obligation.

7.3.1.2   Affordable Rates

"Affordable rates" has been touted as a means to achieve a high degree of subscribership to the PSTN. While widespread availability provides nearly ubiquitous accessibility to the PSTN, the goal of universal service has been monitored in terms of the percentage of homes (and businesses) actually subscribing to service (among other measures). Thus, affordable rates are viewed as a means of accomplishing the objective of ubiquity.

In other situations, the goal of affordable rates has become a public policy that exists independent of the policy of widespread availability. In particular, there have been two predominant effects of the policy of affordable rates:

- Geographically averaged rates that do not reflect the underlying cost differences as they vary by the geographic density of network subscribers, topography and terrain, distance from network switching centers, and other factors;
- Subsidized rates for certain groups of basic local services, particularly residential subscribers' access lines and local telephone usage. The affordable rates policy has also been furthered by a monitoring of the efficiency of a telephone company's operations. Prudency reviews, management audits, and incentives to reduce costs have been motivated in part by the desire to maintain low telephone costs and therefore low associated rates.

Of course, while regulators (and the new telecommunications law) would always want residential basic service rates to be affordable, no one knows what that means. Just as economists need to do a more formal analysis of the cost of UTS, they also need to apply the tools of economics to provide some objective meaning to the term "affordable." Typically, in economics this would mean analyzing the disposable personal income level of households in relation to the amount expended for a given item (e.g., phone service) to determine the proportion of a household's budget spent on telephone service. Historically, on average in the United States, that figure has been about 2% for all services and less than 1% for basic local service (of course, it would be somewhat higher for low-income households). A threshold level of expenditure would then be used as a measure of the affordability of telephone service with respect to an individual household's total consumption budget (e.g., less than 5%).

7.3.1.3 Public Accommodations

LECs often have been required to accommodate others for public convenience. Access to telephone poles by cable television and electric power companies, access to the PSTN by long distance companies and other competing (or even noncompeting) telephone companies, and access to property easements or rights of way are examples of public accommodations.

7.3.1.4 Public Services

LECs have been expected to provide or administer certain services. Emergency services (911), access to operators, public telephones, repair services, annoyance call control, and services required by government agencies (e.g., for national defense and crime prevention) are examples of public services.

The responsibility for those PSOs must be addressed as local telephone markets are opened to competition. In particular, any of the obligations that public policy requires of a designated carrier and that would not be fulfilled voluntarily by competing companies must be identified. Both mandatory obligations and associated funding mechanisms need to be designed and implemented in a manner that is viable in a competitive environment and that provides the least distortion of the benefits of competitive markets.

7.3.2 Economic Principles and Public Policy

Fulfilling PSOs can force loss-making activities on the obliged service provider in two ways. First, an obligation may require that a particular segment of the service provider's business be priced below its IC or below a level that would be required by a competitive firm to pay its underlying shared costs (including common costs or so-called overhead). The underpriced segment of business may be a customer class (e.g., residential), a geographical area (e.g., averaged statewide rates below IC in high-cost areas), an individual transaction (e.g., public telephone access to an operator service for which there is no charge), or any other portion of the firm's business. Second, the firm as a whole may be obliged to engage in loss-making operations even though each of the firm's segments of business is compensatory. For example, the service provider may be obliged to provide a scope of services for a geographical area in a manner that would be less efficient than if the service provider did not have PSOs. While each individual portion of the market may be priced at or above the respective ICs, the service provider may be required to serve an inefficient scope of services.[4] A competing firm that is allowed to select a more efficient scope of services, such as a niche market, may force the prices to be below that which allows the firm with the PSO to cover its common costs. Four illustrations of loss-making operations will clarify some fundamental problems of obligatory loss making in the face of competitive entry.[5]

7.3.2.1 Averaged Rates

Averaged rates across geographical areas may help provide telephone service to high-cost areas at rates below cost, subsidizing the loss with rates above cost in the low-cost areas (see Figure 7.1).

Averaged rates have the obvious benefit of funding widespread availability through an internal cross-subsidy. Averaged rates work only when the providers of the subsidy operate in an uncontested (i.e., not subject to competitive entry) market. If competition is introduced into such a market, the carrier with the PSO will lose the source of subsidy payments from the profitable areas and must find the subsidy funds elsewhere.

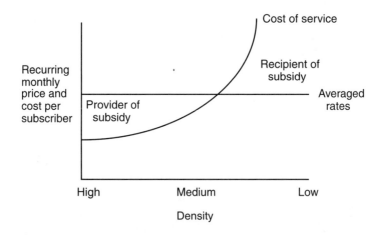

Figure 7.1 Loss-making operation: "average" rates.

---

4. The "scope" of service may refer to a product scope, a geographical scope, etc.

5. The figures illustrate concepts, not quantitative estimates of numeric values. It is also assumed that the average rate is set to exactly cover total cost, including a competitive return on equity.

For example, consider a scenario in which there are exactly 10 end users in each of three density cells (high, medium, and low). First, assume that only one provider, the incumbent LEC, serves this market. Also assume that the cost to each customer in the density cells is $10, $20, and $30, respectively, and the geographically averaged rate is $20 per end user. The cost and revenue implications are depicted in Table 7.1. Notice that no overall subsidy is required; the "subsidy" is internal to the service itself across all density cells.

Table 7.1
Costs and Revenues of Averaged-Rate Scenario

Density	Total Revenue	Total Cost	Subsidy Required
High	$200	$100	$100
Medium	$200	$200	$0
Low	$200	$300	100
Total	$600	$600	$0

The underlying problem arises when competition is introduced to this market. Assume that a new entrant is able to win 20% of the incumbent's high-density-cell customers. If the revenues are lost and the (long-run) costs are saved[6] by the LEC, then a newly revealed overall subsidy requirement of $20 arises ($20 in cost is saved, but $40 in revenue is lost in the high-density cell, meaning that the high-density cell provides only $80 to defray the $100 lost in the low-density cell). As market shares continue to shift over time from the incumbent to the new entrant, the required subsidy will grow.

Averaged rates and the attendant cross-subsidization promote both widespread access and affordable rates in a monopoly environment. However, as competition is allowed and becomes technically and economically feasible, entry occurs in the profitable (primarily low-cost) areas. The result is the incumbent's loss of "providers of subsidy," necessitating increased rates to the remaining (higher-cost) customers. In the end, competition would force deaveraged rates (the competitive rates of the entrant would prevail in low-cost areas and higher averaged rates of the incumbent would occur due to higher average costs of the incumbent among its remaining customers). Thus, maintaining an averaged rate policy in the face of competition will require intervention, fund-

---

6. Here it is assumed that all costs are volume sensitive in the long run. The subsidy requirement would be larger if there were underlying fixed costs.

ing, and/or management of competitive entry. In this scenario, it is not the aggregate subsidy requirement at the moment at which competition is allowed that is important. It is the subsidy requirement that appears as competition develops that needs to be solved.

7.3.2.2 Cross-Subsidies

A similar phenomenon exists with subsidies that are not based on geographical averaging of rates but are the effect of one product or one group of customers cross-subsidizing another. As in the previous example, the subsidy must be viewed over time, not at a single point in time.

The next example pertains to the effects of limits on rate rebalancing (defined as increasing local rates to cost-compensatory levels) with and without cross-subsidies coupled with open entry. The separated circles in Figure 7.2 represent the cost of producing each service (long distance and local) alone, or the stand-alone cost of each service. The cost of producing the services together is less than the sum of the stand-alone costs in the presence of economies of scope. In this example, the stand-alone cost of each service is 100, the IC of each service is 60, and the common cost of the two-product firm is 40. A cross-subsidy exists because: (1) local service is being subsidized (revenues are less than IC); and (2) long distance service is providing the subsidy (revenues exceed its IC plus the common cost).

If entry were prohibited, this cross-subsidy might be sustainable. If competitive entry is open (i.e., there are no barriers to entry), then a stand-alone supplier of long distance service would seek the opportunity to achieve extraordinary profits and displace the incumbent's long distance market. That not only would necessitate raising local revenues to IC but would require revenues to rise to equal stand-alone cost. Of course, neither of these (now) stand-alone companies would survive, since the next opportunity is for a two-product company (costing 160) to displace the two stand-alone companies (together costing 200). The surviving company must, of course, price both long distance and local services so that each product's revenue lies between stand-alone cost and IC.

This example illustrates the reason why competition drives out cross-subsidies. Open entry and cross-subsidies cannot persist together.

Limits on rate rebalancing, especially regulatory requirements to price at or a specified percentage above IC, can lead to inefficient market entry. The preceding example can be extended to illustrate an important source of such inefficiencies. Imagine that local rates are set equal to IC (yielding 60 in revenue). Long distance must now be priced at 100 (stand-alone cost). If another firm with a different scope of services could less efficiently add toll services (say, for an IC of 80), then the alternative supplier would displace the incumbent in the toll market, just as the stand-alone firm did in the previous example. The outcome is unnecessarily high cost of all services. Once inefficient indus-

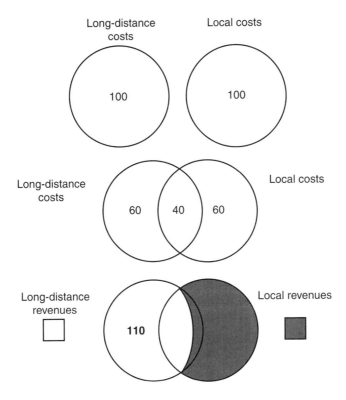

Figure 7.2 Loss-making operation: limits on rate rebalancing with and without cross-subsidies combined with open entry.

tries are in operation, additional inefficiencies often are created by various uses of regulatory and legal processes to preserve the resulting interests.

This latter extension of the example suggests that, for the purposes served by these examples, stand-alone cost is best defined as the IC of entry rather than the cost of entry calculated as if the most efficient entry were the literal stand-alone supply of a service.[7] For example, toll rates (or other nonbasic service rates) historically have been set sufficiently above cost to compensate for local rates (or other basic service rates) that are below cost.

7.3.2.3 Toll Rates

In this example, toll rates have been set to bear an unsustainable majority of shared and common costs of the company. The circles in Figure 7.3 represent

---

7. This is *not* the appropriate definition of stand-alone cost for purposes of detecting cross-subsidies within a firm.

the costs of providing long distance alone and local alone. The overlapping area is the cost common to both local and long distance. As shown, since local service revenues are far less than the cost of providing local service alone, long distance (toll) services bear the majority of common costs.

Although neither local nor long distance is being cross-subsidized in this example (each is priced above its IC), the need for a subsidy will appear as competition enters if rate rebalancing is restricted. To understand this concept more fully, assume that the incumbent provider has revenue sufficient to cover all costs before competition. Like low-cost areas with averaged rates, the toll markets are very profitable (have high contribution) and therefore are attractive to new entrants. As competition enters, the incumbent's costs decline very little due to the largely fixed infrastructure cost required by the widespread availability policy, as shown in Figure 7.4 (the cost decline is ignored in the figure). Revenues erode more rapidly, as shown.

A gap between revenues and total company costs appears. The incumbent must close that gap and recover its total costs if it is to remain viable. The gap can be closed, at least partially, in some combination of three ways:[8]

1. Rebalance rates so that local revenues are increased to fill the gap.
2. Provide external subsidy payments to the incumbent equal to the gap.
3. Structure interconnection charges so the gap is automatically filled by revenues from competitors that grow precisely as the gap grows.

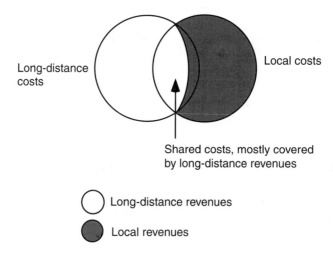

Figure 7.3 Loss-making operation: toll rates bearing unsustainable majority of shared and common costs.

---

8. For this simple example, assume that the firm remains a two-product firm and is operating efficiently given its service obligations.

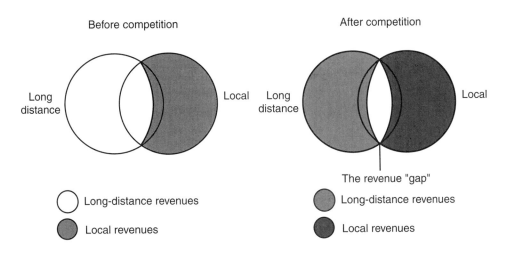

Figure 7.4 Competition enters: incumbent cost decline ignored.

Unlike the previous example, this requirement for PSO funding does not derive from a cross-subsidy (although cross-subsidies exacerbate the size and persistence of the subsidy). Instead, it derives from a combination of the obligation to provide widespread availability of service (thus restricting cost reductions enabled by withdrawing from local service to match the competitors' services and cost) and the policy of affordable rates (which restricts rate rebalancing).

Whether a firm in any of these examples should be subsidized is an issue that requires some attention. Subsidies normally will not be required unless a subset of suppliers has constraints or burdens not assumed by other competitors. Otherwise, subsidy payments will need to be provided from external sources, or restrictions on entry must be used to protect cross-subsidies, assuming, of course, that the obliged or protected supplier constitutes the efficient means of supply.

Consider the example in which both long distance and local services made a positive contribution toward common cost but a "gap" appeared as competition entered (see Figure 7.4). In that case, the constraint imposed on long distance service rates is the incremental entry cost of the competitor, while the constraint on local rates is imposed by regulatory mandate. Were *both* constraints due to the IC of entry, no subsidy should be provided, since the incumbent firm has economies of scope, but the economies are not as great as those of the competing firm(s) that provide the constraints. Notice that the incumbent firm (depicted in Figure 7.4) has lower ICs (represented by the revenues obtainable from the two services). Presumably, other firms could still enter so long as the opportunity to earn a competitive return is included in the IC, as is assumed here.

7.3.2.4 Widespread Availability

The policy of widespread availability by itself can require subsidies even without cross-subsidies and without restricted rate rebalancing (see Figure 7.5).

For the sake of simplicity, imagine that the cost of serving each end user is proportional to the distance over which traffic travels, and a single fixed cost switch can serve two or more users. In that case, the most efficient means of serving traffic between any two locations is through a centrally located CO to which all communication routes connect. A new entrant could serve two of the locations less expensively if it were not obliged to serve the entire geographical area. For example, a new entrant could provide service between any two locations at a lower cost by virtue of a direct connection between the two (locating its CO midway between two end users).

To put numbers to the example (see Figure 7.6), the CO cost is $2 for both the LEC and the competitor, the LEC's incremental access cost is $2.50 per end

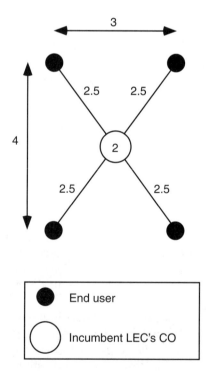

Figure 7.5 Widespread availability.

user, and the competitor's transport cost is $1.50.[9] In this situation, the LEC's total cost is $12, while the competitor's total cost is $5. To remain competitive, the LEC, therefore, cannot price its service above $2.50 per end user. Pricing at such a level, however, would not provide the LEC with enough revenue to cover the costs incurred ($10 versus $12). The LEC would require a subsidy of $2 to continue to meet the widespread availability criterion. While the LEC could not survive in this example without a subsidy to the entire company, it is important to note that there is no cross-subsidy. That is, the revenue received from each end user is (exactly) sufficient to cover the direct IC of service but provides no contribution toward the shared and common costs of the firm.

Table 7.2 summarizes the potential costs and revenues of the LEC under a situation in which no interconnection requirement is imposed on the two carriers. Remember that the cost to the LEC will not change no matter how many of the potential customers it serves, because the LEC continues to have a COLR obligation and therefore must continue to maintain facilities to all four customers. Thus, if the LEC were to lose two customers to competition, the required subsidy would rise precisely by the aggregate revenue (2 × $2.50 = $5) lost.

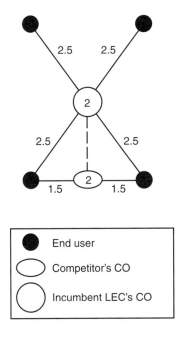

Figure 7.6 Competition.

---

9. Note that the entrant's transport cost to serve a proper subset of the market will always be lower due to the shorter distance between the end user and its CO.

Table 7.2
No Interconnection Requirement

Situation	Price	Quantity Served	Total Revenue	Subsidy Required
LEC retains all four customers	$2.50	4	$10.00	($2.00)
LEC loses two customers	$2.50	2	$5.00	($7.00)

In summary, a regulatory obligation to provide widespread availability can require a subsidy even if no individual segment of the business is being subsidized.

The implied message of these examples is that the problem of preserving PSOs, coupled with subjecting previously franchised monopolies to competition, requires a coordinated, carefully designed set of public policies that reap the benefits of competition and preserve the public policy benefits of the service obligations. All four PSOs must be considered together when determining the "cost" of maintaining the group of obligations described as universal service and assessing how to meet those obligations as competition substitutes for regulation. Well-meaning but poorly chosen policies regarding these matters can result in higher consumer prices, unfair competition, and economic inefficiencies.

7.4 PSO COST-RECOVERY MECHANISMS

The LEC having the PSO must be allowed the opportunity to recover its costs if the PSO is to be preserved. The new law recognizes this and also requires that the cost be funded in a competitively neutral manner. There are several alternative funding mechanisms. Since each must be competitively neutral in its application, the primary distinguishing feature is simply the source of the subsidy itself, in other words, what services are going to be taxed. In economics, an efficient taxing/subsidy-funding system would not exclude any item (i.e., a broad-based income or value-added tax), because doing so would cause economic distortions and sacrifice economic efficiency. Nevertheless, in the real political world, it is typical for one service or a set of services to be taxed while others are not. No doubt, that will be the case for the future funding of PSOs.

7.4.1 Objectives

An economically sound and viable contribution mechanism must satisfy, at a minimum, at least three objectives. First, once the level of funding to be recovered from the sources in question has been decided, it should be assessed in a manner that distorts the efficiency properties of competitive outcomes as little as possible.[10]

The economic efficiency objective requires that contribution assessment be borne by all competitors—both incumbents and new entrants—in a manner that preserves each competitor's relative efficiency in market contests. That is, when the incumbent competes with new entrants, (1) all entrants should pay some funds to the carrier(s) having the PSOs on an equitable and nondiscriminatory basis; and (2) the incumbent's competitive retail services should bear no more or no less of the funding burden than the new entrants' substitute services are required to bear.[11]

Second, it is highly desirable to use recovery mechanisms that are easy to understand and require minimal regulatory oversight once established.

Third, it is important that the subsidy funding (i.e., taxing) system and the specific collection and payment mechanisms within it are perceived as being fair by those paying and receiving funds.

The funding of the various components of the PSOs should be accomplished in stages. In the first stage, the geographic cross-subsidies illustrated in the example in Section 7.3.2.1 should be corrected. A good first step would be to deaverage rates to the greatest extent possible so as to minimize the need for a subsidy flowing from low- to high-cost areas. For example, if prices were set at or above cost across all geographical areas, there would be no appearance of geographic subsidy requirements over time as market shares shift.[12] To the extent that there are public policy or other reasons why rate rebalancing to the full extent of costs is undesirable, the remaining subsidy will need to be calculated, and a funding mechanism will need to be developed.

A second stage, but concurrent with geographical rate deaveraging, would be an attempt to resolve the cross-product subsidy issue illustrated in the exam-

---

10. Establishing the target level of contribution has, in itself, efficiency implications. The choice of raising funds from certain services, (e.g., new information services) rather than from elsewhere (e.g., subscriber connections) affects consumers' (and producers') choices, thus affecting economic efficiency. Because all services are not fully competitive, there is some degree of control over where funds for subsidies can be raised. Both public policy and economic efficiency concerns are likely to be considered when contribution recovery to particular services is targeted.

11. That is, the price floor (minimum price) for the LEC's retail service should include the same contribution that interconnecting and competitive carriers' services include. If market conditions allow, the LEC's actual price charged could be above the price floor, thus providing greater recovery of contribution, but LECs would not be allowed to price below the floor (which includes the contribution). See the discussion of price imputation in Chapter 6.

12. There would, of course, be a temporary need for a subsidy (long-run costs are reduced slowly), but there would not be a long-term need. The temporary need appears because not all the costs incurred are avoidable in the near term. These costs are avoidable only in the long term. "Long term" here is defined as being the point at which the facilities would have been replaced based on engineering economy considerations.

ple in Section 7.3.2.2. This is important because the contribution from competitive services will be unable to continue to subsidize the USO as revenue from those sources is eroded by competition.

Cross-subsidies are measured by comparing the revenues from a segment of the business (e.g., a geographical area or a product) with its ICs.[13] Note that cross-subsidies are masked when segments are aggregated. For example, in the example in Table 7.1, there appears to be no subsidy in the aggregate. The subsidy (in that example, a cross-subsidy in which the high-density area subsidized the low-density area) was discovered at a more disaggregated level. Thus, measurement and correction of the cross-subsidy problem must involve disaggregated IC studies. While it may be difficult or even impossible today to examine subsidies at the level of individual services, it is important to avoid too much aggregation. For example, even the CO level of geographic area is going to be too aggregated to get an accurate measure of subsidy flows among subscribers. Similarly, in terms of products and services, large groups (e.g., all CO services or features) are also too aggregated.

Whatever level of aggregation is selected as practicable, it is the total economic cost (i.e., IC), compared to the corresponding revenues, that are required to measure and eliminate cross-subsidies.

Next, three potential sources of funding UTS are discussed and evaluated. It may be that all three will be required in some combination if both competition and policies promoting public services are to coexist in the new environment.

7.4.2 Alternative Methods

After the subsidy necessary to fund UTS obligations stemming from geographic averaging and the erosion of subsidy funds from competitive services has been calculated, the final layer of funding must be identified. This final layer, or stage, of funding examines the costs associated with the COLR obligation illustrated in the example in Section 7.3.2.4.

After rebalancing rates as far as is politically acceptable to alleviate subsidy needs based on geographic averaging and after eliminating cross-product subsidies to the greatest extent possible, the funding of the remaining universal service subsidy should come from three sources: interconnection charges, taxes, or revenue surcharges. To measure this portion of the subsidy requirement, we need to measure the difference between the revenues from the services associated with PSOs and their respective costs after restructuring rates to

---

13. While revenues *equal to* ICs avoid a subsidy (and therefore a cross-subsidy), a local exchange company needs more revenue to avoid subsidizing groups of products or geographical areas having shared costs. Even more revenue is required to cover the common costs of the company.

eliminate or reduce cross-subsidies. It is important to remember that it is the revenue and costs of the entire family of PSOs that is relevant, not the sum of the individual services' revenues and costs (this subject is taken up later). Even more important to remember is that UTS costs should *not* be compared with competitive service revenues.

The (re)structuring of interconnection charges is a mechanism for funding UTS subsidy requirements. There are strict limits on the use of such a mechanism, however. The limits on the ability to fund PSO costs through interconnection charges are essentially bypass costs. That is, interconnection charges are effectively capped at an amount no greater than the *incremental* cost of bypassing the local exchange company by paying the costs for alternative (i.e., non-PSTN) facilities. As technology advances over time to offer more and lower-cost means of entry and bypass (telephony on cable television, cellular radio, etc.), the IC of bypass is falling. Thus, even if interconnection charges can be structured to entirely fund the currently existing UTS subsidy, it is likely that such charges will soon fail to provide sufficient subsidy funds.

Returning to the example in Section 7.3.2.4 and following from Table 7.2, consider interconnection costs and revenues. If an interconnection agreement between the LEC and the competitor is established such that the $2 interconnection cost is shared equally by both carriers, the cost structure, and therefore the price structure, of the two carriers will change. The LEC's total cost now rises to $13, while the competitor's cost rises to $6. The LEC can now price its service at $3 per customer, but no higher if it intends to remain competitive. As Table 7.3 illustrates, the LEC's total cost and total revenue thus increase by an equal amount ($1), leaving the subsidy required unchanged.

Table 7.3
Interconnection Costs Shared

Company	Price	Quantity Served	Total Revenue	Total Cost	Subsidy Required
LEC	$3.00	2	$6.00	$13.00	($7.00)
Competitor	$3.00	2	$6.00	$6.00	$0

Even if the new competitor is required to pay 100% of the interconnection cost (which is often the case with collocated facilities), a substantial subsidy may still be required. In this example, while such an arrangement slightly reduces the LEC's total subsidy requirement, a substantial subsidy requirement still remains. The competitor's total cost of serving its customers is $7 (see

Table 7.4), thereby allowing the LEC to price its service at $3.50 to each customer. This increase in revenue to the LEC ($2), coupled with no increase in total cost (still $12), reduces the LEC's subsidy requirement to $5.

Table 7.4
Interconnection Cost Paid by Competitor

Company	Price	Quantity Served	Total Revenue	Total Cost	Subsidy Required
LEC	$3.50	2	$7.00	$12.00	($5.00)
Competitor	$3.50	2	$7.00	$6.00	$0

The analysis thus far has assumed that the incumbent LEC retains the COLR obligation and must maintain available capacity sufficient to serve all end users. Now consider the situation that would develop were the LEC to be permitted to withdraw from the COLR obligation and/or transfer the obligation to another company with interconnection requirements between the two carriers (Figure 7.7). In that case, it may no longer be necessary or advantageous for the LEC to continue to maintain facilities to all four customers. In fact, it would be most beneficial for the LEC to situate itself in exactly the same manner as the competitor. That is, the situation depicted in the Figure 7.7 would develop.

The cost structures of the two carriers are now perfectly symmetrical if we assume that the interconnection charge ($4) is shared equally between them. That is, each carrier incurs $7 in total costs.

It is important to realize that the LEC, by withdrawing from its COLR obligation, makes the adjustment to the situation depicted here only in the long run. This abandonment of the existing network and subsequent construction of a new network that interconnects with the old network have two noteworthy implications. First, the economic life of the existing network is affected by competitive entry. It is likely, however, that this new competition was not anticipated and, therefore, not accounted for in the currently employed economic life estimates. The LEC therefore must be able to increase depreciation rates so as to reflect the expected rate of abandonment. Second, it is economical for the LEC to make such a change only when it is time for the LEC to replace its existing plant. It normally would not be economical to make the change immediately. Most likely, the LEC would find that it would be more economical to continue to pay the operating and maintenance expenses of its existing plant serving its two remaining customers than to build all new plant at the initiation of competition.

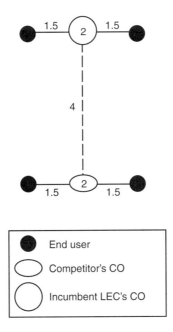

Figure 7.7 LEC permitted to withdraw from COLR obligations

A further complication arises if the LEC is forced to include sunk costs in its price estimates (see Figure 7.8). Under a no-sunk-cost rule (applied to its remaining two customers, not to the lost customers), the LEC's total cost would be $8, while the competitor's cost would remain at $6. This cost differential creates a potential price umbrella under which the new entrant can price (the LEC must price at or above $4, while the competitor can price down to $3). Because the LEC is saddled with recognizing the cost of abandoned plant, not only is there the possibility of a price umbrella, there is also the potential for underrecovery of that abandoned plant.

In summary, the cost rule that provides for a price floor and the decision as to whether the USO does or does not remain with the LEC have to be considered together. That is, the implications of keeping the USO with the LEC or allowing it to be transferred or abandoned, together with the rules governing price floors (e.g., a no-sunk-cost rule), can lead to inefficient outcomes and inefficiently high prices.

Finally, consider the limits on interconnection charges imposed by bypass costs. If the second carrier must interconnect to reach all carriers, interconnection charges in excess of the cost of interconnection can be obtained. Note that the cost of complete bypass as depicted in Figure 7.4 is $14. The LEC could

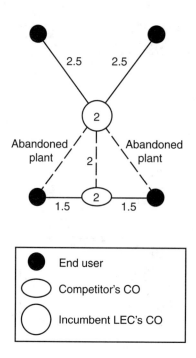

Figure 7.8 LEC under "no sunk cost" rule.

obtain no more than $7 from the new carrier in additional interconnection charges.[14]

In this example, the subsidy requirements of the incumbent LEC can be entirely met with interconnection charges. However, as the costs of bypass fall with new technologies and resulting lower bypass costs, there may be a need for additional funding. Indeed, there may be political or public policy limits well below the bypass limits that require a look to other sources of funding.

To generate the funds necessary to make up the difference between the universal service subsidy requirement and the contributions from interconnection charges, a tax and/or end user surcharge system could be implemented. To ensure the full funding of the remaining universal service subsidy, the taxes should be structured such that there is no practical means of bypassing them. This effectively means that the taxes must be levied on the subscribers to all communications services. As is the case with interconnection charges, however, there exists the possibility that tax contributions will fall over time as technology advances to the point where other types of services can effectively

---

14. The new carrier's cost is assumed to be $1.50 + $1.50 + $2 (switch) + $2 (interconnection cost) = $7. No more than an additional $7 could be collected without giving the new carrier incentive to bypass the incumbent entirely.

be used to bypass traditional types of telephone services. For that reason, a properly constructed system of end user surcharges should be implemented. An end user surcharge could take the form of a conversion of an access charge currently levied on carriers to an access-line charge levied on end users. In that manner, the necessary funds would continue to flow to the COLR, regardless of whether or not that carrier was bypassed by competitors. While the conversion of access charges to end user surcharges preserves much of the present funding structure, end user surcharges could have any of a number of structures.

7.4.3 Revenue Surcharges

Revenue surcharges have been discussed as a mechanism for funding the UTS obligation.[15] Revenue-based surcharges function like a sales tax assessed to all providers of selected services. That is, a certain percentage is assessed to total revenues (sales) of the services in question. This approach has some regulatory precedent. For example, in California, a percentage assessment is applied to the revenue base of all carriers to provide funds to subsidize lifeline services.

Once the level of the contribution recovery has been established and the surcharge rate has been set, the pricing of the interconnection services and the LEC's competitive services is straightforward. Apart from the surcharge, interconnection is priced at cost, and the price floors for the LEC's competitive services also are set at cost.[16] In addition, the revenue surcharge approach has attractive economic properties. In particular, economists have long recognized that outputs, rather than inputs, should be subject to taxation [1]. Because LEC interconnection services are inputs to the competitors' services, it becomes difficult to avoid what is, in effect, taxation of inputs when contribution is recovered through access prices set above cost. In contrast, the revenue surcharge focuses on outputs.

In addition, the revenue surcharge approach has attractive economic properties. In particular, economists have long recognized that outputs, rather than inputs, should be subject to taxation [1]. Because LEC interconnection services are inputs to the competitors' services, it becomes difficult to avoid what is, in effect, taxation of inputs when contribution is recovered through access prices set above cost. In contrast, the revenue surcharge focuses on outputs.

---

15. For a detailed discussion of this type of funding mechanism, see B. Egan, with S. Wildman, "Funding the Public Telecommunications Infrastructure" (*Telematics and Informatics*, Vol. 11, No. 3, 1994).

16. Because of the surcharge, both competitors and the LEC face prices that are above IC.

There are three potential difficulties with this approach. First, the maximum benefit from the approach may require variation of the surcharge rate across competitive services. While that requirement is no more onerous than varying the contribution element across services, arriving at the correct rates could require considerable care. Second, because of its resemblance to a tax, the approach may have some political difficulties. That is, regulatory commissions may be reluctant to assess "taxes." Third, assessing contribution in any form would raise the same objections discussed earlier that competitors should not be forced to recover LEC overheads. Again, such arguments are as invalid in the case of a revenue-based surcharge as they are in the context of a contribution tariff rate element.

7.5 THE STATE OF THE DEBATE

LECs have historically served as the regulated UTS provider including the COLR obligation. Recall that the COLR must stand ready to provide service to any subscriber within the geographic service area on demand, where such demands arrive randomly, without regard to cost. The unique regulatory goal of ubiquitous and affordable basic UTS for all gives rise to some very costly PSOs and therefore gives rise to a contentious debate as to how to calculate the cost and who should pay for it.

Almost all parties to the debate regarding the costs and funding of UTS in a competitive environment agree that: (1) UTS objectives should be met fully at a relatively low cost; (2) UTS objectives should be met with minimum distortion in the economic efficiency of market competition and the desirable incentives it creates for competing firms; and (3) contributions to funding the subsidy required to cover the cost of meeting the obligations be distributed fairly (proportionally) among a broad base of network service providers and/or their subscribers in a competitively neutral manner.

In the current regulatory debate, there are two primary views of how the LECs' UTS obligation should be defined. The LECs, which currently receive basic service subsidies, would ideally like UTS funding to encompass the very broad and costly obligation to serve all subscribers, on demand, at so-called affordable (low or even subsidized) prices. The IXCs, which are funding a large portion of the LEC subsidies via surcharges on long distance usage, want UTS to be defined as a very narrow and relatively low-cost obligation to provide subsidized affordable service only for a small fraction of local service subscribers, specifically those with very low income and those living in high-cost (usually rural) service areas. A third view, which has been presented by some industry observers, is that there should be no subsidy to basic exchange service. This would make any issue of the cost of UTS moot.

7.5.1 Narrow Definition Favored

The absence of an accepted detailed definition of the UTS in either state or federal regulatory jurisdictions has created a lobbying flurry as opposing parties attempt to influence regulators. Pursuant to its broad investigation of universal service in a *notice of inquiry* (NOI) in CC Docket 80-286 (August 30, 1994), the FCC has revealed its preliminary UTS policy views in its *notice of proposed rulemaking* (NPRM) (July 13, 1995).[17] It is fair to say that the FCC favors the narrow (low-cost) view, which has been the trend in every state that has recently revisited the issue. The political reason is obvious: Narrow interpretations reduce the total cost and required subsidy amounts and therefore reduce the regulatory heat from industry pressure groups that are otherwise saddled with paying the subsidies, not to mention that the costs and required subsidies themselves have been growing rapidly, which is why most regulators have had to go over the entire issue again.

Most states are leaning toward reforming the system of subsidies in favor of targeting them to low-income and high-cost subscribers. Almost all parties to the debate believe that, going forward, it is no longer sensible or perhaps even feasible to subsidize broad segments of local service subscribers without regard to their individual ability to pay cost-compensatory rates. Rather, future subsidies should be targeted to those subscribers in need. But it is important to note that this view has its dissidents, namely, small rural telephone companies and some consumer groups that represent those on the receiving end of the subsidies. But, compared to the political heavyweights who are the major industry players, the small telephone companies and consumer groups eventually will not be able to carry the day.

7.5.2 Defining UTS

Regardless of the emerging industry consensus to limit the definition of UTS by targeting it to small groups of subscribers, there is still no widely accepted definition of UTS. Even as a concept, UTS is in a constant state of flux. The new law has ensured that this situation continues by stating that UTS is by definition an evolving concept. Therefore, the costs of UTS should be analyzed within a framework flexible enough to adapt the study results to any definition that ultimately may be adopted by the Joint Board, as mandated by the new law.

Since it will be some time (if ever) before the states agree on firm definitions of advanced services for purposes of setting universal service policies

---

17. Pursuant to the Telecommunications Act of 1996, the FCC will be revisiting a number of the issues and will expand its entire investigation of universal service considerably. See the first FCC action pursuant to the new law establishing a Federal-State Joint Board: FCC 96-93, CC Docket No. 96-45, March 8, 1996. This and the other FCC proceedings that are referred to can be found at URL http://www.fcc.gov.

pursuant to the new law, it is most useful to deal with the existing definition of basic universal telephone service, also referred to as POTS.

Four primary questions must be considered in a UTS cost study. First, what constitutes basic service? (In this area, at least, there seems to be agreement among the major industry factions.) Second and third, regarding the main body of the dispute: What should the per-subscriber cost of the subsidy be (and how is it calculated)? What is the total number of subscribers to be subsidized? Fourth, how should we treat historical costs vis à vis prospective costs?

Definition of Basic Local Exchange Service

Generally speaking, in the context of UTS, there is little dispute among LECs, IXCs, and other LEC competitors as to what constitutes a static definition of basic local exchange service on a per-subscriber basis (POTS). The contentious issue is how many subscribers should receive a subsidy.

Both sides tend to agree that universal service implies that every basic service subscriber should have a dial tone line with access to emergency services and perhaps additional minor features such as touch tone and local usage. For example, AT&T maintains that, for purposes of determining basic local service subsidies, only residential service is relevant (some believe that single-line business service may be included). AT&T includes as part of residential service voice-grade single-party dial tone with touch tone and local usage, access to repair service, telecommunications relay service, directory assistance (411) and 911 emergency service, and a white pages directory listing. MCI's definition of basic universal service is nearly identical. In numerous state and federal regulatory proceedings, the LECs tend to agree with the IXC's narrow view of targeted universal service subsidies on a going-forward basis (while continuing to maintain that the IXC's preferred treatment of historical costs is incorrect). From time to time, a few other interest groups, especially consumer advocates, have promoted an expanded concept of universal service, including ISDN and VDT. Because these views will increase the costs of UTS and the required subsidies, from an economic perspective they should not become popular (to date, few states have adopted them). However, the new law calling for advanced services may change that.

7.5.3 The Primary Remaining Dispute Over UTS Costs

The remaining dispute is concentrated in two areas: (1) the per-subscriber cost of the subsidy and (2) the total number of subscribers who should be subsidized. Associated with these major disputes is whether or not the per-line and total subsidy amounts (defined as basic service revenue minus basic service cost—usually a negative number) should be funded at a level based on historical (i.e., average) cost estimates or going-forward (i.e., incremental) cost esti-

mates. In either case there is an underlying issue as to which costs are legitimate to include. Even assuming that the interested parties could find consensus
on costing methodologies, there is still the ever present dispute over who
should pay the subsidies.

7.5.3.1   Per-Subscriber Cost and Subsidy

The numbers most often quoted in the trade press as the nationwide cost of the
UTS range from a low of only $4 billion per year to a high of over $20 billion.
There are about 110 million residential access lines in the United States, so the
subsidy estimates vary from $3 to $15 per month per subscriber line.

   The numerous attempts to quantify the amount of the subsidy covering the
costs of the UTS have fundamentally missed the mark. The problem is that existing studies have utilized a costing framework that is not suited to answering the
relevant questions for the current and prospective market environment. Most estimates of subsidy requirements focus only on the cross-subsidy flows from toll
to local service and are based on estimates of direct (i.e., incremental) costs of
providing toll and carrier access services. This is fine as far as it goes, but it does
not directly address the full cost of UTS. Studies that rely on ICs or "average" ICs
were designed to answer questions of service pricing, marketing, and line-of-
business contribution (profitability). Such studies, which are useful for determining regulatory pricing floors and current service-specific contribution flows
(see Chapter 6 on costs for pricing and profitability calculations), are inherently
not suitable for determining the cost or subsidy requirements for UTS.

7.5.3.2   The Cost of Running a Local Telephone Company

Questions regarding the costs and subsidy requirements of the UTS are questions of *total* (i.e., average), not *incremental*, costs. That is, in a competitive
marketplace, what is the total cost of running a local telephone company?
There are two indisputable features of future markets for telecommunication
services: that competitive market entry will expand and that price cost margins
for most services will be lower as a result. If it is agreed that a fundamental
feature of competitive markets is that prices for network services will be driven
toward their respective costs, then clearly those services facing competition
eventually will not be able to continue to provide the existing levels of subsidies flowing toward covering a portion of the costs of providing basic local
service. Market competition is the natural enemy of cross-subsidy. Therefore,
in the presence of market competition, the relevant questions regarding funding
of the UTS going forward are: (1) what is the *total* cost of the UTS obligation;
and (2) how should it be funded (including *who* should contribute)?

   Regarding the first question, there is also the issue of what portion of
going-forward funding of the UTS represents historical costs to be recovered

from future revenues, as opposed to truly future costs to be recovered from future revenues. Regarding the latter question, the issue is the equity, economic efficiency, and sustainability of the funding mechanism. These issues will be investigated within the following framework of analysis.

7.5.3.3 Number of Subscribers That Should Be Subsidized

Given that total cost (average unit cost) is the appropriate metric for gauging the UTS, there remains the initial issue of defining the number of residential subscriber lines to which this metric applies. Given also that major players have taken strong public positions that UTS refers only to that narrow portion of the subscriber base represented by low-income and high-cost subscribers, the analysis should be structured so that its results are not particularly sensitive to the regulatory outcome of this issue. At any rate, it is reasonable to assume that, regardless of the short-term resolution of this issue, in a future world of rate rebalancing (which all major LECs support), the focus will inevitably be only on subsidies for low-income and very high cost subscribers. For that reason, both the low-income/high-cost and the total subscriber scenarios must be evaluated.

7.5.4 Definition of COLR

In the current environment, a portion of the average residential subscriber's access line and local usage costs is recovered from nonbasic services. This "total subscriber scenario" will be considered to reflect the cost of the COLR obligation. Therefore, the definition of COLR used herein is as follows: *the total cost to a LEC of providing, on a regulated common carrier basis, basic local telephone service to all households, on demand, where such demands arrive randomly from the public at large.* In common parlance, that definition is translated as the total cost of running a local phone company (i.e., stand-alone cost). This includes the cost of deploying spare capacity in the public network to be ready to serve the randomly arriving demands in a timely fashion in accordance with whatever service standards are required to be met by regulators.

Under this definition of COLR, the cost of the narrowly defined UTS is a subset of the total COLR cost. That explains why the terms often are used interchangeably. In practice, the term *universal service* is often used loosely to refer to the costs associated with the COLR obligation. Unfortunately, even in academic literature, the stated or implied definitions of UTS and COLR are mixed.

7.5.5 Costing Principles for UTS and COLR

Viewed from the perspective of incremental economic analysis, the relevant cost question, either historical or going forward, for determining the costs of the

UTS (including COLR) obligations is this: What is the total cost of building and operating the LEC's POTS network with the UTS obligation minus the cost without the obligation? Of course, the answer is the total cost of serving unprofitable subscribers since, over the long term, no one would voluntarily serve customers at a loss.

In a monopoly environment such as that which has historically existed, this is the same as taking the total stand-alone cost of the LEC's POTS network and subtracting the total POTS revenues. The balance is both the net cost of the UTS obligation and the subsidy requirement. Of course, LECs historically have been vertically integrated enterprises offering both basic POTS and nonbasic services. Unfortunately, this has all been done in a monopoly environment. So the answer to the relevant cost question for UTS (keeping in mind that it is in the context of the historical monopoly environment) may be found by applying the following economic question: Given the total costs of the vertically integrated (i.e., multiservice) POTS enterprise, what costs are saved (not incurred) by eliminating (not providing) nonbasic services?

Notice that this question is not the same one that would be applied in a competitive environment, whether in a historical or a prospective context. The reason is that, in a competitive environment, it is not legitimate to include either the direct costs or the revenues of otherwise competitive services in the calculation of total costs or subsidy requirements for UTS.

7.5.6 Historical and Prospective Costs

Just as the number of lines in service and costs and rates of providing basic telephone service change over time, the costs associated with UTS obligations vary as well. In fact, in the federal (interstate) jurisdiction, the explicit subsidy costs and contributions change annually, along with the three primary sources of funds: the *carrier common line charge* (CCLC) applied to long distance usage, the high-cost fund charge (also a surcharge placed on long distance service), and lifeline service (e.g., the FCC's Link Up plan).

In the state jurisdictions, the net revenues or financial contributions from services that are priced above their ICs are used to offset a portion of the LEC's residential basic exchange service costs. Therefore, even when viewed in the context of the current period, revenues from current sales of services are always recovering the historical costs of past investments, including capital costs in the form of annual depreciation charges. Thus, contrary to the claims of various parties to the subsidy debate, especially the IXCs, there can be no issue or confusion as to whether or not current and future contribution flows are designed to subsidize a portion of the costs of a LEC's *historical* investments. The only real issues are: (1) what amount of contribution should be expected from each of the nonbasic services; and (2) what is the contribution shortfall (revenue minus cost) from basic residential service?

UTS cost studies should include all historical costs of the PSOs so that, if and when the policy debate converges on a relatively narrow and targeted subsidy amount, the regulatory IOUs associated with an LEC's fulfillment of historical UTS obligations are not overlooked. For example, an LEC that accepts the risks associated with price cap regulation, which effectively breaks the link between tariff rate levels and costs (see the discussion in Chap. 6), may not be able to petition regulators for net revenue increases to compensate for competitive losses of contribution to continue to recover the costs of UTS obligations. Indeed, the IXCs have been claiming in state and federal regulatory proceedings that one of the risks that LECs accept under incentive regulation (i.e., price caps) is that competitive losses may make it impossible to recover capital investments, including those associated with UTS.

In summary, UTS can refer to either: (1) the prospective costs of continuing to provide basic exchange service to a relatively narrow portion of the residential subscriber base (i.e., targeted to low-income and high-cost subscribers); or (2) the current costs of UTS based on past expenditures applied to a relatively broad portion of the subscriber base (i.e., the COLR obligation). There can be no doubt, however, that the latter reference represents the full costs of UTS. Within the narrow definition of UTS, there is no conceptual dispute in any quarter that low-income and high-cost subscribers groups deserve some level of subsidy. Of course, whether narrowly or broadly defined, there will still be a great deal of controversy over the exact level of subsidies required going forward, which can only be addressed by a long-term costing effort using highly granular data.

7.5.7 Public Policy and Fairness Considerations

To determine which portions of a vertically integrated telecommunications enterprise require a subsidy to be sustainable as an ongoing and financially viable business operation, it must be determined if: (1) any portions of the business are sustainable as a stand-alone business entity; or (2) natural monopoly characteristics of the underlying cost structure would not allow such sustainability in the presence of competitive market entry.[18]

The analytical framework for determining the amount of long-run costs, subsidy requirements, and contribution flows must reflect underlying realities. The most important reality is that there are fundamentally only two types of local telephone company subscribers, those providing net subsidies ("subsidizers") and those receiving subsidies ("subsidizees"). From a public policy perspective, whatever the historical political and economic motivations, regulatory institutions have created the situation we observe today regarding

---

18. See Appendix B for a discussion of the basic economics of natural monopoly used in this context.

which certain PSTN subscribers, or groups of subscribers, are net subsidizers or subsidizees.

Unless we are willing to make value judgments as to whether competition policy for the future should favor historical subsidizers or subsidizees, the fact remains that, if the subsidy system and the contribution flows within it are to be reformed so that the system can accommodate the future increasingly competitive market paradigm, we must be able to evaluate the sustainability of the old system and the subsidy requirements of the new competitive system. That requires a study sufficiently granular or disaggregated to capture individual subscriber service cost characteristics.

If we can reliably evaluate the costs of serving individual subscribers and geographic collections of subscribers, we can also evaluate the following:

- Whether the cost structure is characteristic of natural monopoly and therefore deserving of a subsidy;
- Natural monopoly notwithstanding, whether (and to what degree) tariff-rate averaging for basic local service is sustainable in a given area that is vulnerable to competitive entry;
- Where universal service may be reliably and profitably provided under a free market featuring open entry (this may still require so-called lifeline service for low-income and high-cost subscribers).

In addition, the disaggregated cost study will be able to shed some light on the going-forward subsidy level and the best way to reform the funding mechanism that generates the required subsidy.

Reference

[1] Diamond, P., and J. Mirrlees, "Optimal Taxation and Public Production," *American Economic Review*, Vol. 61, 1971, pp. 8–27.

Improving the Rural Telecommunications Infrastructure

8

8.1 INTRODUCTION

Advanced (digital) telecommunications technology has the potential to dramatically improve the quality of life and the rate of economic development in rural America.[1] Public access to advanced telecommunications technology need not imply that users have to be physically located in proximity to urban areas, where most information and production are generated. But while technology adoption in communication networks continues at a rapid pace, increased market competition among telephone network operators forces them to invest where the money is—in dense urban and suburban areas. Thus, while a modern and effective telecommunications infrastructure is crucial for rural economic development, its financing raises a multitude of difficult public policy issues.

The analysis in this chapter examines the rural telecommunications infrastructure, focusing on technological developments and the costs and financing of network modernization. While there has been considerable hype in the industry and trade press about digital information superhighways (as if we can all just sit back and wait for "it" to happen), a look at the facts would lead to a more pessimistic view, especially for rural areas of the country.[2]

---

1. For background reading on the relationship between rural development and telecommunications infrastructure see E. Parker et al., *Electronic Byways: State Policies for Rural Development Through Telecommunications* and *Rural America in the Information Age: The Communications Policy for Rural Development* (Lanham, MD: University Press, 1989). See also *Rural America at the Crossroads: Networking for the Future* (U.S. Congress Office of Technology Assessment, OTA-TCT-471, April, 1991).

2. For a dose of healthy skepticism (or even cynicism) on the prospects for deployment of information superhighways in a competitive market place see B. Egan, "Building Value Through Telecommunications: Regulatory Roadblocks on the Information Superhighway" (*Telecommunications Policy*, 1994).

Some recent technological developments provide exciting prospects for improving rural telecommunications. In particular, new digital wireless technologies may speed up the development of an advanced network infrastructure in rural areas. However, to fulfill the promise offered by digital rural radio service, the government needs to allocate more spectrum to support its development.[3] Because the cost characteristics of wireless technologies are not nearly as sensitive to physical distances as the cost of wired networks, they are well suited for rural applications. Still, these technologies most likely will not be deployed in rural areas until well after they have appeared in dense urban and suburban markets. The reason, as usual, is simple economics: in dense urban markets, the consumer convenience and cost advantages of wireless subscriber telephone connections over wired ones are enormous.

Federal and state governments are considering and beginning to implement pro competition policies in all aspects of telecommunications. As competition in urban and suburban markets for local and long distance telecommunications heats up, alternative-service providers will invest in the latest technologies in pursuit of profits and competitive advantage. Under the old monopoly system, rural telephone systems were able to offer prices and a level of service quality that often were close to those of urban and suburban subscribers (at least for basic residential service). But such service required massive cross-subsidies from the relatively dense urban and suburban areas to cover a portion of the costs of service. In the face of competition, that situation is no longer sustainable, competition being the natural enemy of cross-subsidy. In recognition of the problem of maintaining rural subsidies in the presence of competition, the Telecommunications Act of 1996 strongly reaffirmed the public policy rationale underlying rural telephone network cross subsidies, namely, to ensure that all Americans have access to similar services at similar prices. This objective is outlined in Section 254 (b) of the Act, under UNIVERSAL SERVICE:

(3) ACCESS IN RURAL AND HIGH COST AREAS

Consumers in all regions of the Nation, including low-income consumers and those in rural, insular, and high-cost areas, should have access to telecommunications and information services, including interexchange services and advanced telecommunications and information services, *that are reasonably comparable to those services*

---

3. For more information on wireless technology developments and public policy regarding wireless network infrastructures, refer to Chapter 4.

provided in urban areas and that are available at rates that are reasonably comparable to rates charged for similar services in urban areas [emphasis added].

So far, regulators and policymakers have not seen fit to assess a proportional subsidy requirement on new competitive service providers or their subscribers. Assuming that regulators choose to maintain the status quo of competitive entry where no proportional subsidy funding requirement applies to entrants, it is inevitable that service gaps will be exacerbated. It therefore remains to be seen if the new telecommunications law can truly stem the erosion of rural subsidies to the point where rural subscribers will continue to enjoy price and service levels similar to those enjoyed by their urban and suburban counterparts.

For the rural telecommunications infrastructure to continue to be on a par with that of urban areas, the government will have to adopt competition policies that can sustain subsidy flows from low-cost to high-cost areas. It is hoped that those policies also will be competitively neutral so that no network provider, entrant or incumbent, is disadvantaged in the process. That twofold requirement poses a difficult (perhaps impossible) challenge for technology and competition policy, the result being that the government will likely have to retreat from its stated goal. It could do so and still keep face simply by redefining universal service and affordability much more narrowly than it has in the past. By targeting subsidies to the relative handful of truly low-income and high-cost rural network subscribers, the government could drastically reduce subsidy requirements and still maintain affordability for those who really need subsidies to stay connected to the public telecommunications network.

Much of rural America is served by small independent telephone companies. There are more than 1,300 local telephone companies in the United States, the top 10 of which serve over 80% of subscribers. The rest serve areas with a relative handful of subscribers. Historically, financing for the modernization of rural company network facilities has come from a combination of the local tariff rates charged by the rural telco and cross-subsidies derived from: (1) rural company charges to interconnecting toll carriers; and (2) revenue-sharing arrangements with larger local telephone carriers that serve a relatively dense area with lower-cost (i.e., higher-profit) subscribers. Increased competition has added considerable uncertainty to the traditional revenue flows derived from those sources.

While direct competition for telephone subscribers may be long in coming to many rural areas, the competitive erosion of cross-subsidies currently provided by toll calling, business, and high-profit residential market segments is surely going to proceed rapidly. Naturally, competitive network operators do not want to pay any subsidies for rural development. At the same time, small telephone companies want the government to ensure that rural network infrastructures and individual subscriber service in rural areas are affordable and equivalent to those available in urban and suburban areas.[4]

The reality, of course, is that the outcome for the future will be similar to that of the past. Namely, rural network infrastructures will lag behind those in urban areas in terms of advanced service capability. While rural subscribers generally have been able to obtain single-party basic telephone service that is on a par with urban subscribers, that has not been true for advanced telephone services like digital data and other business services. With competition, the service gap will naturally expand to include new digital services for residential subscribers. In an increasingly competitive environment, the real risk is that the service disparity will become much worse. To prevent the erosion of rural subsidies from newly competitive services, a number of federal and state initiatives are under way with the goal of preserving subsidy flows. Those initiatives are usually presented under the rubric of so-called universal service objectives. At the federal level, the FCC has a current investigation underway to better target subsidies to rural areas needing them.

Federal initiatives aside, the best way to establish rural objectives for a network infrastructure is at the state level. Telecommunications depreciation policy, basic rates, and economic development planning for local telephone service are set at the state level; each state determines its objectives, timetables, and financing requirements. An important gap in telecommunications infrastructure planning exists in most states, especially regarding coordination with the important transportation and energy sectors. Synergies between telecommunications network providers and public power grid operators in the area of fiber optic transmission have, thus far, been underutilized. Increased cooperation in that area is critical. The same is true, but to a lesser extent, in the case of the transportation sector. For example, telecommunications technology for monitoring and locating purposes is currently used to complement rail, air, truck, and delivery services traffic. In the future, information technology will be used much more intensively in the transportation sector—emergency service vehicles and automobiles will be equipped with digital telecommunications gear for navigation functions, and traffic control engineers will use it for traffic control, routing, and emergency services. The early beneficiaries of more cooperation among sectors are rural education, health care, and other professional services.

---

4. For a recent discussion of the goals and public policy concerns about continued subsidies of rural telephone companies, see the testimony of Margot Smiley Humphrey on behalf of the National Rural Telecom Association before the U.S. Senate Commerce Science and Transportation Committee, May 19, 1994. See also National Association of Development Organization (NADO) Research Foundation, "Telecommunications and Its Impact on Rural America" (April 1994).

8.2 RURAL NETWORK TECHNOLOGY TRENDS

The following analysis indicates that where terrain permits, fiber optics will continue to be the technology of choice for all shared network facilities. At times, however, local conditions may call for microwave radio trunk transmission lines instead of fiber. For dedicated subscriber loop plant, several alternatives are available (once again, depending on local terrain and the spatial distribution of individual subscribers), including coaxial cable, copper wire, and digital radio. It is important to note that due to significant variations in local demographics and topography, some aspects of the overall analysis and their corresponding conclusions may not apply in many specific rural areas, although they are relevant for broad public policy considerations.

In fact, the most important conclusion that can be drawn from this analysis is that technological solutions must be tailored to specific circumstances, taking into consideration topology, terrain, subscriber demand, and spatial distribution. A cookie cutter approach to technology deployment, while easier from a network standards perspective, usually is not the least-cost method for optimizing the network for local supply and demand conditions or for planning future network upgrades.[5] Indeed, flexibility in network deployment strategies is the key to successful low-cost investment. That means that flexible standards must be developed by both wireline and wireless network equipment manufacturers to allow efficient interconnection between networks and a high degree of connectivity between end users.[6]

The cost of advanced rural communication network infrastructures is substantial. In a future competitive market environment, it may not be possible to finance such construction without significant increases in subscriber rates unless a new and stable source of subsidy funding is adopted by regulators.[7] Assuming a construction interval of 10–20 years—a normal time span for turning

---

5. This is not to say that fundamental network planning should not pay close attention to life-cycle costs of network equipment, including all its software dimensions. Planned and unanticipated network software upgrades for network switching, monitoring, control, and service functions often represent a large share of total network costs, and the cost savings associated with compatible or single-supplier purchases of software right-to-use fees must be considered.

6. Of course, while nonproprietary flexible network software and interface devices may lower the costs for telephone companies that purchase from competing vendors, such standardization can reduce profits of vendors that make money on sales of proprietary hardware and software systems. When rural telcos are faced with purchasing in an environment of competing proprietary network systems, more often than not a least-cost strategy is to select a single vendor so that life-cycle costs are minimized.

7. This is contrary to the conclusions reached in my prior study of this issue nearly five years ago when the industry's regulatory regime was still based on monopoly supply, and the threat to cross-subsidies for rural areas was much smaller. See B. Egan, "Bringing Advanced Telecommunications to Rural America: The Cost of Technology Adoption" (*Telecommunications Policy,* February 1992).

over telephone plant—one estimate of the cost of digital service is about $1,000 per subscriber.[8] That amount per subscriber would endow rural subscribers with digital communication capability comparable to narrowband ISDN service. While that service might suffice for residential subscribers using home computers or other devices, the narrowband service capability may not meet the communication requirements of business customers. As subscriber demand for multimedia services develops, fiber optic technology or other suitable high-capacity transmission media may be necessary.

Archiving digital multimedia communication capability in rural areas is a costly proposition, averaging $2,000–$5,000 per rural subscriber (much more in sparsely populated areas and areas with rough terrain). Broadband communication facilities would allow consumers to enjoy high-quality interactive multimedia services, video telephony, and entertainment video services, potentially allowing a single subscriber to use more than one communication activity simultaneously. For example, a broadband telephone connection may allow a subscriber to access an online database while viewing a movie, reading, or listening to the news. The cost of such functionality is high because it requires new alternatives for subscriber loop plant to replace traditional twisted-pair copper phone lines.

Where possible, existing coaxial cable television loops could be interconnected to a fiber backbone of shared network facilities to provide broadband multimedia capability. Elsewhere, FTTH (or fiber "near" the home) is required. Current satellite and microwave radio would not be the best option for most service applications because bandwidth limitations and delay times make these technologies unsuitable for a multimedia real-time environment. However, both radio and satellite are useful for infrastructure development in some applications. Satellites, for example, are preferred for delivery of distant video programming and may be interconnected to the wireline network infrastructure. But the use of satellites for voice service or other real-time two-way communications will likely be minimal.[9]

---

8. This estimate is a broad average and depends heavily on embedded subscriber loop plant characteristics. For example, where digital switching is already available and the subscriber line is short (less than 18,000 ft) with no signal repeaters or amplifiers, the average cost of a digital upgrade is one-third of this amount, or $300. In older plant (about half the embedded base), the per-subscriber costs are much higher (about $2,500) due to digital switch replacement and rehabilitation of nonfilled cable plant which generally will not support digital service.

9. The round-trip transmission delay for two-way satellite service is 250 ms, which usually results in poor-quality voice conversations, although some researchers believe this problem could be somewhat mitigated by the use of advanced electronics. In cases like rural Alaska, where customers never had a high-quality wireline option for voice service, satellite is more readily acceptable. However, the costs for voice satellite service in thin rural markets can be very high, even when transponder capacity is leased from others (thereby removing upfront manufacturing and launch costs from the calculation). The delay does not present a serious problem for data transmissions.

That could change, however, with the future deployment of LEO and MEO digital satellite systems.[10]

Likewise, microwave radio is useful and cost effective in many situations where fiber is not practical, such as over rough terrain or water. Many of the existing microwave facilities are useful for providing advanced telecommunications because they are already digital and may feature high bandwidth and capacity for new service applications. The FCC-approved *basic exchange telecommunications radio service* (BETRS) is the primary application of microwave radio technology for local service and is expected to be the preferred alternative when wireline service is not feasible. But cases of wireline services not being feasible to transmit basic local service are rare, and rural radio service, as currently defined by the FCC, is not being widely deployed as an alternative to traditional wireline service in rural areas.[11]

The FCC could change that by assigning new spectrum for high-powered rural radio systems. High-powered digital radio systems for fixed telephone service are cost effective in rural applications compared to wireline systems, but only if there is enough spectrum and only if the FCC relaxes system power restrictions to allow for large macrocell radio coverage areas (e.g., 15–30 mile radius) featuring maximum sharing of available spectrum within a single base station area. Foreign countries, especially those with nascent network infrastructures, are deploying new digital wireless systems as an alternative to traditional wireline connections.[12] The U.S. government's recent focus for radio spectrum policy has been on new convenient low-power cellular and advanced paging and cordless telephone services, which, while ideal for pedestrian and mobile applications in congested urban environments, are not cost effective or feasible in rural settings.

For distribution of basic local service, both satellite and microwave will generally be limited to relatively high cost applications.

---

10. For more information on new digital satellite systems see G. Gilder, "Ethersphere" (*Forbes ASAP*, October 10, 1994). The entire Gilder series of articles featured in *Forbes* sections over the last three years provides a thought-provoking discussion of future telecommunications technology trends, especially regarding new digital wireless systems.

11. For example, in a recent investigation of rural radio service, the Oregon Public Utility Commission concluded: "BETRS is not now a viable system. There are too few BETRS systems in operation. No additional BETRS systems are planned for Oregon. BETRS appears to have significant drawbacks in terms of relatively high maintenance and investment costs. These drawbacks have resulted in low use of BETRS in Oregon" (Oregon PUC Staff Discussion Paper, "The Economics of Wireless and Wireline Telephone," draft, April 20, 1995).

12. See A. Paulraj, "Wireless Local Loop for Developing Countries—A Technology Perspective" (*1994–95 Annual Review of Telecommunications, International Engineering Consortium*, Chicago, 1995). Two North American manufacturers, Motorola and SR Telecom, are each deploying (or plan to deploy) many such systems; see J. Gifford, "Wireline Local Loop Applications in the Global Environment" (*Telecommunications*, July 1995) and "Rural Network Possibilities" (*Interlink 2000*, August 1992).

8.3 WHAT IS RURAL?

There is no standard definition of rural telecommunication subscribers; however, some general observations should be made. Government data indicate that about a fourth of all households in the United States are in nonurban areas, or *metropolitan statistical areas* (MSAs). Nonmetropolitan counties are those with no urban areas greater than 50,000 in population, but there are many possibilities for classification errors. For example, there could be metropolitan areas close to the border of adjacent non-MSA counties, or there could be many neighboring towns with populations less than 50,000 each. It is potentially misleading for policymakers to use such data for policy purposes without adjusting them for classification problems.[13]

The Telecommunications Act of 1996 provided the following definition of "rural local telephone companies":

> The term "rural telephone company" means a local exchange carrier operating entity to the extent that such entity—(A) provides common carrier service to any local exchange carrier study area that does not include either—(i) any incorporated place of 10,000 inhabitants or more, or any part thereof, based on the most recently available population statistics of the Bureau of the Census; or (ii) any territory, incorporated or unincorporated, included in an urbanized area, as defined by the Bureau of the Census as of August 10, 1993; (B) provides telephone exchange service, including exchange access, to fewer than 50,000 access lines; (C) provides telephone exchange service to any local exchange carrier study area with fewer than 100,000 access lines; or (D) has less than 15 percent of its access lines in communities of more than 50,000 on the date of enactment of the Telecommunications Act of 1996.

It is important to distinguish "remote" subscribers from simply "rural" subscribers. "Remote" refers to those whose access to the telephone network is difficult due to physical "remoteness" caused by either extreme distance or terrain. While remote subscribers with no telephone service might represent a socially deserving segment of the general population, for public policy purposes they should be separated from the general body of rural subscribers. Public policy must be able to focus on upgrading communication infrastructures for those customers already hooked up to the network regardless of policies for reaching customers who are not only rural, but physically remote. Otherwise, policy debates over the subsidies required to provide service to remote nonsub-

---

13. For a more detailed discussion of the problem, see U.S. Congress OTA, *Rural America at the Crossroads*, pp. 36–38.

scribers can derail progress in technology adoption for the vast majority of rural subscribers. Furthermore, the available evidence is that remoteness is neither a particularly common problem nor one that requires much total subsidy to solve. Those pockets of truly remote subscribers that do exist will be served most economically by new digital satellite communication networks.

There are few truly remote subscribers relative to the base of all rural subscribers. One estimate puts the number of remote customers at 183,000, or only about 1% of all rural subscribers.[14] Fortunately, for actual telephone statistics and data on rural subscribers, a wealth of information exists for small independent telephone companies from industry trade groups such as the *United States Telephone Association* (USTA), the *National Telephone Cooperative Association* (NTCA), and an agency of the U.S. Department of Agriculture that for many years was known as the *Rural Electrification Administration* (REA). The REA's areas of responsibilities were recently combined with other areas, and the new agency is called the *Rural Utilities Service* (RUS). RUS provides investment and financial data for almost 900 small telephone companies serving about 6 million subscribers in very thinly populated markets. Thus, for the purposes herein, the RUS data will be considered representative of rural subscribers. While many other data sources will be used in this analysis, the basis for most statistics will be the RUS data.[15] Depending on one's particular view as to the absolute number of rural telephone subscribers in the United States, for broad policy analysis, the per-subscriber results based on RUS data simply may be increased by an appropriate factor to arrive at universe results.[16]

Beyond the distinction of rural versus remote, there is also an important distinction between existing and new customers. Costs of technology adoption may be sensitive to the fact that the necessity of starting from scratch in some areas renders moot the issue of whether to use some of the existing facilities in a network upgrade. For most subscribers, a network upgrade must consider the embedded base of technology to ensure a cost-effective construction decision. Keeping in mind the distinctions between rural versus remote and existing versus new subscribers, this analysis concentrates on the cost of network upgrades

---

14. See Parker et al., *Electronic Byways*, p. 67. Parker's book classifies about 20 million households as "rural" on a base of about 92 million households in the United States. Other estimates of remote subscribers appear in FCC Report No. DC-1066, CC Docket 86-495, "New Radio Service (BETRS) Established to Improve Rural Phone Service" (December 10, 1987).

15. The source of 1993 financial and investment data for small telephone companies is REA, "1994 Statistical Report, Rural Telephone Borrowers" (U.S. Department of Agriculture).

16. Since the top 10 telephone companies serve roughly 80% of an estimated 100 million U.S. households, and considering that the 900 RUS companies serve only about 6 million subscribers, it follows that most "rural" subscribers are served by large telcos. (Recall that about one-fourth of U.S. households are in non-MSA "rural" areas.) However, the RUS data and statistics for small companies probably provide the best average characterization of rural areas.

for existing subscribers—the vast majority. Remote and new subscribers should be considered separately.

8.4   FINANCIAL PROFILE FOR RURAL TELEPHONE COMPANIES

There are over 1,300 telephone companies in the United States, about 900 of which are borrowers in the federal government RUS financial assistance program. The top 53 LECs, which report annually to the FCC, account for about 90% of the approximately 150 million access lines in the United States.[17] The seven RBOCs alone account for about 70% of all telephone lines; adding GTE and Sprint accounts for nearly all of the top 53's 90%. Thus, all but the top nine telephone holding companies are quite small by comparison. However, despite the huge differences in the scale and scope of the operations among U.S. LECs, when comparing statistics for average per-line financial results between large and small companies, the data are surprisingly similar. One reason is that, while the larger LECs may enjoy the low average per-line costs of serving large metropolitan areas and spreading fixed network costs over a large subscriber base, they also serve a considerable number of rural service areas. Similarly, while small rural LECs may serve much less dense areas overall, they too serve relatively dense towns within those rural areas. Furthermore, larger LECs tend to have a scope of operations very different from that of smaller LECs, including investments in regional toll service network facilities and specialized and business services.

Tables 8.1 and 8.2 provide financial benchmark data for key operating ratios, costs, and revenues for large and small LECs.

8.5   OPERATIONS, INVESTMENT, AND EXPENSES

A comparison of FCC and RUS data for large and small LECs indicates that large LECs enjoy substantial capital and labor productivity advantages due to their large scale of operations and dense subscriber base. For example, large LECs support, on average, about 30% more telephone lines per employee than small LECs.

Average annual expenses per line are $607 for small LECs and $446 for large LECs. However, those amounts include annual depreciation charges per line, which, due to the small LECs' larger investment in physical plant per line,

---

17. Statistical data for non-RUS companies are based on the FCC, *Statistics of Communications Common Carriers* (U.S. Government Printing Office, 1993–1994 Edition). LECs with annual revenues exceeding $100 million are included in the report.

Table 8.1
LEC Financial Benchmarking (1993)

Companies	Access Lines per Employee	Total Operating Revenue per Line	Regulated Operating Revenue per Line	Basic Area Revenue per Line	Total Toll and Network Access Revenue per Line	Total Operating Expense per Line
Large LECs (53 reporting to FCC)	293.87	$605.36	$543.10	$193.44	$270.54	$446.23
Small LECs (833 reporting to REA)	228.22	$799.10	$733.97	$198.74	$535.23	$607.73

Table 8.2
LEC Margins and Investment Benchmarking (1993)

Companies	Operating Margin per Line	Total Plant in Service per Line	Net Plant in Service per Line	Depreciation Reserve	Sheath Meters of Copper	Lines per Switch
Large LECs (53 reporting to FCC)	$159.13	$1,736.64	$1,049.44	39.57%	38.33	7,742.14
Small LECs (833 reporting to REA)	$191.37	$2,532.95	$1,478.29	41.60%	177.94	1,499.3

would be expected to cause annual capital-related expenses to be higher. Since depreciation expense requires no cash outlay, operations expense net of depreciation provides a better measure of relative expense performance. Net of depreciation expense, small LECs annual expense per line is $450 and that of the large LECs is $330.

Even though small LECs have 40% more investment per subscriber line, the annual network-related expense ($128) is almost the same as for large LECs ($120), and annual customer operations expenses were $70 per line for small LECs and $84 for large LECs. Corporate operations expense (i.e., overhead) per line for small LECs is $120 and for large LECs is $70. This cursory analysis of average expense data reveals that small LECs are quite efficient relative to their larger LEC counterparts when the ongoing network and business office operations are considered. That is especially significant when we take into account the conventional wisdom that important production cost economies are associated with larger scale and scope of network operations. Overhead expense performance for smaller LECs relative to larger LECs is not good. But corporate overhead involves expenses that are more easily "reduced" on a per-line basis by spreading them over more access lines.

8.6 REVENUE AND OPERATING MARGINS

Figure 8.1 portrays major sources of revenue and expense for small LECs in average percentage terms, and Table 8.3 provides some indication of the variability of per-subscriber revenue and expense among individual firms. The data presented in Table 8.1 showed that annual revenue per line for small LECs is $799 per year, or $66 per month. Corresponding amounts for large LEC revenue is $605 per year, or $50 per month.

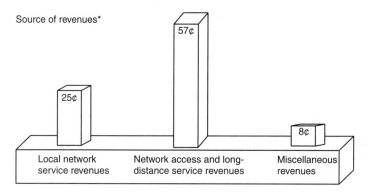

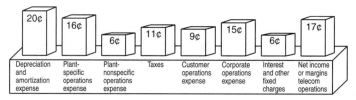

* Adjusted for uncollectible revenues.
** Excludes other operating income and expense.

Figure 8.1 Revenues, costs, and margins for small LECs (1993).

Table 8.3
Variability of Per-Subscriber Revenue and Expense Among Individual Firms

Criterion	Composite Revenue and Expense Ratios of Reporting Borrowers, Average Amount per Subscriber	Per-Subscriber Medians for Revenues and Expenses of Reporting Borrowers	Per-Subscriber Middle Ranges for Revenues and Expenses of Reporting Borrowers
Number of borrowers	883	883	883
Revenues			
Net operating revenues	$799.10	$798.30	$659.09–1,056.86
Local network service revenues	$198.74	$157.14	$123.75–208.47
Network access and long distance service revenues	$535.23	$556.70	$432.37–798.42
Miscellaneous revenues	$88.28	$74.40	$49.02–94.68
Uncollectible revenues (debit)	$3.15	$0.98	$0.00–3.10
Expenses			
Total operating expenses	$607.73	$611.73	$499.52–824.57
Plant-specific operations expense	$128.25	$113.13	$78.35–165.47
Plant-nonspecific operations expense	$44.83	$32.57	$18.84–49.87
Depreciation and amortization expense	$157.46	$150.09	$116.78–202.77
Customer operations expense	$70.51	$78.44	$59.36–101.98
Corporate operations expense	$120.79	$138.41	$97.54–227.20
Other operating income and expense (cr)	$0.39		
Federal taxes	$48.21	$35.58	$1.86–74.04
Other taxes	$38.08	$37.20	$22.35–58.08
Net operating income or margins	$191.37	$194.51	$134.27–264.54
Interest on funded debt	$50.93		

Table 8.3 (continued)

Criterion	Composite Revenue and Expense Ratios of Reporting Borrowers, Average Amount per Subscriber	Per-Subscriber Medians for Revenues and Expenses of Reporting Borrowers	Per-Subscriber Middle Ranges for Revenues and Expenses of Reporting Borrowers
Allowance for funds used during construction (cr)	$0.74		
Other interest expense	$2.34		
Total fixed charges	$52.53	$51.52	$31.96–81.60
Nonoperating income and expense	$15.72	$22.21	$5.84–52.07
Extraordinary items	$0.36cr		
Jurisdictional differences	$0.04		
Nonregulated net income	$6.00	$0.92	$0.30–8.33
Total net income or margins	$160.22	$171.93	$109.42–261.58

Basic local monthly service charges per line are similar for both large and small LECs, at about $16 per month. Thus, regulation continues to achieve the social objective of rate parity between rural and nonrural areas for POTS. The quality of POTS service is similar between large and small LECs; RUS companies report that 98.5% of residential subscribers have single-party service (the remainder have shared-party-line service).

These average revenue numbers reflect both business and residence lines. The FCC reports that 64% of access lines for large LECs are residential, while the RUS reports that small LECs have 82% residential lines. Throughout the United States, business basic local service rates are higher than residential rates. Therefore, the basic rates for residential service for rural subscribers are somewhat higher than those for large LECs after accounting for the higher ratio of business to residential lines.

Table 8.2 showed operating margins per line for small LECs of 24% of revenue ($191.37 per year). For large LECs, the corresponding margin is similar, at 26%. Thus, for now, the cash flow performance is similar for both large and small LECs.

The most important difference in the revenue streams of small and large LECs is that a whopping 67% of small-LEC revenue is derived from toll and toll carrier access services, while, for large LECs, the number is 45%. Rural telephone subscribers spend almost twice as much on toll service per dollar of household income than urban customers. Relative to large LECs, small LECs provide very little toll service directly and instead share in the use of the toll network facilities of interconnected large LECs and IXCs. This is a harbinger of future problems for small LECs that have little hope of increasing their toll operations. Large LECs, on the other hand, especially the RBOCs, have much to gain in the future when the government removes restrictions into the huge interLATA toll market. Carrier access charges and toll settlements paid from larger telephone companies to smaller ones increase the ratio of toll and carrier access revenues. As competition in the industry for toll and carrier access services escalates, this important revenue support for small telephone companies is increasingly at risk. The fact that some very high cost rural telephone companies depend on toll subsidies for their very existence represents a special problem for the future. For such companies, average loop costs can easily run 2 to 10 times the overall rural average.

8.7 FINANCIAL TRENDS

Whatever the prospects for the financial future of rural LECs, the trend for the last five years is certainly a healthy one. For the time period 1989 to 1993, RUS LECs achieved an 8% increase in per-line revenue and operating margins. Basic service revenue for RUS LECs increased over the period by 8% and toll and network access revenue increased by 11%. Those figures are impressive, considering that the corresponding FCC data for large LECs indicate percentage reductions in revenue per line (–10%) and operating margins per line (–18%).[18]

Furthermore, investment in rural networks is proceeding apace. From 1989 to 1993, the per-line investment for RUS LECs increased by 9%. The depreciation reserve ratio (an indicator of the rate of capital replacement) has steadily increased, albeit slowly, from 38.1% to 41.6% (a 9% increase). Large LECs have done somewhat better on average; depreciation reserve ratios rose considerably, from about 34% to about 40% (a 16% increase). Thus, the rural LECs' rate of capital recovery increased by only one-half that of the large LECs over the last five years. However, the large LECs had started back in 1989 with a depreciation reserve percentage far below that of the rural LECs, and are only now catching up.

That having been said, the rural LECs are now at risk of stagnating and falling behind. Large LEC depreciation rates for 1993 were 7.1%, compared to

---

18. Caution must be exercised in the reporting of trends in RUS data because annual data apply only to the companies that borrow money from the RUS, and the mix of companies changes from year to year. For example, from 1989 through 1993, the time period covered by this study, RUS borrower companies numbered 803, 897, 902, 899, and 883, respectively. Thus, total and average per-subscriber financial results are not directly comparable from year to year.

only 6.2% for the small LECs (about the same as it was for 1989). In 1993, the large LECs invested in capital additions at a rate of 7.5% of the total plant in service, indicating that almost all the financing was generated internally from depreciation charges. No comparable estimate of total capital additions over time is available for RUS companies, because the exact number of companies that borrow (and report) those data to RUS varies from year to year.

8.8   RURAL TELEPHONE PLANT CHARACTERISTICS AND COSTS

Based on RUS company cost characteristics, one broad-gauge estimate of the total cost of providing rural telephone service in the United States is $19 billion per year. That total assumes that all 22 million non-MSA subscriber lines are classified as rural and that they generate an average annual cost of $871.08 ($72.59 per month).

There are significant differences in the physical characteristics of rural versus urban telephone plant. RUS companies' markets are very thin, averaging only four subscriber lines per square mile of area served and only six lines per route mile of telephone transmission plant. For large telephone companies, the average density of subscriber lines is greater by an order of magnitude.[19] Large LECs have five times more lines per switching office and almost five times fewer transmission facilities per line than small LECs (measured by sheath meters of copper cable; see Table 8.2). The average length of subscriber connections to the LEC exchange switch is about 10,000 ft for large LECs and about twice that for small LECs. However, the net result is that the average investment and expense per subscriber line are only about 40% higher for the small LECs (Table 8.2).

Figure 8.2 shows a breakdown of small LEC total capital expenditures by major category of plant. Eighty-five percent of small LEC capital investment is represented by switching plant (31%) and cable and wire facilities (54%). Large LECs have 82% of total investment in switching (38%) and cable and wire facilities (44%). For both large and small LECs, the remainder of the investment is primarily in land, building, and support assets.

The average access line (loop) length for RUS companies is 20,330 ft, which is significant, considering that lines longer than 18,000 ft usually require special treatment to ensure high-quality basic service. The main problem is the attenuation of the analog signal, which may require boosting by the use of repeaters and amplifiers, passive reduction of attenuation losses by loading coils,

---

19. Detailed data for subscriber loop characteristics for both Bell and REA companies are available in B. Egan, "Bringing Advanced Telecommunications to Rural America: The Cost of Technology Adoption" (*Columbia Institute for Tele-Information*, Research Working Paper No. 393, Columbia Business School, October 1990), Table 4.

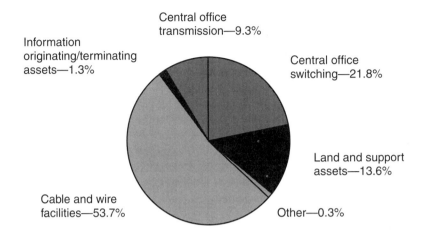

Figure 8.2 Capital expenditures of small LECs by plant type.

or both. Such loops pose a problem for the narrowband digital and new broadband services that require relatively high quality circuits for error-free digital transmission. However, the mode loop length is less than the average for RUS companies. Consequently, 55% of the loops are less than 18,000 ft. The majority of RUS company loops are actually nonloaded, but many still receive treatment of some kind to improve transmission and signal quality. In contrast, about 90% of RBOC loops are less than 18,000 ft, and a large majority of those are nonloaded with an average length of only 7,500 ft.

On average, there are about 7,400 access lines per telephone company exchange (switch) in the United States. *Bell operating companies* (BOCs) have about 12,000 lines per exchange [1]. Non-Bell *independent companies* (ICOs) have only about 3,000 lines per exchange. For 1993, the RUS reports an average of only 1,223 lines per exchange.

Average statistics regarding costs and network operations can be misleading when an individual LEC or specific geographic region is considered. Caution must be used before average statistics are ascribed to any company or group of companies. An examination of the RUS data for individual companies indicates some highly skewed distributions. Figures 8.3 through 8.5 illustrate the high variability in small-company network characteristics, including the number of exchanges, the number of subscribers, and the average exchange size. For example, Figure 8.3 shows that the average number of exchanges per small LEC is 5 with a standard deviation of 8.5, the majority of the companies having only 1. Figure 8.4 shows that the average number of subscribers per company is 6,341 with a standard deviation of 14,000, and most companies have under 1,000. Figure 8.5 shows that most RUS companies have between

Frequency (borrowers)

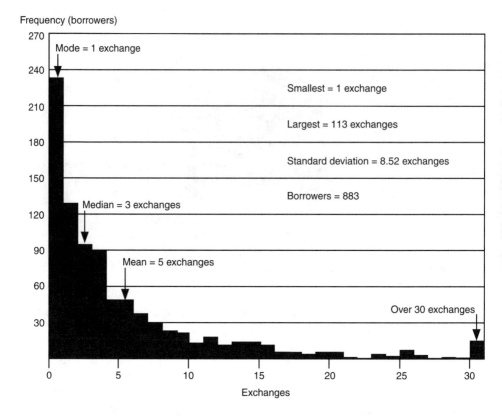

Figure 8.3 Distribution of number of exchanges per small LEC (1993).

200 and 400 subscribers per exchange, with an average of 1,223 and a standard deviation of 1,499. A considerable number of companies have over 2,800 subscribers per exchange.

Indeed, even within a single rural exchange area there are substantial differences in the physical characteristics of subscriber connections. That means that it is not only misleading to ascribe average company or exchange statistics to individual companies or exchanges, but that it is also problematic to apply average loop characteristics of a single exchange to individual subscribers. This has enormous implications for public policies that are trying to accurately target funding assistance to those subscribers who are truly in need.

Figure 8.6 is a stylized example of a representative local exchange area for a rural telephone company. The average exchange comprises about 1,200 households, with a relatively dense downtown area containing 65% of the total lines in the exchange area. The remaining 35% are located in the rural surrounding area of the exchange. The typical rural exchange, as shown in Figure 8.6, has 768 households in the downtown area at a density of 256 subscrib-

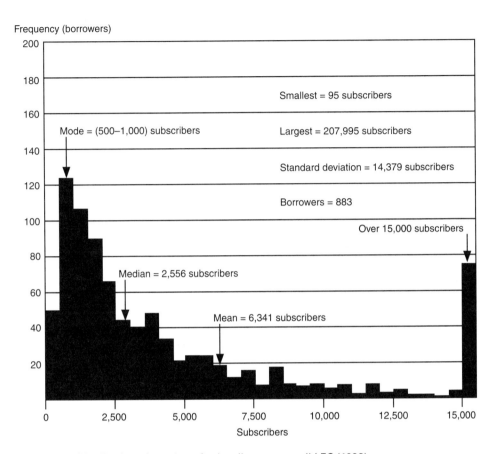

Figure 8.4 Distribution of number of subscribers per small LEC (1993).

ers/sq mi and 440 rural households with an average density of 6 sq mi. This example of a "typical" exchange shows that it is the rule rather than the exception that costs vary widely for individual subscriber connections even within the same exchange area.

To illustrate the impact of subscriber density on the average cost per subscriber for rural LECs, Figure 8.7 provides cost estimates for the average urban and rural subscriber in the stylized exchange presented in Figure 8.6 and shows how such costs vary with subscriber density. The overall average per-subscriber cost is $2,200. For the urban zone of the exchange the average cost is $800, and for the rural zone it is $6,000. As expected, the difference in cost is due primarily to the placement of longer loops for the rural subscriber.

A further examination of the variability of rural loop costs among small LECs (Table 8.4) provides a breakdown of total investment per subscriber for three density bands: 0–10 lines/km)=, 10–100 lines, and 100–500 lines. The

Frequency (borrowers)

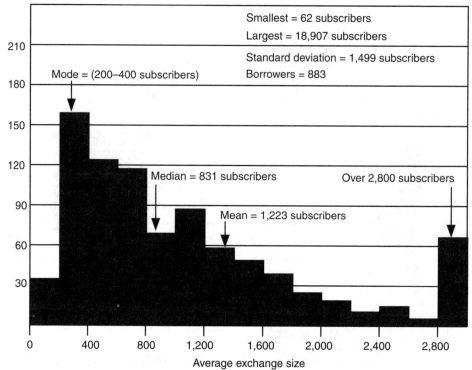

Figure 8.5 Distribution of average exchange size per small LEC (1993).

per-subscriber cost in the lowest density band (0–10/km) is about one-third higher than for the second (10–100/km) and three times higher than the highest density band (100–500/km), with the average investment being $2,055 per line. Even within each density band, it would be misleading to ascribe the average cost result to any one company. For example, there could be drastic differences in topology and terrain that would dramatically affect costs but that do not show up in these data (one company may serve a relatively flat area with sandy soil, while another's area might be hilly or mountainous and featuring solid rock). The spatial distribution of subscribers in a single exchange area could be exactly the same for both companies and yet the per-subscriber costs for each could vary by an order of magnitude or more. The bottom line is that local conditions matter a lot.

Table 8.5 provides further support for the need to consider local conditions in the assessment of average-cost characteristics. Table 8.5 displays 1993 statistical correlations between key publicly available measures of subscriber distance and density and the investment and expense costs per line actually ob-

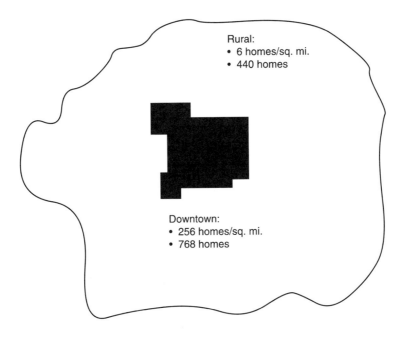

Approximately: 65% of subscirbers in downtown area
35% of subscribers in rural area

Figure 8.6 Rural telephone company local exchange area.

Table 8.4
Density Band Cost Results for Small Rural LECs

Subscriber Density Range	Total Plant in Service per Line
0–10/sq km	$2,667.91
10–100/sq km	$1,893.15
100–500/sq km	$882.83
All LECs	$2,054.7

served for 886 RUS companies. The subscriber density measures that were cor-related with average cost per line were subscribers per route mile of cable, sub-scribers per square mile of serving area, and subscriber lines per switch. The very low values of the standard correlation coefficients demonstrate that there is no significant relationship between density measures and costs. Yet, it is well

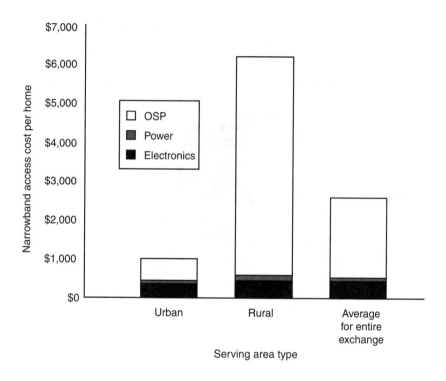

Figure 8.7 Narrowband access model costs per home.

known that local factors like terrain notwithstanding, the primary engineering cost driver in local telephone networks is the distance of subscribers from the exchange. The second set of correlation coefficients is based on a positioning of all the observed values for each variable in rank order from highest to lowest and correlating the rank-ordered vectors. The very high rank correlation coefficients do indicate significant relationships, but now they have no meaning for any given company since the ranking of variable values were made without regard to which company the values belonged.

Kentucky is considered one of the most rural states in the United States, and Table 8.6 is presented to show how small LECs' average costs and revenues may vary within any given state. Sixteen rural Kentucky LECs borrowed from the RUS in 1993. Table 8.6 provides operating and financial statistics for each of them. The weighted average revenue and cost per line and network density for the combined Kentucky rural LECs (next-to-last row in Table 8.6) are fairly close to those for the national averages (last row of Table 8.6). This illustrates exactly how varied the rural LECs actually are.

Table 8.5
Correlations of Average Costs and Subscriber Density

		Standard Correlation Matrix			
Criterion	*TPS/L**	*TOE/L*†	*S/RM*‡	*S/SM*§	*L/S*‖
TPS/L	1.00	0.50	−0.09	−0.11	−0.15
TOE/L	0.50	1.00	−0.01	0.09	0.3
S/RM	−0.09	−0.01	1.00	0.87	0.18
S/SM	−0.11	0.09	0.87	1.00	0.49
L/S	−0.15	0.30	0.18	0.49	1.00

		Rank Correlation Matrix			
Criterion	*TPS/L**	*TOE/L*†	*S/RM*‡	*S/SM*§	*L/S*‖
TPS/L	1.00	0.73	0.65	0.84	0.98
TOE/L	0.73	1.00	0.99	0.93	0.73
S/RM	0.65	0.99	1.00	0.92	0.67
S/SM	0.84	0.93	0.92	1.00	0.88
L/S	0.98	0.73	0.67	0.88	1.00

Note: There are 886 observations, where each observation represents a different company.
* TPS/L = Total plant service per line.
† TOE/L = Total operating expense per line.
‡ S/RM = Subscribers per route mile.
§ S/SM = Subscribers per square mile.
‖ L/S = Lines per switch.

Conventional wisdom (at least to the layperson) is that rural telephone companies serve sparsely populated regions with few or no urban areas. That is not true. The available data make it clear that inferences for any given company based on the average statistics for the group could be grossly misleading. Similar data are available for small LEC revenues and expenses. Those data provide an important message for policymakers and regulators who may be tempted to develop competition policies and rural subsidy requirements based on average statistics for costs and revenues. There is no such thing as an "average" rural company and no such thing as a meaningful average measure of the subsidy requirement.

Table 8.6
Kentucky REA Borrower LECs Financial Benchmarking (1993)

Company	Access Lines	Access Lines per Employee	Total Operating Revenue per line	Regulated Operating Revenue per line	Basic Area Revenue per line	Total Toll and Network Access Revenue per line	Total Operating Expense per line	Operating Margin per line
South Central Rural Tel. Coop. Corp., Inc.	21,923	249.13	$653.10	$592.56	$258.87	$333.69	$457.29	$149.04
Mountain Rural Tel. Coop. Corp., Inc.	12,028	245.47	$597.55	$527.42	$146.41	$381.01	$365.10	$206.59
Peoples Rur Tel. Coop. Corp., Inc.	6,059	195.45	$571.16	$528.22	$158.70	$369.52	$388.58	$137.30
Ballard Rural Tel. Coop. Corp., Inc.	5,571	232.13	$607.97	$567.81	$110.02	$457.79	$372.83	$126.39
Foothills Rural Tel Coop. Corp., Inc.	11,897	321.54	$553.63	$482.70	$157.38	$325.31	$327.47	$189.79
Brandenburg Telephone Company	20,860	221.91	$560.96	$496.41	$139.28	$357.12	$410.09	$165.05
West Kentucky Rural Tel. Coop. Corp., Inc.	15,039	268.55	$562.02	$499.74	$165.79	$333.95	$414.50	$100.08
Continental Tel. Co. of Kentucky	78,597	727.75	$720.93	$669.05	$249.89	$419.16	$529.68	$115.70
Duo County Tel. Coop. Corp., Inc.	9,849	266.19	$629.76	$563.52	$194.25	$369.27	$410.39	$203.41
Alltell Kentucky Incorporated	20,046	435.78	$572.73	$466.17	$164.76	$301.41	$424.00	$126.88
Logan Telephone Cooperative, Inc.	5,671	283.55	$687.92	$609.91	$211.16	$398.75	$481.17	$148.02
Harold Telephone Company, Inc.	5,145	135.39	$635.59	$606.51	$190.14	$416.37	$471.17	$128.11
Thacker-Grigsby Telephone Co., Inc.	6,411	188.56	$607.20	$567.84	$96.87	$470.98	$415.99	$196.59
Leslie County Telephone	6,986	268.69	$666.55	$580.33	$190.26	$390.07	$538.30	$82.88

Table 8.6 (continued)

Company	Access Lines	Access Lines per Employee	Total Operating Revenue per line	Regulated Operating Revenue per line	Basic Area Revenue per line	Total Toll and Network Access Revenue per line	Total Operating Expense per line	Operating Margin per line
Salem Telephone Company	1,807	258.14	$462.93	$409.27	$131.34	$277.93	$455.92	$(54.48)
Lewisport Telephone Company	1,146	143.25	$626.96	$538.23	$112.12	$426.11	$504.80	$41.86
Weighted Average—Kentucky Companies		427.72	$640.58	$578.17	$199.23	$378.95	$457.08	$137.19
Total for all REA Borrower Companies	9,487,560	228.22	$880.23	$907.83	$351.17	$556.66	$662.81	$217.42

8.9 NETWORK MODERNIZATION

Notwithstanding the differences in individual company costs, at a broad policy level, the average statistics for loop length, transmission electronics, and investment are useful for evaluating the average and the total cost of rural subscriber loop upgrades. There is a great disparity between the tasks confronting large and small LECs to upgrade their loop plant to ISDN compatibility. Although bridged taps limit the ability of loop plant to support new digital service, this is no longer a serious problem for RUS companies.

In terms of digital network switching and intelligent network (i.e., switches equipped for SS7) facilities, small LECs compare favorably to large LECs. Table 8.7 provides recent data on digital network facilities for Bell, other large LECs, and smaller independent companies.

As the economies of scale derived from digital and fiber optic technology continue to lower the *incremental* per-subscriber costs for advanced telephone services, the *total* costs associated with converting subscriber lines to narrowband and broadband digital service remain high or even prohibitive. Digital subscriber lines will allow rural subscribers to take advantage of new information-age services including online computing, database, information and transaction services, remote monitoring, and advanced facsimile and data services. Those primary near-term applications for advanced rural telecommunications will enable subscribers to telecommute and to improve their productivity in the office or the home. Eventually, broadband digital service will become possible, ultimately providing for bandwidth on demand for a wide range of multimedia applications for everything from high-resolution images and high-speed graphics to video telephony and full-motion (e.g., entertainment) video.

Basic narrowband digital service begins with upgrading rural network functionality. Initial upgrades will support only low-speed data and voice service. Expanded network capability will support higher data rates from 56 Kbps service up to 144 Kbps full ISDN service. This is the same modernization scenario scheduled for urban and suburban network upgrades, except that rural areas face some special challenges due to longer loop lengths. In both urban and rural areas, business customers may require broadband services, while most residential customers probably will be satisfied with narrowband capability for advanced voice and data telephone services. If residential demand for integrated multimedia services takes off, narrowband network upgrades could be leapfrogged by the provisioning of broadband network connections capable of simultaneously supporting traditional telephone and new multimedia services. That scenario would be very expensive and particularly risky in light of the fact that the most cost-effective alternatives for such applications are terrestrial wireless and satellite networks. It is especially risky for rural LECs to deploy broadband subscriber connections due to the very high sunk costs in-

Table 8.7
Telecommunications Infrastructure Upgrades in Large and Small Telephone Companies (1992)

LEC	Percentage of Digital Access Lines	Percentage of SS7 Access Lines
Bell companies		
Ameritech	74	55
Bell Atlantic	63	95
U S West	45	54
Large Independents		
Centel	100	72
Century Telephone	82	13
GTE	83	30
TDS Telecom	94	0
Small Independents		
NECA interstate access tariff participants	91	15

volved in the face of uncertain demand and certain competition from technological alternatives.

Not only is the broadband network infrastructure expensive, but the additional subscriber premises equipment cost must be factored in. New terminal equipment is currently very expensive. Even the basic digital set-top converter box used to manipulate and control telephone and digital television signals coming into the house is very expensive. Early production units will retail at around $500–$700 each.

A major problem with narrowband digital service network upgrades, as with next-generation broadband services, is that there are no significant demand drivers, primarily because network services, almost by definition, require two-way end-to-end connectivity. Yet physical network upgrades are gradual processes in which more and more customers obtain access to the new technology over a period of many years. It takes a long time to implement widely available interconnectivity, which is the factor that will provide the demand-pull for further technology adoption. After all, what good does it do to have advanced telecommunications equipment in your home if the people with whom you want to communicate do not have similar capabilities?

Thus, developing and deploying advanced digital telecommunication networks is a difficult and costly proposition. Narrowband digital service, in the form of ISDN, has been in the implementation stage in highly populated areas for almost a decade now, and still there is no residential service and only very limited access to business service. With widely available residential ISDN service not expected until late this decade, it is clear that even more advanced network upgrades will be delayed for both physical and financial reasons.

8.10 BUSINESS SUBSCRIBERS

The rapid development of an advanced communication infrastructure for rural America will depend on how easy it is for businesses to access the technology. Businesses consider telecommunications capability an important factor in their location decisions. To the extent that businesses will have advanced services available to them, rural areas may gain more consideration as a viable alternative to urban and suburban locations. Furthermore, as telecommunications capability improves in rural areas, demand-pull will begin to stimulate further technology adoption as businesses and their various suppliers and customers make use of more efficient network facilities. However, exactly what constitutes advanced telecommunication for businesses is an unsettled issue.

Relatively large businesses in rural areas, whether in the service or the manufacturing sector, often require broadband communications capability to maximize operating efficiency and keep up with their urban and suburban counterparts. Broadband in that case refers to digital transmission speeds of 45 Mbps and higher.[20] At such speeds, high-quality data services and video telephony are possible. Such transmission speeds are much greater than the narrowband ISDN service that is gradually being deployed. Broadband service generally requires coaxial or fiber optic cable for subscriber connections, while narrowband service may be provided over more traditional copper facilities. Microwave and fiber optic transmission technologies are nominally capable of supporting both narrowband and broadband services, but, as already explained, fiber optic cable is expected to be the dominant medium for shared network facilities in the future, even in rural areas.

Since fiber optic and coaxial cable subscriber connections not only allow for future broadband telecommunications but also simultaneously provide for high-quality narrowband services, there is some question as to whether incurring the costs of narrowband ISDN on copper facilities is cost effective in the long run. Some analysts believe that early deployment of broadband facilities is

---

20. Investments in digital signal processing equipment for compression can significantly reduce the transmission capacity requirements for broadband to the range of 10 Mbps.

the better long-term strategy and suggest completely bypassing the deployment of narrowband digital service on copper.

Rural economic development depends partially on attracting businesses that require efficient telecommunications. Thus, the focus should be on getting fiber optics deployed in the public network as far downstream as possible, so that business customers have the option of accessing the network for high-speed service applications. It will not be necessary to subsidize business access to the fiber optic public network, but it is important that businesses have a cost-effective means by which to build or lease their own access lines to a high-speed digital public network, since that option usually exists in urban and suburban settings. This can be accomplished through an aggressive statewide plan for a fiber optic network infrastructure.

8.11 RESIDENTIAL SUBSCRIBERS

The deployment of advanced rural telecommunication facilities for residential subscribers should be viewed in several stages. Dedicated coaxial and fiber optic access lines generally are not required to support the demands of residential customers for known services. Indeed, most of the copper loops in the "downtown" portion of rural exchanges, like those in the "typical" rural exchange described earlier, are already short enough to cost effectively upgrade to narrowband digital service. The larger problem is that subscribers in the rural portion of the same typical exchange require expensive loop rearrangements and improvements to reduce or eliminate loop electronics on longer loops, thereby allowing for high-quality digital service.

Furthermore, since the late 1970s, many rural LECs have pursued a plan to upgrade rural loop transmission quality and achieve economies in loop provisioning by deploying remote terminals that were placed between the central exchange and the subscriber. This upgrade strategy was endorsed and encouraged by the REA's guidelines for borrowing companies. In effect, by investing in the deployment of RTs at specified locations called *serving area interfaces* (SAI) and through the placement of *subscriber loop carrier* (SLC) systems, rural LECs were able to lessen their investment in the loop transmission facilities dedicated to individual subscribers, while improving loop transmission quality by making the subscriber connection shorter. But, as can often happen, savings in one generation of network upgrades may be costly in transitioning to the next generation (which may not have been anticipated).

It turns out that the deployment of new ISDN and broadband digital network capability is somewhat easier in an environment of dedicated subscriber connections. The placement of shared remote loop electronics makes it difficult to upgrade, on demand, any given subscriber's line for ISDN or broadband service. No smooth and cost-effective migration from POTS to ISDN is possible in these situations, so if future subscriber loop upgrades are to occur in a timely fashion, RTs might need to be retired early. Unfortunately, this situation is typical not just of rural areas and small LECs but also of many areas served by large LECs (which also deployed a number of RTs).

This discussion provides some critical insight for sound fundamental network planning. Most experts agree that households of the future will no longer be satisfied with POTS and instead will demand their own choice of services and service suppliers, meaning that networks must be designed flexibly. In other words, not all households will want or be able to pay for ISDN or broadband service, and, in any event, not all subscribers would want it all at the same time. Thus, a cash-flow-oriented fundamental network plan would try to accommodate the structure of future demand.

8.12 NETWORK UPGRADE COSTS

Figure 8.8 illustrates an advanced digital rural subscriber connection. The basic loop architecture is similar to today's average rural POTS loop except for a few features. Assuming that the basic POTS loop meets the maximum length for high-quality digital service (e.g., 12,000–18,000 ft) and that the serving CO already houses a modern digital switch, the placement of sophisticated electronic equipment located in the three boxes between the subscriber premises and the CO enables the subscriber to use a range of new digital services.

Upgrading the loop plant of rural telephone subscribers for digital service presents a financial dilemma. A high percentage of existing subscriber loops cannot support an acceptable level of digital transmission, even for existing basic (analog) services. Most rural loops are engineered to support analog voice at 3–4 kHz and very low speed data service up to 9.6 Kbps. To attempt more than that risks intolerable errors in transmission. Thus, the motivation for upgrading the rural loop plant is that current bandwidths will not support the use of many new digital service applications.

It would be misleading to conclude from the data on rural company loop investment that the upgrade problem is simply solved over time by replacement of investment through rapid depreciation. Increased cash flow from depreciation, an important source of funds for new loop plant, also implies rate increases for current subscribers or increased subsidies from others. In addition, the new loop plant is nominally more expensive than the old, even with technological advances, because of price inflation.

Generally, the main problem with upgrading rural subscriber loops to digital service is the presence of load coils. The coils must be removed by cutting them out and replacing the cable at the load coil point. Alternatively, if suitable, loop carrier or remote switching terminal equipment may be installed. Normally, that would be all that is required in the physical loop upgrade. How-

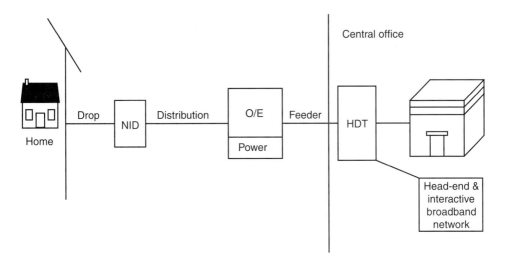

Figure 8.8 Advanced digital rural subscriber connection.

ever, some rural telephone companies still have old nonfilled cable in their loop plant. Such cable may not support high-quality digital service even at low speeds if moisture has penetrated the cable. Nevertheless, analog voice is acceptable on nonfilled cable. The financial requirements for upgrading gel-filled cable rural loops for digital service are not too much of a burden for current telephone company construction budgets over a reasonable time period. However, for nonfilled cable loops, the costly and aggressive rehabilitation program may require external financing. The process of replacement will speed up, because the remaining nondepreciated useful life of nonfilled cable is relatively short (it was last installed in the early 1970s).

8.13 NARROWBAND DIGITAL SERVICE

Figure 8.7 presented the costs for current narrowband rural LEC loops. The estimated cost of upgrading existing rural loops to provide for ISDN service is only about $100–$300 per subscriber (again assuming that the loop is qualified in terms of length and electronics). For nonqualified loops (featuring load coils, nonfilled cable, etc.), the *average* cost ranges from $500–$2,000 per subscriber;[21] some customer loops will be even more expensive to upgrade, as in

---

21. It should be kept in mind that all reported per-line cost results make no assumptions about the demand side of the equation; if they did, the costs would be higher. For example, the cost numbers presented do not include any costs associated with either CPE and terminals, set-top boxes, or network service software and programming services provided by the LEC or another vendor.

the case in which spatial distribution of subscribers was not conducive to sharing outside plant facilities. One goal of the upgrade (e.g., deloading rural loops) could be very expensive when there is no cost-justified possibility for shortening the dedicated portion of the subscriber loop through the use of a ISDN-compatible *remote subscriber terminal* (RST) or DLC system. The current state-of-the-art loop architecture assumes that a fiber trunk connects an RST to a digital host CO (see Figure 8.8).

8.14 BROADBAND AND MULTIMEDIA SERVICE NETWORKS

Based on a broad-based analysis of existing (1992) RUS company cost structures, the monthly cost of deploying a digital multimedia network in rural areas is estimated to be between $92–$132 a month per line, depending on the time period for deployment (10–20 years) [2].

Whereas rural network upgrades for narrowband digital service are based on maximum loop lengths of 18,000 ft from the switching node, higher bandwidth and power requirements of switched broadband networks require a smaller serving area featuring loops of only 6,000–12,000 ft, depending on the services contemplated and the specific network design. This raises costs considerably. For example, reducing a maximum serving area distance from 18,000 ft to 6,000 ft means that nine network nodes are required versus only one.

The digital loop diagram in Figure 8.8 indicated where electronics can be installed to allow subscribers to upgrade service for broadband capability like entertainment video service. Recalling that the downtown area of the rural exchange might well be within the 6,000-ft limit, this situation certainly favors that area over the outlying rural area in any upgrade decision. Figure 8.9 provides an estimate of a rural LEC's broadband loop upgrade, HFC. Initially, HFC systems allow for simultaneous two-way narrowband voice and data telecommunications and one-way (downstream) video service. Assuming that the maximum number of households served per HFC network node is 480, Figure 8.9 shows how per-subscriber costs might be expected to vary as subscriber density varies (i.e., as one moves out from the downtown area toward the rural areas of the exchange). Subscriber-access connections in the dense downtown area can be upgraded to broadband service for $1,000, while serving subscribers in the outlying rural areas of the exchange can cost up to $10,000. The illustrative costs are for network subscriber connections only and do not include the costs of upgrading other network and nonnetwork functions, including sophisticated broadband network system hardware and software and programming services. One estimate is that those upgrades could add another $400–$1,500 per subscriber. To expand the HFC network capability to support interactive multimedia and two-way broadband services (e.g., video telephony) would cost even more.

Another possibility for providing digital multimedia telecommunications to rural areas is to upgrade the existing rural coaxial cable systems with fiber optic trunk lines and interconnect to the PTSN. A truly integrated broadband system usually requires incurring per-subscriber costs similar to those already discussed for telco network upgrades. There are other (even more sophisticated) methods of providing multimedia services to the home, but the costs of those alternatives generally are equal to or higher than the HFC network upgrade.[22] FTTH systems are touted as being the ultimate in broadband telephony featuring high-quality bandwidth on demand with capacity for any conceivable service. The costs of such systems for rural applications currently are so high as to not even be seriously considered by rural LECs. However, that conclusion in no way deflects from the great potential of fiber optic trunk network systems (i.e., FTTH, FTTC) in rural settings to support subscriber demand for sophisticated narrowband digital service [3].

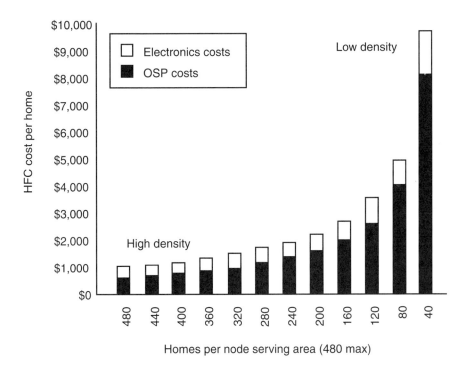

Figure 8.9 Cost per home versus serving area density broadband hybrid fiber coax system.

---

22. For a discussion of such systems and the types of services for which they might be used, see R. Henry, "Video and Broadband Services in Rural America" (*1994–95 Annual Review of Communications*, International Engineering Consortium, Chicago, 1995).

8.15 WIRELESS ALTERNATIVES

For situations in which it is simply too expensive to use wired loop architectures, there are several alternative choices, including satellite, point-to-multipoint radio, and cellular radio. These alternatives must be evaluated on a case-by-case basis, including an estimation of the cost of an efficient connection to the public wireline network.

Digital wireless technology certainly has the potential to become a cost-effective replacement for fixed wired telephone service for everything from POTS to broadband service. In particular, new digital wireless cable networks already are competing with traditional wired cable in urban areas, and it is widely believed that the new digital cellular PCNs will be providing a cost-effective alternative to both fixed and mobile cellular telephone systems in service today at competitive prices [4]. Rural areas, however, pose a special problem for successful deployment of cellular systems for fixed telephone service. The main problem is that radio base station equipment and tower siting costs and maintenance are not cheap. Without sufficient subscriber density in the radio coverage area, it is difficult to spread the fixed costs over a large enough subscriber base to reduce per-subscriber costs to the point where rates charged to subscribers are affordable and still cost compensatory.

Another related problem is that, while new digital PCN systems using small cells (microcells) are optimized for low-power operation in urban settings (i.e., dense market areas), they usually are not cost effective in rural areas because of the sparse number of subscribers who can share a single base station. Due to the distances involved in a rural setting, the permitted power levels for transceiver base stations need to be much higher than in urban cellular markets, or it is not possible to reach enough subscribers to make the investment worthwhile. Too many low-power antenna sites would be required to cover rural areas in a cost-effective manner. Current microwave radio systems for rural telecommunications (BETRS) are very expensive to deploy and operate and tend to be cost effective only in the thinnest rural markets or where terrain will not permit wired subscriber connections [5]. Even though these wireless systems cost less than the wired alternative, both may be prohibitively expensive for subscribers. The network operator requires large subsidies from regulatory agencies to cover the costs of the system and still charge reasonable and affordable rates to its local service subscribers.

Many recent articles have touted the virtues of using wireless access as a cost-effective substitute for wired access in rural areas [6]. Hatfield, Paulraj, and others show that in the thinnest markets (e.g., 0–100 subscribers/sq km), fixed microwave radio (i.e., BETRS) systems may be cost effective to deploy. Other authors show that cellular systems using large cells (macrocells) may be cost effective in many rural markets, including downtown areas. Figure 8.10 provides a broad-gauge look at the relative cost effectiveness of macrocell and microcell wireless access systems versus wired access.

Raw cost efficiencies aside, much of the problem with deploying wireless networks in rural areas lies more with the long head start and continuing inertia of wired service and the ingrained preferences of telephone company managers and engineers for the old (and well-understood) way of doing things.[23] A second important problem to overcome is the current federal rules governing the provision of digital cellular service in rural areas and the limited radio spectrum frequency that has been licensed for use by rural radio systems.[24] Currently, rural cellular service must be provided under restrictive conditions imposed by the government on RF use. Rural cellular providers must share radio frequencies with existing high-power paging services, which causes interference problems. Channelization schemes used by current urban area cellular radio licensees are not optimized for use in rural areas. Without channelization schemes optimized for rural areas, the slice of RF spectrum allocated to rural wireless service cannot be used efficiently, which causes per-subscriber costs to be higher than they would be otherwise. Government restrictions on base station power levels, usually based on an urban market model, are also too low for rural areas, where there is much less chance of RF interference from competing sources. The combination of power restrictions that are too low and radio carrier channels that are too narrow prevents a single radio channel to be shared cost effectively in a rural setting.

If the government would allocate sufficient dedicated RF spectrum (e.g., 20 MHz) and increase allowed power levels, cellular equipment manufacturers would be able to produce and network operators would be able to use state-of-the-art digital access techniques such as TDMA and CDMA to increase efficiency by broader sharing of wide carrier channels (e.g., spread spectrum). That would allow for rural cellular service to become a cost-effective replacement for expensive rural wired POTS access arrangements and could reduce the costs of broadband network upgrades.

For subscribers in rural areas that are truly remote (perhaps even unserved), new digital satellite systems offer the best hope of obtaining high-quality digital telephone service. Many systems being launched beginning in 1996 will provide coverage over the entire continental United States. Initially, prices for usage on these systems will be very high, and some subsidies may be required to make them available. One of the main reasons why satellite service

---

23. For a detailed discussion and some piercing commentary on this issue, see G. Calhoun, *Digital Cellular Radio* (Norwood, MA: Artech House, 1988) and *Wireless Access and the Local Telephone Network* (Norwood, MA: Artech House, 1992).

24. See the discussion of the government's role in B. Egan, "Economics of Wireless Communications Systems," and of the FCC's most recent decisions on rural radio service (BETRS) in the FCC report "New Radio Service (BETRS)."

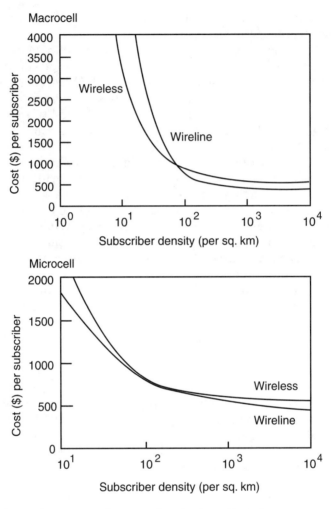

Figure 8.10 Cost effectiveness of macrocell and microcell wireless access systems versus wired access.

has not been viewed as potentially competitive with wired service is the annoying (and heretofore unavoidable) delay time associated with voice transmission on the system uplink and downlink segments (280 ms). Many of the new digital satellite systems have finally overcome this quality differential by using LEO satellites, which feature only a fraction of the delay time.

The recent rapid introduction of digital satellite television using DBS service demonstrates that rural areas will be able to benefit substantially from new digital satellite transmission services. The cost of this technology is not distance sensitive (and usually not usage sensitive); therefore, rural subscribers may finally obtain equivalent service at equal prices to their urban counterparts, even in a competitive market setting [7].

8.16 INFRASTRUCTURE DEVELOPMENT, SUBSIDIES, AND UNIVERSAL SERVICE

For now and in the foreseeable future, federal, state, and local governments will play a key role in developing the rural telecommunications infrastructure. Indeed, regulators are largely responsible (along with the industry) for creating the current complex web of industry cross-subsidies, which are the very lifeblood of many rural systems and which allow rural POTS subscribers to enjoy a level of service and prices that are at par with urban and suburban subscribers.

As the industry makes the transition from a monopoly structure to a competitive one, rural subsidies clearly are at risk. The federal government remains concerned about rural issues, and the new federal legislation aimed at furthering competition in the industry contains numerous important provisions to protect subsidies for universal service, including both low-income subscribers and rural high-cost areas.

8.17 THE TELECOMMUNICATIONS ACT OF 1996

The sweeping new telecommunications law signed by President Clinton on February 8, 1996, contains a host of provisions for protecting the interests of rural telephone companies and their subscribers. The most important provision in the law is the pronouncement contained in the section on universal service principles:

> ACCESS TO ADVANCED SERVICES—Access to *advanced* telecommunications and information services should be provided in *all* regions of the Nation [emphasis added].

The following passages are from the same section of the law:

> QUALITY AND RATES—Quality services should be available at just, reasonable, and affordable rates.

> ACCESS IN RURAL AND HIGH COST AREAS—Consumers in all regions of the Nation, including low-income consumers and those in rural, insular, and high-cost areas, should have access to telecommunications and information services, including interexchange services and advanced telecommunications and information services, that are

reasonably comparable to those services provided in urban areas and that are available at rates that are reasonably comparable to rates charged for similar services in urban areas.

By specifically including advanced services in its universal service mandate, the government explicitly supports the rural industry's most sought-after result: that rural areas will have access to comparable services at comparable prices to those available in urban areas. The language of the law is, almost verbatim, the position taken by the rural telephone company lobby in the raging debates over the legislation and represents a huge legislative victory for the rural telephone companies. By implication, that represents an important legislative loss to large telephone companies and other businesses that lobbied against continued urban-to-rural-area subsides. The Act codifies for the first time what was only implied by past government social and regulatory policies, namely, that rural (i.e., high-cost, low-demand) subscribers should get similar services at prices similar to urban (i.e., low-cost, high-demand) subscribers.

The law also provides for a broad-based sharing of the costs of meeting universal service in rural areas with the following funding provisions.

EQUITABLE AND NONDISCRIMINATORY CONTRIBUTIONS—All providers of telecommunications services should make an equitable and nondiscriminatory contribution to the preservation and advancement of universal service.

SPECIFIC AND PREDICTABLE SUPPORT MECHANISMS—There should be specific, predictable and sufficient Federal and State mechanisms to preserve and advance universal service.

TELECOMMUNICATIONS CARRIER CONTRIBUTION—Every telecommunications carrier that provides interstate telecommunications services shall contribute, on an equitable and nondiscriminatory basis, to the specific, predictable, and sufficient mechanisms established by the Commission to preserve and advance universal service.

The law contains general provisions supporting the availability of access to advanced telecommunications services for schools, libraries, and hospitals on favorable prices, terms, and conditions.

ACCESS TO ADVANCED TELECOMMUNICATIONS SERVICES FOR SCHOOLS, HEALTH CARE, AND LIBRARIES—Elementary and secondary schools and classrooms, health care providers, and libraries should have access to advanced telecommunications services...

In the case of health care the law specifically mentions rural areas:

> HEALTH CARE PROVIDERS FOR RURAL AREAS—A telecommunications carrier shall, upon receiving a bona fide request, provide telecommunications services which are necessary for the provision of health care services in a State, including instruction relating to such services, to any public or nonprofit health care provider that serves persons who reside in rural areas in that State at rates that are reasonably comparable to rates charged for similar services in urban areas in that State.

Beyond the various provisions promoting universal service in rural areas, the Act, in Section 707, authorizes a new Telecommunications Development Fund to:

> support universal service and promote delivery of telecommunications services to underserved rural and urban areas. The source of funds for this special provision is to be from proceeds of a portion of the FCC's auctions of radio frequency spectrum.

8.18 RUS GUIDELINES FOR BORROWER COMPANIES

The RUS provides low-interest loans for network upgrades to the majority of small rural LECs. As the primary source of public funding for rural telephone network upgrades, the RUS's published guidelines for network upgrades have important national standing for infrastructure policy. The most recent guidelines were adopted March 15, 1995, and the major provisions are as follows [8]:

1. Every state must have a modernization plan to improve the rural telecommunications network and must submit it for RUS approval, (the plan may be drafted by the State regulatory agency or the borrower company itself);
2. The plan must provide for the elimination of party line service;
3. The plan must provide for the availability of services for improved business, educational, and medical services;
4. The plan must encourage and improve computer and information highways for subscribers in rural areas;
5. The plan must provide for rural subscribers to receive: conference calling, video, and data rates of at least 1 Mbps;
6. Uniform deployment schedules in rural and nonrural areas;
7. Expeditious deployment and integration of emerging technologies;
8. Affordable tariff rates for medical and educational services;

9. Reliable powering for POTS service including alternative power sources during electric utility power outages;
10. In the "short term" all new telecommunications network facilities shall be constructed so that: all single party service subscribers have access to lines capable of speeds of at least 1 Mbps, all switching equipment must support custom calling features, and E911;
11. In the "medium term" (6 years after plan approval) all new facilities must be capable of: transmitting (motion) video signals, and E911;
12. In the "long term" all plans should be able to accomplish: elimination of party line service, universal availability of digital voice and data service (56–164 Kbps), offer service at transmission speeds of no less than 1 Mbps, and offer video service.

Needless to say, the new RUS guidelines and the formidable network capabilities they require have sparked considerable controversy.[25] Suffice it to say that while the RUS has laid out the rules for approving loan applications for network upgrades, it is far from clear that there is enough government funding available to pay for such substantial upgrades, and it is equally unclear whether nonborrower companies would otherwise plan to make such upgrades. A consistent state infrastructure upgrade policy would ideally be based on industry consensus, but with the RUS setting the least common denominator is at such a high level it may not be possible to reach an industry consensus. Keeping in mind that many large LECs also serve rural areas (in fact most rural areas of the United States are served by large LECs), until the large LECs (which cannot borrow from the RUS) concur with the RUS proposals (not a likely proposition), it will not be possible for states to implement consistent network infrastructure upgrade plans.

It appears that the RUS rules were written for an era of continued monopoly provision of telephone service. That model is outdated in light of other federal and state initiatives that promote market entry. Specifically, large and small LECs, whether or not they are RUS borrowers, recognize that the future competitive environment means that local network upgrades involve considerable market risk and that there is no longer any good prospect of recovering all the investment from tariff rates on captive customers. One obvious provision that is lacking in the RUS's rules is that infrastructure investment plans should meet the fundamental test of market viability (i.e., there is no business case called for).

---

25. After the RUS released its proposed interim rules on December 20, 1993, there was an overwhelming response, including concerns regarding the new requirements for state modernization plans. More than 39 parties commented. A good summary of those comments appears in the Federal Register, Vol. 60, No. 29, 7 CFR Part 1751, February 13, 1995.

That having been said, the RUS should be applauded for its vision, which, rather than a mandate, is a reasonable goal toward which to strive. The problem is that, without large LEC concurrence, it will never be implemented on a large scale. The RUS probably sensed that when they called for state PUCs, which regulate all LECs, to coordinate and submit their own infrastructure upgrade plans in accordance with RUS minimum requirements.

It will be up to the state and federal governments to try to coordinate their respective roles regarding telecommunications infrastructure policy, a complicated task that has barely begun.[26]

8.19 PUBLIC POWER GRID

The use of a dielectric transmission medium, such as fiber optics, provides an unprecedented opportunity for inexpensive infrastructure development by taking advantage of newly found synergies that combine existing electric utility distribution infrastructures with those of telecommunications. Construction costs of fiber optic facilities may be substantially reduced by utilization of public power grid rights of way and pole or conduit facilities. Since optical transmission is not susceptible to electromagnetic interference caused by power lines, fiber cables could use the distribution plant offered by the statewide power grid by purchasing or leasing facilities from rural electric utilities. Such inexpensive fiber deployment may even include lashing the fiber cable to the electric utility ground and phase wires, which often run along the tops of towers and poles. There are many possibilities. Products already on the market include a fiber cable that utilizes the metallic ground wire for strength. The ground wire in the cable supports the requirements of electric utilities, and the fiber communications capacity may be resold to others for telecommunications service.

Many large electric utilities already operate major private communication networks and benefit greatly from real-time monitoring of the power grid loads and consumer usage. Smaller rural electric companies also could save money by self-providing telecommunications services, for load management, monitoring, internal communication and the like, which they currently obtain from interconnection the rural LECs.

Rural electric cooperatives serve geographically large and thin rural markets, which often span many independent telephone company exchanges. Because they cannot justify stand-alone internal communication networks, small electric utilities must rely on many rural telephone companies and pay relatively high tariff rates. The sharing of power company facilities with local tele-

---

26. This is also the conclusion reached in the Congressional study by B. Egan, "Economics of Wireless Communications Systems," pp. 45–46.

phone companies can provide economies for both, providing a win-win situation. In addition, some large businesses choosing to locate in rural areas are often able to get sufficient power, while advanced communications capability is lacking. If the shared infrastructure were available, businesses might be more likely to locate and expand in rural areas. Safety communications for fire and alarms are other new service applications that place only nominal bandwidth requirements on the communications infrastructure. There seems to be a natural synergy here for rural communication infrastructure development, but one that is underexploited. The electric power industry tends to be conservative, but many firms are now examining novel arrangements with communication service providers.

8.20 TOLL SERVICE

Rural telephone companies, which have long wanted to enter directly into the lucrative toll market, could begin to take advantage of the revenue opportunities that a fiber optic infrastructure could provide, not to mention the possibilities for providing the new data and video services that fiber optics could support. This is important if it is true that traditional large telephone company toll subsidies enjoyed by small rural telephone companies will eventually disappear due to increasing competition. Small rural telephone utilities may pool traffic and interconnect with the fiber optic backbone trunk network to efficiently—and profitably—provide high-quality toll voice and data services. Fiber optic backbone networks may also allow rural subscribers to purchase digital services and access remote databases of enhanced service vendors.

8.21 STATE PLANNING

The process of rural telecommunications infrastructure development is an evolutionary one that will occur gradually as advanced facilities become available. For that reason, it is important that the process begin as soon as possible. State telecommunications planners must take on the role of coordinating network interconnection and development activities, exploiting potential synergies for the benefit of all subscribers. In the early stages, such coordination will concentrate on surveying all communication facilities, public and private, and evaluating short-term and long-term interconnection and compatibility potential. At first, microwave and satellite network facilities will be evaluated, along with existing coaxial cable network facilities, to determine interim infrastructure possibilities. The long-term focus will be on migrating to a more efficient infrastructure based on digital fiber optics and radio technology. The goal will be to share network facilities whenever it is cost effective to do so, and guide the replacement of older network facilities with advanced facilities, stressing network compatibility along the way. Without compatibility, interconnection of communication networks will be inefficient or even impossible, and potential synergies will be lost or diminished.

The rate of development of rural telecommunications infrastructures may depend largely on demand drivers. There are some logical ways to pursue network technology adoption, paying close attention to demand patterns in the current infrastructure. For example, secondary and tertiary schools, libraries, hospitals, and regional airports tend to be among the heaviest consumers of information and telecommunication services in rural areas. Public power utilities and other rural infrastructure firms, including occasional large manufacturing or service companies, also represent logical node points for rural networks. Existing telephone company switching offices, combined with the aforementioned locations, represent demand drivers and potential network hub sites, providing for efficient communication infrastructures. This set of candidates for network node (hub) points should allow for a number of alternative deployment scenarios for state telecommunications planners to consider. Hubbing allows the economies of satellite, microwave, and fiber transmission to be used cost effectively in relatively thin markets, thereby maximizing the NPV of the rural construction program.

8.22 REGULATORY ISSUES

Planning for an advanced rural telecommunications infrastructure raises many regulatory and public policy issues. Prominent among those issues are who should own and control the infrastructure and how it should be financed. Obviously, there are no "right" answers to such questions, but some general economic principles may guide the thinking on the issues. First, private ownership and control generally are preferred to public ownership and control, for reasons of the operating efficiency incentives that competition provides.[27] Second, government must have a proactive role as an overseer, enabler, and planner. As discussed previously, private network development may help support infrastructure development in a win-win situation in which net revenue opportunities accrue to both private and public industry participants through efficient interconnection and compatibility. State government may be most helpful in the role of identifying how public and private communications network activities can complement each other and strengthen the overall infrastructure.

As a rule, an infrastructure approach does not imply centralized ownership or control. It does imply cooperation among the various players, however,

---

27. Many rural electric utility cooperatives are very successful operations; therefore, publicly owned and operated arrangements are not necessarily worse than strictly private ones.

and that is the enabling role of government: bringing together the players and encouraging infrastructure development. Much more can be done than we are observing today. Most states have not yet placed sufficient emphasis on telecommunications infrastructure and its role in economic development. New technologies that are just beginning to be deployed have very low unit costs once demand thresholds are met, but they do have very high up-front capital costs. For that reason, an infrastructure approach to planning, which maximizes capacity sharing through a hubbing network architecture, holds great promise for dealing with the problem of thin rural markets. For example, even in states like Kentucky, which feature an inordinate number of rural areas, many existing locations could generate enough traffic demand to justify a fiber, radio, or satellite hub, depending on the specific demand application(s) required.[28] Eventually, fiber hubbing will dominate as the technology of choice for most new shared network applications, while radio, coaxial, and copper cable will be used for dedicated short-haul subscriber plant. Satellite and microwave radio will be utilized wherever wireline facilities cannot be deployed cost effectively, especially in physically remote applications.

Finally, a host of important pricing issues are associated with recovering the costs of advanced telecommunications infrastructure development. Two primary issues are: (1) broad toll-rate averaging; and (2) toll-to-local service subsidies. Trends in both areas are troubling for rural telephone companies and will no doubt become the subject of extensive public policy debates. A full treatment of these issues is beyond the scope of this chapter, but a few observations deserve discussion.

Increasing competition in toll services and the absence of regulatory rules for retail tariffs of competitive toll carriers are slowly eroding the broad rate-averaging rules, which have been in effect for many years. The effect of rate averaging is to subsidize relatively low use subscribers (and those in thin rural markets) relative to high-use subscribers and those in dense markets. Volume discounts offered to heavy toll users (especially businesses) by competing toll service providers already have been undermining traditional rate averaging. The toll subsidy that generally flows from larger telephone companies to smaller ones is also going to decrease as competition continues to drive prices down. The best solution here probably is to target subsidies more carefully toward those companies that need them most, rather than bestowing them on entire classes of small companies, as is currently the case. The FCC, Congress, and state regulators are already moving in that direction.

---

28. For a more detailed discussion of the existing Kentucky infrastructure for power, transportation, and telecommunications, see B. Egan, "Bringing Advanced Telecommunications to Rural America," Section 6 and Appendix.

8.23 RURAL TELEPHONE SERVICE SUBSIDIES AND UNIVERSAL SERVICE

Based on RUS data as a representative proxy for high-cost rural areas of the United States, the total rural subsidy is estimated to be about $5 billion per year based on revenues of $14 billion and costs of $19 billion [2]. That means that rural telephone rates would have to rise, on average, about 35% to pay the full costs of providing rural POTS service, an increase in average monthly tariff rates of about $19. A per-line average subsidy calculation is valid only in a monopoly environment, where it is considered legitimate to equate the total cost of USO (and corresponding subsidy) to be the average cost (subsidy) per subscriber multiplied by the number of subscribers.

In a competitive market, presumably some of the more-profitable subscribers of the incumbent monopolist will migrate to a competing market entrant, thereby altering the average subsidy calculation for the remaining PSTN subscribers. This critical fact has been lost on existing efforts to quantify the going-forward subsidy requirement. *In a competitive environment, there is no such thing as an "average subsidy."* Of course, a total subsidy requirement does still remain, it just is no longer equal to the monopoly average subsidy per subscriber multiplied by the number of PSTN subscribers. Without exact data on the costs, revenues, and profit (subsidy) contribution of each and every telephone subscriber, the concept of an average subsidy in a competitive environment is meaningless.

Specifically, the broad average subsidy figure derived from monopoly-era data is substantially understated in the face of competition. Competitive entrants will invariably pursue only profitable LEC subscribers, like those who reside in relatively dense downtown areas of the rural exchange, larger business subscribers, and those subscribers who purchase a lot of nonbasic services. In a competitive environment, it is the sum of the individual PSTN subscribers' total costs and revenues that determine the total subsidy required to cover the amount by which costs exceed revenues.

In other words, in a competitive market, the cost of serving any given subscriber is not even half the profit or subsidy picture. What really matters to a competitive entrant or incumbent telco is demand (the expenditure side of the profit equation). From a demand-side perspective, only profitable subscribers provide positive contributions toward supporting the low average tariff rates for the high-cost, low-profit (or no-profit) subscribers. Without individual subscriber data on costs and corresponding revenue, it is not possible to estimate what the cross-subsidies would be in the rural LEC exchanges, but it is well known that a small fraction of total subscribers are responsible for generating the most revenue and providing an inordinate amount of funds to support the costs of the remaining majority. It is easy to visualize that the rural subsidy to truly high-cost unprofitable subscribers ultimately will need to be much greater than the average $19 per month that each rural subscriber now receives. (It will probably be at least twice that amount.)

Thus, current subsidy and universal service policies that focus only on the supply or cost side of the equation will be seriously flawed. Unfortunately, this is the approach of all current government initiatives.

To accurately determine the cost of an LEC's PSOs, including USO and COLR,[29] it is necessary to compare the total PSTN investment and expense cost of the LEC to the cost of the (hypothetical) PSTN if there were no such obligation to serve or be ready to serve. Indeed, those conditions define the environment and context for determining the cost of the LEC's PSO.

Thus, the cost of the PSO must be the cost the LEC would incur with the obligation (i.e., current PSTN costs for POTS) versus what that cost would be with no such regulatory obligation. That is equivalent to asking which customers would have been served anyway (the answer, of course, is those that could be served profitably). Assuming, for purposes of discussion, that subsidies should be targeted to cover a portion of the costs of providing POTS for certain residential PSTN subscribers, for example, high-cost, low-income (low or no profit), and that all other subscribers should directly pay the cost of their POTS service, then it is clear what the proper cost "object" is when quantifying the cost of the PSO and, in turn, the corresponding subsidy requirement.

If competition and subsidies for small LECs are to coexist, some general economic principles should underlie the subsidy funding mechanism. The mechanism should be: (1) fundamentally fair for consumers; (2) competitively neutral for competitors; and (3) sustainable for the long term.

Furthermore, pursuant to a host of recent federal and state investigations of universal service, it appears that almost all parties to the debate have agreed on certain aspects of the various plans and proposals for future reforms, namely:

- All charges for telephone service should be cost compensatory except for a narrowly defined and highly targeted set of basic service subscribers (e.g., low-income, high-cost).
- Subsidy funding requirements should be shared by a broad base of market players, and the funding mechanism should be administered by an independent third party.

Beyond the need to define exactly what constitutes the "benefited" service and subscriber group(s) deserving of subsidies, some primary issues remain unresolved: (1) What is the cost of the current and future USO and COLR obliga-

---

29. The COLR obligation refers to the cost of having to stand ready to serve everyone in a given geographic area on demand, where such demands arise randomly from the public at large, with no opportunity to exit the market.

tions, and how should the cost be determined? (2) Exactly how should the subsidy-funding mechanism work? For example, how should the subsidy fund be administered? Who should administer the fund? How should funds be collected and distributed within any given geographic area—locally, regionally, or statewide?

A cost study method must be developed that establishes both the historical and going-forward costs of the USO.

8.23.1 Historical Perspective

The historical cost of an LEC's PSO is simply the cost of providing POTS service to residential subscribers. When that cost is compared to POTS revenues, the difference is the net cost (subsidy requirement) of the obligation to serve.

In the current local monopoly environment, the funding of the subsidy amount comes primarily from markups of prices over costs (i.e., cross-subsidies) for LEC business services and residential non-POTS services. In the new competitive environment, all carriers must be required, in a competitively neutral way, to share in the funding of the costs of PSOs.

The reason that the historical cost of an LEC's PSOs must be considered is that monthly subscriber bills now and in the future simply represent the time payment plan that regulators (on behalf of PSTN subscribers) and the LECs agreed to under the precompetitive regulatory regime. Thus, in the postcompetitive regulatory environment, the LEC must still be enabled to reimburse its historical investors. Had the LEC (and its investors) known that the regulatory regime would change, eliminating its opportunity to compete at the margin and still recover the cost of its regulatorily imposed historical PSOs, the LEC would not have continued to invest in the network infrastructure that served customers and locations that the LEC viewed as too risky and without good prospects for cost recovery.

In a historical context, LECs, as regulated common carrier monopoly providers of POTS, were obligated to serve (or be ready to serve) all subscribers in a given service area on demand at broadly averaged "affordable" prices for POTS. Thus, to know the historical cost of the obligation to serve, one would need to know the cost an LEC would have incurred without the obligation to serve versus the actual (observable) cost incurred with the obligation. The average subsidy requirement for rural telephone subscribers in the United States was previously provided. This broad-gauge level of analysis provides regulators with a ballpark estimate of the total and average per-subscriber subsidy. It is important to keep in mind, however, that this broad-gauge estimate cannot be used to evaluate the cost (subsidy requirement) for any given subscriber or location because the available information is not granular enough to make that determination. The average is the net total subsidy requirement from all POTS subscribers, the sum of those subscribers who are both subsidizers and subsidizees. Nevertheless, the estimate does establish the total (average) amount of the subsidy funding requirement in the historical monopoly environment. That value may serve as a lower bound of the per-subscriber subsidy required in the new competitive environment (recall that competitors would tend to strip off the more profitable subscribers, which would tend to raise the average subsidy requirement for the remaining PSTN subscribers).

The total and average subsidy amounts determined in stage 1 of the costing process may begin to be deaveraged into density cells by an examination of the available data for the density characteristics of an LEC's POTS subscribers. This stage of the analysis is somewhat easier because it is also possible to examine the PSTN investment and expense costs of smaller, rural (less dense) LECs and compare those figures to the same embedded accounting data for larger (more dense) LECs.

8.23.2 Incremental Perspective

The definition of the going-forward cost of an LEC's PSOs is the same as the historical one, except that the costs would be based on the total incremental cost of the obligation instead of the total historical cost. Thus, the cost calculations would reflect incremental instead of embedded technology and business practices.

8.23.3 Long-Run Service Incremental Cost Versus Total (Average) Cost

Some parties to the subsidy debate have asserted that the cost and funding of LEC PSOs should be based only on the incremental cost of POTS, not the total or average cost. Unless the *long-run service incremental cost* (LRSIC) study methodology is carefully constructed to approximate the total cost of the obligation, it is flawed and thus unjustifiable from an economic perspective. This theoretical and practical distinction is important because it is the total (average) cost of a business enterprise that must be covered by total (average) revenues for the firm to be sustainable. Open market entry and competition will, over time, force rates for all services to be driven toward their costs (unless subsidies continue to be provided by government mandate).

From a market perspective, a subsidy system that relies on continued price cost margins on competitive services to fund a portion of the costs for residential POTS service is inherently unsustainable. Indeed, any LEC and, in turn, LEC customers, who must incur the costs of regulatorily imposed PSOs not similarly borne by market entrants (and their subscribers) will be disadvantaged in the market place. The end result is a shift in consumer welfare from the LEC's customer base toward the entrant(s) and its customers.

To avoid such discrimination and inadvertent shifts of wealth between subscriber groups, regulators should adopt a consumer-friendly and competitively neutral subsidy-funding mechanism before competitive entry forces a type of de facto rate deaveraging to occur in which one group of subscribers must pay for regulatorily imposed burdens and one does not.

8.23.4 Summary

Fundamentally, there are only two types of POTS subscribers: (1) those low-cost subscribers in relatively dense areas served by relatively short loops who have relatively low local calling rates (subsidizers); and (2) those relatively high-cost subscribers served by relatively long loops who have relatively high local calling rates (subsidizees). Of course, even in high-cost (long-loop) geographic locations, there almost always will be some highly profitable subscribers who purchase a lot of non-POTS services. For that reason, a proper cost analysis must be conducted at the individual customer level. In other words, to know whether any particular subscriber is a net subsidizer or subsidizee requires monitoring individual subscriber characteristics. Obviously, that is not practical, even though it is the only way to guarantee that subsidies flow to subscribers who otherwise would not be able to obtain POTS service (or at least that quality of POTS that other "profitable" PSTN subscribers receive).

Because LECs were historically monopoly providers of residential POTS, it is straightforward to calculate the total cost of any given LEC's PSO by examining the embedded accounting cost data for PSTN investment and expense and converting those data to an annual or monthly per-subscriber amount.

The results from density-cell analysis may be further disaggregated to customer-specific analysis levels via a computerized cost proxy model that calculates the costs for each subscriber location.[30]

8.24 SUBSIDY-FUNDING MECHANISM

The mechanism for collecting funds to pay the cost (subsidy requirement) of LEC PSOs must be competitively neutral with respect to incumbent LECs, which had—and likely will continue to have—such obligations, and competitive entrants, which do not. In other words, the contributions toward covering the subsidy costs should be shared equitably by all telecommunication service providers. To be sustainable, competing service providers must not be able to

---

30. Such a model was recently used by U S West to calculate costs by U.S. Census block groups in U S West, "Targeting High Cost Funding to High Cost Areas Using U.S. Census Block Groups" (draft, October 28, 1994). Even at this level, however, there is no recognition of variations in individual subscriber costs and differences in individual subscriber expenditures or income, a potentially more important determinant of individual "need" for a subsidy.

avoid payments by totally bypassing the PSTN. Many such funding mechanisms have been proposed by industry groups and most involve a revenue surcharge mechanism.[31]

8.25 FINANCING ALTERNATIVES

The costs of deploying efficient communication infrastructures are high compared to any historical measure of the costs of technology adoption. The reason is that technological trends are moving toward lower ongoing usage costs and higher up-front capital costs. Digital network equipment has few moving parts and features very large scale capacity relative to older-generation network equipment. As such, the new equipment is more cost efficient from a maintenance and repair expense perspective, but it also is more capital intensive and is typically purchased in greater lumps, because it is well suited for large-scale operations. The same tends to be true of fiber optic transmission equipment, although for many network applications fiber soon will be cost effective even relative to the older-generation copper and coaxial cable costs. The bonus with fiber optics is not only its very high capacity but also its high quality and reliable service compared to metallic and radio technologies. Nevertheless, up-front deployment costs for fiber optics are substantial, and every effort to cost effectively introduce it is important.

Telephone rates are the obvious first choice for financing advanced rural network infrastructures; indeed, most of the financing must come from that source. Fortunately, under the traditional finances of telephone utilities, it appears likely that internal cash flows will fund much of the infrastructure deployment costs. But, as has been pointed out, these traditional internal cash flows are at risk because of increasing competition and the advance of technological alternatives.

Borrowing is the next alternative to consider. The RUS and others provide subsidized loans to rural telephone companies. Without government assistance, rural telephone companies would have to go to other capital markets that offer less attractive terms. Unlike large telephone utilities, many rural companies are already highly leveraged. That is not bad in and of itself, but it does affect the propensity of lenders to approve more funds on favorable terms. Regulators also may become concerned about the level of business risk that leverage implies, even though ratepayers may benefit from the lower average cost of debt capital relative to equity finance.

---

31. Several proposed mechanisms vary in detail, but all follow the general mechanism described in B. Egan and S. Wildman, "Funding the Public Telecommunications Infrastructure" (*Telematics and Informatics*, Vol. 11, No. 3, 1994).

RUS and some other lender practices are basically sound for financing advanced rural telecommunications infrastructures because they operate within an incentive structure that tends to give the right signal to borrowers to make good investments (not to mention the new aggressive network upgrade guidelines). RUS uses equity-based financing and loans that are usually "self-liquidating." The proposed investments of borrowers must meet general technical guidelines for acceptable and approved equipment purchases. This system prevents speculation and abuse of government loan funds. Even though the RUS program is a loan subsidy program, only the interest rate discount is truly subsidized, a relatively small portion of the entire loan and repayment sum. The loans are self-liquidating from revenues and cash flow from telephone rates. Overall, this approach seems socially efficient because it allows the private sector to determine the market requirements and opportunities for sound investment decisions and requires the borrower to have a substantial equity stake. The only government role is to provide an inexpensive source of funds, technical support, and monitoring.

Direct subsidies, especially of the current untargeted variety, are much worse and are not socially efficient. The current flow of toll-to-local, urban-to-rural and large telco to smaller telco subsidies, is generally inefficient because it is not based on need; instead, it is based on a grand formula for broad rate averaging and revenue sharing. In fact, some of the vast sums of money in the toll revenue pool now divided among telephone companies through the use of a broad formula could be used to increase the RUS's loan authority or could be distributed based on bona fide financial need. Whenever subsidies are not targeted, there are potentially wasted resources. The introduction of basic telephone lifeline service based on a needs (income) test is a good example; it has proved to be much more socially efficient than a blanket subsidy for all local service subscribers (especially those who can afford it). As the financial data provided earlier indicate, many small rural telephone companies have very healthy cash flow situations and do not really need subsidies.

Direct government subsidies for rural telecommunications should be discouraged because the investments funded presumably will generate some level of on-going subscriber revenues. Those revenues should be included in any loan repayment formula, even if the repayment is only a partial one.

8.26 CONCLUSIONS

Perhaps one of the most important policy conclusions from this analysis is that there is a notable difference between the costs of local network upgrades for the base of existing rural telephone subscribers and the costs of serving new and physically remote subscribers. From a public policy perspective, the latter group must be treated as exceptions requiring special cost subsidies; otherwise, policy for the masses could fall victim to debates over subsidies for the few.

A second important conclusion is that the existing body of rural subscribers is currently being served cost effectively and profitably, and a timely digital network upgrade is a reasonable proposition without necessitating large rate increases. Such a proposition, however, assumes that the government, as it promotes open entry and competitive markets, will continue to require a broad sharing of the costs of maintaining universal service.

A third important conclusion is that small rural LECs cannot be classified according to average costs and subsidy requirements. There is too much variability in the costs of serving subscribers based on differences in local market conditions and geography. Furthermore, considering the wide differences in individual subscriber expenditures for telephone services, there is even too much variability among subscribers within the same geographic location to be able to target subsidies accurately. That presents a major problem for policymakers who want better targeting subsidies to those companies and subscribers who really need it. Average statistics just will not do. There is no such thing as an average rural LEC or an average subscriber, which has tremendous implications for state and federal government subsidy and competition policy. If open market entry is allowed, existing subsidies will dry up—period. Therefore if the government is serious about having both competition and subsidies, the current system is in dire need of reform. The alternative is the natural market solution, which is to drive out cross-subsidies. That may not be so bad considering that the existing subsidies may not all have been justified in the first place. Nevertheless, if it remains a public policy objective that rural areas of the United States should be able to continue to obtain a comparable level of basic telephone service (e.g., POTS however defined) at prices comparable to that in urban areas of the United States, then the subsidy system must be reformed.

Finally, the key to rapid adoption of advanced technology for rural subscribers is to take an infrastructure approach to the problem. Such an approach implies significant coordination and monitoring of public and private network investment and business activity, preferably at the state level. Specifically, in the current environment, there appear to be significant lost opportunities for the realization of public benefits and potential synergies from cooperation of the energy, transportation, and telecommunications sectors.

The infrastructure approach could go a long way toward solving this problem and actually follows from the technology itself. First and foremost, new telecommunications technologies can be very efficient, but that efficiency depends on two critical factors, which often are nonexistent in rural areas of the country: (1) economies of scale; and (2) end-to-end service capability. The first factor operates on the supply side of the equation and simply says that technologies such as digital fiber optics require relatively large scale operations to achieve the low unit costs that are ultimately available. End-to-end service operates on the demand side of the equation and simply says that, unless advanced network functionality is adopted on a very wide scale, demand drivers will be unable to speed up the technology adoption process. It is no good to have ISDN service capability or broadband digital, for that matter, unless the other party to the call also has it. *Thus, the critical issue for efficient technology adoption in rural telecommunications is sharing of network facilities, both to achieve scale economies and to stimulate demand drivers.*

Fiber optics is generally the most cost-effective technology for shared network service applications. New digital wireless technology has tremendous potential for reducing the cost of dedicated subscriber connections. Fiber is not cost effective for dedicated (nonshared) customer connections. Most businesses, especially large ones, share network facilities among a number of telephones and therefore may cost effectively adopt fiber technology before residential customers. However both businesses and residences must share facilities as much as possible to take advantage of the superior economies of scale that fiber exhibits relative to competing technologies.

Another important advantage of fiber optics is that it can support new broadband services like video telephony, multimedia services, and very high speed data service. It is not necessary that demand for broadband services precede fiber optic technology adoption because fiber is also very cost efficient for simultaneously transmitting narrowband services. Sharing and multiplexing allows fiber to become cost effective even when only narrowband service applications are used.

An infrastructure approach to rural telecommunications technology adoption should maximize the possibilities for sharing, thereby stimulating investment in those technologies offering the greatest cost efficiencies. The bonus with adopting digital fiber optic technology early on is that the network will be robust with respect to almost any conceivable demand scenario that ultimately develops.

References

[1] "Statistics of the Local Exchange Carriers 1994—For the Year 1993," USTA, July 1994.
[2] Weinhaus, C., et al., "Redefining Universal Service: The Cost of Mandating the Deployment of New Technology in Rural Areas," *1994–1995 Annual Review of Telecommunications*, International Engineering Consortium, Chicago, 1995.
[3] Deutscher, D., "Rural Fiber Network in Service," *1994–95 Annual Review of Communications*, International Engineering Consortium, Chicago, 1995.
[4] Egan, B., "Economics of Wireless Communications Systems in the National Information Infrastructure," U.S. Congress Office of Technology Assessment, August 1995.
[5] Oregon PUC Staff Discussion Paper, "The Economics of Wireless and Wireline Telephone," draft, April 20, 1995.
[6] Hatfield Associates, "The Cost of Basic Universal Service," draft, July 1994; Paulraj, A., "Wireless Local Loop for Developing Countries—A Technology Perspective," *1994–95 An-*

nual Review of Telecommunications, International Engineering Consortium, Chicago, 1995; and Egan, B., "Economics of Wireless Communications Systems in the National Information Infrastructure," U.S. Congress Office of Technology Assessment, August 1995, pp. 45–46 and 95–98 and Chap. 9.

[7] Murphy, B., "Rural Americans Want Their DirecTV," *Satellite Communications*, March 1995.

[8] 7 CFR Part 1751, RUS Telecommunications System and Design Criteria and Procedures.

Public Policy and Institutional Considerations

9

9.1 BROADBAND MULTIMEDIA NETWORK INFRASTRUCTURES

Convergence, the compelling buzzword of the information age, conjures up images of combining many different communications media using the common communications medium of digital signaling. Convergence potentially allows for effortless interactive communication among people and machines, where voice, text, and images are all carried on an integrated digital information superhighway. The implications of having a powerful new communications infrastructure for creating added value for businesses and households are staggering.

In a global economy that is increasingly information intensive, almost everyone agrees that an advanced telecommunications network infrastructure is vital for economic growth and value creation. Investments in IT generally, and telecommunications technology in particular, have been primarily responsible for the rapid increases in service sector productivity. Spillover effects serve to bolster other sectors like manufacturing, which also rely to some extent on information technologies. The result is a better standard of living. The same phenomenon occurs at the level of individual households when they invest in time-saving information technologies. Indeed, but for a raise in take-home pay, the adoption of IT in the home is one of the few ways in which a household's members can improve their lot in life both in quantity (productivity of time spent working at home) and quality (more leisure time).

Based on the economic discussion and analysis in earlier chapters, if a truly advanced public network infrastructure is deployed, it is highly likely that the mass market will finally begin demanding interactive multimedia applications and that those applications eventually will become as commonplace in the household as talking on the telephone and watching television. However, it is unlikely that the financial commitments necessary to make the infrastructure a reality will be justified by the demand for a single killer application the way that spreadsheet applications were largely responsible for PC purchases taking off. It is also unlikely that any single network technology, wired or wireless, will dominate the overall multimedia scene. It would not, however, be surprising for the mass market to prefer a single multimedia technology or delivery system for many, if not most, specific applications or activities (much like infrared wireless technology has totally dominated the mass market for remote control of stereos and television sets). Unlike the passive technologies used for watching television and listening to stereos (and associated remote control devices), interactive multimedia technologies will spawn numerous genres of novel applications, many of which will be customized to the specifications of individual users.

That raises at least two important questions for would-be interactive multimedia network operators:

- Will the demand for multimedia services continue the remarkable growth rates of the WWW over the past few years, eventually overcoming the demand for traditional voice and data telecommunications?
- What network technology will be most successful in delivering interactive multimedia services to the mass market of residential subscribers?

The answer to the first question is, "No, of course not." Nothing can sustain that kind of growth over an extended period of time, and the Web is not about to dominate two-way voice and data telecommunications (which will someday include digital video telephony). That being said, interactive multimedia applications on computer networks like the Web are still in the early stages of their growth cycle, and their phenomenal growth can be expected to continue for years to come.

As to the second question, it is safe to say that the future of interactive multimedia is not going to be dominated by a single huge computer network like the Web. Neither the Web nor the entire Internet, for that matter, is actually a service. Nor is it an application. Rather, the Net and the Web represent a general-purpose computer-based network platform that can be used for innumerable end-user services and applications. Admittedly, the widespread availability and affordability of Web access have allowed it to flourish, but that is only because of the ubiquitous fundamental network platform that supports it, namely, the telephone network. While for now the Web may be the only game in town as far as public networked multimedia is concerned, it will not be alone as an interactive multimedia delivery system; over the long term, it may even lose its dominant position. Chapters 1, 2, and 3 explained some of the reasons for this conclusion, not the least of which is the Internet's first-come-first-served network design.

The most successful commercial applications of interactive multimedia technology in the future will require alternatives to the Internet's rather unorganized system structure or, at the very least, fundamental changes to make the

Internet more suited to supporting high-quality commerical applications. Commercial success in a competitive market requires that a multimedia network feature value-based pricing and service delivery. In that regard the Internet's legacy as a "free" forum for a wide variety of interactive communications is both a blessing and a curse. It is a blessing for getting public demand for Internet access and usage to reach a critical mass and a curse for those trying to make money when there is no way to prioritize messages according to their value. Indeed, to this day within the Internet user community and governing bodies themselves (assuming they have any real say in setting global Internet system design and service policies), debates are raging about how the Internet could or even should be commercialized. Some user factions hold the chaos of the current "free" public Internet sacred, while others want to restrict access so they can make money selling services to people who value those services the most. Successful commercialization requires implementing a hierarchical value-based pricing and service delivery system design. If the Internet cannot accommodate that fundamental change, then other competing service delivery systems that are yet to emerge will.

Besides the obvious requirement that enough valuable programming and information must be created to attract consumers, two technological linchpins are necessary for the realization of a ubiquitous, widely used, and affordable interactive multimedia network infrastructure. Substantial progress must be made in the production cost and performance of consumer multimedia devices and terminals and in the corresponding high-speed public broadband network, including subscriber-access connections. Both of these technological developments must occur if interactive multimedia services are ever to find a true mass market.

Chapter 2 discussed technological trends and market plans that should substantially reduce the cost and increase the performance of subscriber equipment in the next few years. But that by itself will not be enough. Chapter 3 discussed the longer-term problem of achieving the target cost (low) and speed (high) of the individual subscriber network connections that feed information to multimedia terminals and devices located in the household. The initial cost of a household's broadband network connections are very high and will remain so for some time to come.

Perhaps even more important is that, in a competitive network environment featuring rapidly changing technologies, the construction of an expensive and long-lived nationwide network system is a risky proposition for any network operator. Yet that is the very infrastructure that needs to be in place before mass market demand can really take off. Thus, for infrastructure network development to proceed apace, close cooperation and joint service arrangements between network operators and service providers may need to occur so that each has a shared interest and responsibility in ensuring that the underlying network infrastructure gets built.

Unfortunately, public policymakers have yet to gain a serious appreciation of the problem. Technology notwithstanding, substantial regulatory barricades must be removed or, at a minimum, reduced before construction on the information superhighway can proceed with any speed. The policy tool of regulation is a blunt instrument for implementing competition policy. Even when regulatory rules are designed with the explicit intention of increasing the number of firms in a local market and the level of competition among them, once they are in place, two undesirable consequences often result: (1) the rules themselves take on a life of their own well beyond what was originally intended to jump start competitive market entry; and (2) the rules often hinder rather than stimulate infrastructure development. Chapter 6 provided a comprehensive discussion of the numerous problems associated with the use of asymmetric regulatory rules to try to implement competition policy.

9.2 ROADBLOCKS ON THE INFORMATION SUPERHIGHWAY

Throughout the world, a flurry of government activity has been promoting investment in advanced telecommunications network infrastructures. Because public sector budgets are strained, new government policies and regulations aim at stimulating private sector investment. But because businesses worship at the altar of cash flow, investment incentives and the regulations that are supposed to provide them must be directly related to the pursuit of high profits and competitive advantage. Established regulatory processes, on the other hand, continue to focus on constraining earnings and market power and therefore represent a formidable institutional roadblock to increased private investment. Squeezing the profits of incumbent network operators, in the name of protecting consumers, harms competition.

The recent experience in the United Kingdom provides a good example of the problems caused by the government attempting to regulate the profits of incumbent infrastructure network operators while at the same time pursuing a competition policy aimed at increasing market entry and infrastructure investments. Such inherently conflicting policies can ultimately restrict competitive market entry. The United Kingdom has the longest history of any country with incentive regulation during a period in which the government was trying to stimulate competitive entry. An article in *The Economist*, "The Great Telephone Paradox," describes price cap regulation in the United Kingdom [1]:

> For consumers, all this is wonderful: average call charges have fallen by more than 50% since price caps were introduced in 1984, and are now well below those in other European countries...But, complications arise from the other aim imposed on Mr. Cruickshank [the U.K.

regulator]: to encourage other companies to build networks that can compete with BT [British Telecom, the incumbent].

The incentive for them to invest is that, as nimble and efficient firms, they can undercut BT and still make a handsome profit. But each time Mr. Cruickshank forces down BT's prices, its competitors' potential profits also decline…The less profitable the British market, the fewer companies are likely to want to invest in competing with BT.

Today, BT's only major competitor in the United Kingdom, Mercury Communications, is not doing very well in expanding into the market, and BT has announced its intention to buy Mercury's parent company, Cable and Wireless.

Government authorities throughout the world need to avoid the temptation to overregulate the introduction of competitive entry into telecommunications service markets. That is the best way to ensure that a rapid deployment of the information superhighway will not fall victim to a fundamentally flawed regulatory process.

9.2.1 The Regulatory Process

The problem with regulation lies not with the economic theory of regulation (see Chapter 6) but in the political process of regulation as it is practiced. In most developed countries, it is the process of overregulating competition and market entry that is sinking, or at least significantly delaying, the development of an advanced public communications network infrastructure.

In the practice of regulation, policymakers have always been faced with the fundamental dilemma of balancing economic efficiency (which could mean raising some basic service prices) while protecting "captive" consumers by keeping prices regulated at artificially low levels. That explains much of the regulatory schizophrenia observed in the real world, where regulatory policy is aimed at introducing competition and, at the same time, providing subsidies to large groups of network subscribers. Of course, free market competition and cross-subsidies are natural enemies, so it is no wonder that regulating to achieve both at once can be a frustrating (and costly) process.

The new U.S. telecommunications law is a case in point. The law begins with a grandiose pronouncement that deregulation and competition are the best means to stimulate private sector investment in public network infrastructures, but the law itself imposes a plethora of rules and regulations restricting the market activities of network operators. At the same time, it also promotes huge cross-subsidies in the name of protecting a host of disadvantaged or otherwise deserving groups, including low-income and rural consumers, new competitors, schools, hospitals, and libraries.

If the government truly wants the private sector to fund the development of high-technology industries and infrastructures, then regulations must, at a minimum, offer rewards for investment that are commensurate with the risk of undertaking it. Government policy initiatives should focus on implementing regulations that improve—or at least do not harm—private market incentives to make infrastructure investments. In addition, if a current regulation is not proactive in furthering that objective and does not otherwise serve a specific higher public policy objective (e.g., income redistribution), then it should be eliminated.

Still, for political reasons, government authorities responsible for telecommunications policy remain much more preoccupied with the possibility that some companies may obtain monopoly power and high profits than with the potential for stimulating investment in network infrastructures via market-based policies and incentives. Herein lies the most pitiful irony of the situation: The most fundamental drivers of private market investment incentives are exactly the opposite of those that government authorities are prepared to adopt in the name of infrastructure development.

The solution to the problem is clear. Policymakers should eliminate restrictions on market entry, profits, and the scope of operations of all market players. "Managed competition," like "efficient regulation," is an oxymoron. Regulators, by trying to maintain both regulation and competitive entry, end up with the benefits of neither and the costs of both. As discussed in Chapter 6, until market entry creates a situation of workable competition, direct price regulation of incumbent monopolists is the best tool for protecting basic service subscribers. Everything else should be deregulated—period.

It's an unfortunate reality that deregulation policy initiatives never seem to result in any genuine deregulation. Genuine deregulation implies eliminating rules and not replacing them with others. While the new telecommunications law clearly states that deregulation is a goal of public policy, it goes out of the way to protect existing regulations, regulatory structures, and processes. Chapter 6 provided ample proof of the new law's problems in that regard.

As is so often the case with bureaucratic processes, fixing current regulatory flaws will likely be a painstaking process. The first step is to make the public aware of the legislation's flaws. The second is to call on the carpet the policymakers who are placing their personal agendas ahead of the good of their constituents. Taking policymakers to task for their actions has fallen largely on the academic community and the popular press. Almost every other political entity (including industry and consumer advocacy groups) is too embroiled in the policy debate itself to be capable of objectively furthering the goals of deregulation.

Just as the procompetitive rhetoric of the new federal legislation manages to grab the headlines, most significant decisions of federal and state regulators are coupled with press releases trumpeting their programs for adopting procompetitive policies. In fact, there has been so much positive press about the procompetitive policies of government regulators that one might expect to see a flood of benefits flowing to households across America. In reality, consumers are experiencing little more than a trickle of benefits compared to those that could occur under genuine deregulation.

Historically, regulatory policies have had two purposes: (1) to provide consumer benefits, which, in the real world, can be defined as more service choices and competitive prices; and (2) to emulate the forces of competition by mandating lower prices in situations in which markets themselves fail to provide such discipline. Such was the case for monopoly public utilities.

Largely due to wondrous advances in digital technology, there are now many ways for people to communicate with one another, rendering obsolete the old monopoly model of public utility regulation. The mandate for modern regulation, therefore, is clear. It is supposed to promote, or at least not hinder, competitive market entry so that market forces can do their thing: provide more choices at competitive prices. The best way to achieve that objective is through deregulation. Unfortunately, even some so-called consumer advocate groups (mostly based in Washington, D.C.) still espouse the notion that regulation is synonymous with consumer benefits.

By now, most regulators know better. But even among those enlightened policymakers who know that deregulation and competition are truly synonymous with consumer benefits, things seem to go awry in the implementation phases of competition policy.

Take the case of the FCC. In speech after speech, Reed Hundt, the current chairman of the FCC, states succinctly that deregulation is the way to promote competition and achieve consumer benefits. Most state regulators parrot those remarks, and academic economists could not agree more. But actions speak louder than words, and the FCC's actions impeach all the rosy rhetoric. Before discussing the supporting evidence provided by specific regulatory decisions, a brief digression on the nature of the regulatory process itself would be useful for a better understanding of why it is so difficult to render good decisions for the public at large when so much is at stake for powerful private interests.

Regulatory decisions, like the new telecommunications legislation, are boring and do not make for a very good read. They are mostly a forum for bureaucrats, high-priced lawyers, and industry consultants. Even though modern telecommunications technology and regulation are difficult for laypeople to understand, political machinations are not. Not only is the latter well understood by the public, but it is generally viewed with a cynicism that borders on disgust.

Historically, the regulatory process can best be characterized as an exercise in monopoly turf allocation. Mark Fowler, the former chairman of the FCC, aptly characterized the federal regulatory and legislative process as "soft corruption" [2]. In short, the regulatory process itself is a market subject to its own unique forces of supply and demand. As long as the government, at taxpayers' expense, is willing to supply regulation, there will be a demand for it. As market supply grows, so will demand, and vice versa. Even when regulatory agencies try to do the right thing to promote fair competition and consumer benefits, lawmakers can derail them by threatening budget cuts or job security. A classic example of political meddling occurred in the early 1980s, when the FCC tried to rebalance telephone company tariff rates to better align them with underlying costs. Congressional pressure prevented that from happening.

Competition is not an orderly process, and regulators tend to deregulate in very small increments in an effort to control the transition. That is precisely the problem. The process should be reformed so that regulators can take the economic high road of market discipline and promote competition among firms that is truly based on the merits and the business acumen of the players themselves.

Once the regulatory approval process gets hold of it, every new (de)regulatory rule that the government proposes to implement (which might at first genuinely represent an effort toward deregulation) provides another opportunity for one or another competitor to gain a regulatory advantage. Regulators become nothing more than referees in prize fights between competing business interests. The new telecommunications law is again an excellent case in point: It has become an absolute field day for private interests to gain an edge by convincing regulators that their own interpretation of each and every passage in the law is best. Anyone who doubts that that is the case needs to read the lead news article that appeared in the *Wall Street Journal* on March 29, 1996, "The FCC Is Besieged as It Rewrites Rules in Telecommunications: Executives, Lobbyists Cram the Agency's Schedule and Offer Lots of Advice." That article provides ample support for the conclusions herein about the political economy of the regulatory process. Of course, no one should blame private interests for pursuing their own agenda. The problem is solved by closing the door on regulatory opportunism by passing laws that truly deregulate. Gaming a market system is vastly more difficult than gaming a regulatory system.

A cursory review of recent decisions in federal and state (de)regulatory reform proceedings reveals that regulators are, more often than not, promoting and protecting business interests rather than improving the lot of consumers. What is, or at least should be, particularly galling for consumers is how blatant these decisions are in their preoccupation with promoting private business interests at consumers' expense. One could get rich receiving a dollar every time the term "competitive safeguard" appears in decisions to deregulate some aspect of the telecommunications business. The term can (unfortunately) be taken at face value. Regulators are safeguarding competitors, not consumers. By comparison, the terms "consumer benefits" and "consumer safeguards" rarely appear; when they do, they are almost always used in the context of protecting one or more firms trying to gain a regulatory advantage in the marketplace. In other words, consumers are generally not being considered, except in their role as customers of one or another competing firm; when a firm wins a regulatory advantage, its own customers, rather than consumers generally, benefit. This is a subtle but important distinction when the banner of consumer benefits is being waved. It also explains the outcome of many regulatory decisions. The bottom line is that (de)regulatory decisions are full of provisions that safeguard the proprietary and financial interests of would-be competitors that want to ride on the political gravy train of regulatory rent seeking.

Regulators should try to encourage competition the old-fashioned way, by letting the players have at it. Reed Hundt is adamant in his public pronouncements that the FCC is moving full speed ahead on deregulation. That is probably true compared to past administrations, but, from the perspective of consumers, it still falls far short.

9.2.2   The Growth of Regulation at the FCC

While the discussion in this section applies to both state and federal regulation, it is beyond the scope of this book to cover state regulatory trends in any detail. However, while it is much easier to critique the FCC's regulatory performance, that should not be construed as support for creating more state controls, since the regulatory track record in many states is even worse [3].

Let us assume for the moment that, as Mr. Hundt asserts, the current FCC has an unprecedented track record for deregulation. Why then do regulatory budgets and staffing requirements continue to expand? The usual explanation is that there are high costs associated with formulating and implementing new detailed rules and (de)regulations governing the new competitive market environment, including the costs of administering and enforcing the new policies. The real answer is that, just as with tax laws, all the detailed provisions and complex rules spawned by new (de)regulatory policies, including the new federal law, create huge costs for administration and litigation. Creative business minds are constantly taking advantage of the myriad of regulatory loopholes and angles. Who's measuring the benefits or the costs to consumers?

A revealing example of the strange coincidence of deregulation leading to increased regulatory staff and budgets can be seen in a recent letter exchange between the FCC and the chairman of the Senate Commerce Committee, which is responsible for FCC oversight.

The exchange started with a letter from the FCC to the chairman of the Senate Commerce Committee requesting continued agency funding. It included a specific request not to cut funding levels. The following excerpt is from that letter:

> Our concern over reducing the FCC's funding from $175.7 million to
> $166.2 million is heightened by the prospect of the potential immi-

nent enactment of a telecommunications reform bill. This bill would completely overhaul communications regulation for the first time in more than sixty years. It is expected that this reform bill will create a host of new responsibilities for the Commission, including the completion of scores of major rulemaking proceedings. [Note: The FCC's budget is much more than that indicated here; according to the FCC chairman, only about one-third of its total budget is funded from tax revenues, the rest is from industry generated funds.]

The following excerpts came from the Senate committee chairman's lengthy reply:

> The suggestion that the FCC will not be able to implement the Telecommunications Act of 1996 (the Act), unless current funding levels are increased is troubling...
> Indeed, the legislation contains a number of provisions that will reduce spending by the FCC...
> For 1996, the FCC requested approximately $224 million to support 2,300 employees. That is roughly 2/3 more than the budget of just three years ago...Since 1992, FCC expenditures have risen at a compounded average annual rate of about 15.2 percent, while those of the communications industry have grown only 10.4 percent...
> [P]assage of the Act also has empowered the FCC to consider initiatives in four key areas which will more than adequately deal with today's fiscal austerity imperatives...
> First, the Congress authorized regulatory forbearance, an authority which the Act specifically grants to the FCC. At present, the FCC places an array of requirements on industry (e.g., compliance with archaic accounting and depreciation rules and very extensive reporting requirements), which, in turn, generate high demands on the FCC's scarce resources...Elaborate and costly regulatory requirements such as those associated with the FCC's so-called "video dialtone" rules, which the Act eliminates, also warrant immediate assessment. The telephone carriers today are typically "capped" at both the federal and state levels as you know. Any risks to ratepayers as a consequence of these and other reforms, therefore, are small...
> Second, I encourage the FCC to follow the leadership of experts including the Vice President in seeking to privatize functions to the maximum possible extent...Third, I encourage the FCC to terminate programs which overlap with or duplicate those of other federal entities...Another example of overextending the FCC's regu-

latory reach is the list of 17 items recently released by the FCC's Wireless Bureau...The FCC's rush to regulation seems excessive and this proceeding would appear to be a solution in search of a problem...Finally, I also understand that you plan to conduct a rulemaking with respect to the broadcast license renewal provisions of the Act. This is another example of a provision which does not lend itself to agency discretion.

Whether and to what extent this reply reflects good economics (it certainly does) or partisan politics (possibly—the chairman of the FCC is a Democrat and the chairman of the Senate committee Republican), it is clear that the practice of deregulation does not necessarily imply a displacement of regulation in favor of a reliance on market forces.

Congress must have realized by now that regulators cannot be trusted to deregulate. It is against their nature. By placing the FCC in charge of implementing the new rules of the road, Congress left a huge regulatory roadblock on the information superhighway. Only a short time ago, the FCC's budget was on the Republican Congress's chopping block. That was a pro-consumer idea whose time had come. Now the FCC's bloated budget is likely to expand even further (there are plans to expand financing the budget from industry sources). The FCC is incapable of reining itself in. It should be abolished in favor of a much smaller and more efficient organization that will regulate only those services that cannot be reasonably entrusted to the forces of the free market.

For many years now, most industry economists have had a cost-effective and efficient prescription for genuine pro-consumer deregulation: (1) protect captive monopoly customers by setting ceilings on prices for services that are not subject to competition; (2) enforce nondiscriminatory access and interconnection;[1] and (3) deregulate everything else. The savings in regulatory costs would be enormous. There may be some other minor areas of concern that regulators must continue to address, like handling consumer complaints, but in large measure these three steps are all that is really needed. Interestingly, in principle (i.e., on paper), most regulators would agree. The problem is, as Mr. Fowler put it, that "soft corruption" has a firm grip on the (de)regulatory process; simple rules that should be easy and inexpensive to administer suddenly become enormously complex and very expensive, with little regard for the ultimate net benefit for consumers.

---

1. Similar to protecting the captive monopoly customer, this usually means adopting a simple uniform price ceiling for telephone network access and interconnection and monitoring the results, so that service quality, terms, and conditions for carrier access and interconnection are equal (the "equal access" doctrine).

9.2.3   Selected Broadband Policy Failures

There are many regulatory barriers to entry and investment for prospective broadband infrastructure providers. A case in point is the regulation of both cable television and telephone companies. In the case of cable television companies, policymakers in Congress in 1992 had a clear choice between the rule of competition and the rule of regulation, and they chose the latter. Instead of adopting a simple price ceiling to protect captive customers of basic cable service suppliers and letting telephone companies or others into the market, Congress and the FCC embarked on a costly regulatory journey. The result is that there is still a cable monopoly, with roughly the same prices and no better service.[2] Fortunately, a mere two years after its enactment, this policy has effectively been reversed by the new law. One can only assume that it did not take long for Congress to recognize its earlier poor choice of regulation over competition.

Consumers probably all have their little horror stories to tell about how they were personally affected by the government's reregulation of cable television. Allow me to relate mine. It started off nicely enough. Like most others, I too got a dollar or two off my monthly basic cable bill. But I also got (for reasons only the FCC staff and industry insiders understand) three NBC stations, all with identical network programs, in the first nine channel positions on my television dial. (It all has something to do with forcing local cable operators to carry over-the-air signals broadcast by network affiliates in the region.) This is progress? Considering the abundant programming available and the scarcity of channel capacity, I am absolutely sure that, if left to their own devices, cable network operators would never voluntarily make their customers pay for duplicate channels in some of the most attractive channel slots.

In the case of the FCC's (seemingly endless) review of so-called price cap regulation of telephone companies and attendant cost allocation schemes,[3] the most recent "comprehensive" review promised to establish yet another layer of regulatory oversight and, of course, more complexity for monitoring and cost allocation. In a particularly daft move, the FCC proposed to adopt cost-based

---

2. The FCC has declared victory and walked away. But it is not clear at all that cable prices adjusted for quality of service and price per channel adjusted for quality had gone down or that subscribership had risen as a result of what is unambiguously an expensive and burdensome regulatory implementation process. For some detailed background data and analysis see T. Hazlett, "Regulating Cable Television Rates: An Economic Analysis" (Working Paper Series No. 3, Institute of Government Affairs, University of California, Davis, 1995).

3. In fact, the FCC's price cap plan, like most state plans, retains some cost allocation rules and earnings regulation and, in turn, all the bad incentives and economic distortions attributed to them.

regulation of VDT service,[4] a new service in which, by definition, neither telephone companies nor anyone else has any market share or market power. Fortunately, this is also being eliminated by the new law, but the fact that it was even considered is indicative of the size of the problem of regulatory inertia.

Even though the latest proposed rules for regulating VDT service were preempted by legislation, other terribly onerous rules had been in existence for years, effectively stifling the offering of this new service by telephone companies. A number of regulations forced telephone companies into protracted and expensive litigation to meet both the stringent requirements of the FCC and intervenor complaints. Naturally, the incumbent monopoly cablecos happily obliged the regulatory process by contesting every telco application. The result was deadly. Many frustrated telcos simply gave up, and VDT, like so many exciting new service concepts before it (e.g., videotex and information service gateways), has yet to become a reality. Why is the FCC making it so hard for consumers to obtain new services? We are not talking about the possibility of ignorant consumers purchasing life-threatening drugs here. The simple answer lies in the process of soft corruption that Mr. Fowler described.

On the wireless scene, just when it appeared that the FCC would aggressively expand the amount of radio frequency for new digital services via spectrum auctions and liberal flexible-use policies, the effort is falling short. The amount of spectrum associated with so-called broadband spectrum licenses is woefully inadequate to support the deployment of a mass market broadband telecommunications system. As a result, wireless access systems will be relegated to supporting only narrowband telecommunications (see Chapter 4). The FCC should allocate more spectrum to broadband services and lift restrictions on the amount of spectrum that any one licensee may acquire so that new broadband digital radio services can be brought to the mass market.

While the FCC has already (finally) adopted an important flexible-use policy granting wireless network operators the freedom to choose what services will be provided over their slice of spectrum, the policy fails to be fully implemented. The FCC continues to protect incumbent wireless operators from competition by barring new wireless operators from using their spectrum for broadcast or point-to-point radio services (among others). In particular, the FCC fails to apply its flexible-use policies to provide market opportunities for existing wireless operators that might wish to add market value by using their slice of the spectrum for services with high consumer demand. The obvious case in point is the FCC's disallowance (over its own internal staff's recommendations) to allow fallow UHF spectrum to be used for new telephone services. UHF spectrum is adjacent to that for cellular phone service and could readily be con-

---

4. In fairness to the FCC, many such rules are the result of Congressional interference or "soft corruption." However, when consumer benefits are at stake, little comfort is gained from that excuse.

verted to cellular service. Use restrictions placed on cellular systems and other broadcast satellite and terrestrial wireless cable television systems should be lifted to allow for two-way services.

The largest and potentially most wasteful regulatory maneuvering of all is occurring in the area of digital wireless telecommunications. In its plan to implement a nationwide policy for advanced digital television, the FCC has proposed setting aside a huge amount of RF spectrum in a frequency band that is coveted by many potential wireless service providers. The problem is that the FCC proposes to give the spectrum away to incumbent broadcast networks without first considering the foregone revenues it could generate from the private bids using spectrum auctions and, more important, without even considering the value of the spectrum for other uses and whether there is any substantial consumer demand for high-resolution TV. Based on the demand studies conducted to date, there is certainly no consensus on the value to consumers of new high-resolution digital TV pictures, especially considering that the television sets themselves could cost as much as a new car. Nicholas Negroponte, director of the MIT Media Lab, when asked about viewer preferences and willingness to pay for higher resolution pictures responded: "People don't give a damn about resolution" [4].

The bottom line is that the FCC's own policies continually put it in a position of choosing winners and losers among technological and market alternatives, a position it maintains it does not want to be in.

9.3   HYPE AND REALITY

Beyond the political and regulatory rhetoric, industry hype in the United States regarding the private sector's willingness and ability to invest in a ubiquitous and affordable broadband network infrastructure is also vastly overblown.

More often than not, private businesses are simply hyping new technology to signal their potential rivals that they will aggressively deploy the latest, most advanced multimedia network infrastructure in the hope of beating competitors to the punch. At the very least, this would give a competitor pause before making a hefty and potentially redundant investment in network infrastructure. At the same time that firms are "committing" to vast network upgrade plans and clamoring for deregulation and free market competition, all the major industry players continue to lobby for government protection of their traditional markets. The government should have learned by now from their past well-intentioned efforts to manage competition that loopholes in the rules are the norm, not the exception. Private enterprise always seems to be able to exploit the regulatory process to gain a market advantage.

Interestingly, though the market signaling that is catching all the news headlines may constitute credible threats from players with deep pockets, it is likely that the first network operator to seriously take the plunge into mass market investments will have the most to lose financially. Technological progress and the attendant proliferation of private network alternatives are making it increasingly difficult to bet on the financial prospects for capital recovery of public infrastructure investments. Whether the potential to lose money is there or not, the government should encourage businesses to take the risk by freeing up market entry.

Throughout recent history, a number of integrated network projects became spectacular failures, but at least they were only troubled broadband market trials, which can be abandoned with relatively minimal cost. Some of the more prominent failures include:

- The highly touted FTTH trial by BellSouth in Heathrow, Florida—a very expensive experiment;
- The failed GTE integrated network project in Cerritos, California;
- The U S West/AT&T/TCI integrated network trial in Denver;
- Most recently, the Time Warner "full-service network" broadband trial in Orlando, which the press has been reporting as having problems.

More limited and less expensive trials of interactive TV technology have also failed to generate any significant consumer interest.

As advances in digital technology cause the convergence of voice, data, and video telecommunications, it is also causing the complete commoditization of public digital network distribution systems. The common carrier public network platform for multimedia services will have to compete with a host of alternative private network delivery systems. These contract carriers are proliferating, fulfilling demand in a significant number of market niches and siphoning off important revenue streams from the integrated (and commoditized) common carrier network providers. Examples of cost-effective private network technologies include several wireless alternatives such as private two-way satellite systems, new digital cellular networks, antenna-based "wireless" cable television networks, and two-way broadcast and packet data radio networks. Private wireline networks that use digital fiber optic and coaxial cable and various hybrid networks that use combinations of wireline and wireless technologies also abound. Integrated network operators will also get some stiff competition in many market segments from truly portable media such as read/write CDs, digital video and audio tapes, and other computer-based communication alternatives. These also constitute communication "networks" that will capture a portion of the multimedia revenue streams that might otherwise be available to integrated public network operators.

With such a cornucopia of private competitive network alternatives on the horizon, it is interesting to contrast that with the convergence rhetoric of the major industry incumbents. With the proliferation of buzzwords like "seamless nationwide," "high-speed digital," and "information superhighways," it seems as if consumers can just sit back and watch the information revolution happen. Yet trends in investment rates in public network infrastructures, beyond the existing programs to upgrade core networks for digital service upgrades, are stagnant. The core network refers to the trunk or shared capacity of public networks, which has already been undergoing rapid digitization and capacity expansion for many years and still continues to do so today. The rest of the (noncore) public network facilities are not all shared and, in fact, are usually dedicated to specific households and business locations. These are referred to as local access network facilities or loop plant. It is the latter on/off ramps that are far and away the primary cost issue for mass deployment of the so-called information superhighway; they are also far and away the slowest to be upgraded for high-speed digital service.

The likely explanation is that the industry remains convinced that direct government involvement is the worst thing that could happen, and the players are keeping their money in their pockets until something proactive is done to ensure them that their infrastructure investments can be made without regulatory interference. Given the track record of regulation in promoting the development of new markets, that seems to be a logical conclusion. But a truly effective policy mix will inevitably require some government intervention, not the least of which would be to tear down current barriers to competitive market entry. With the new legislation as a start, the government is finally stepping up and taking some responsibility for national competition policy. Nevertheless, a lot more needs to be done to rein in regulators as they try to micromanage the transition to free market competition. The risk, as always, is that private industry will continue to delay risky infrastructure investments.

9.4 REFORMING GOVERNMENT POLICYMAKING

A historical examination of traditional institutional processes for governing the telecommunications industry around the world could best be described as a study in the political economy of monopoly turf allocation. While the details of the process might vary by country, the general result of the process is the same. In all cases, communications-market turf in any given geographic region is allocated to monopoly providers for telephone, cable television, broadcast, and, in many cases, newspapers. The monopoly provider paradigm has long been questioned, and, by now, most countries have authorized at least limited entry of competitive service providers within a given domestic media market. But such entry is hardly open to all comers, and more often than not the new paradigm is one of regulated competition rather than just plain competition. Even with the new law in the United States, policymakers and the state and federal regulatory authorities responsible for implementing it have still failed to authorize any significant cross-market ownership so that telephone, cable, and broadcast companies may diversify their media businesses in any given geographic area. To the government's credit, the new law does begin to relax rules limiting common ownership of broadcasting stations and newspapers, but that is hardly enough to stimulate investment in an integrated broadband network infrastructure capable of serving mass market demand for interactive multimedia services.

There are exceptions, such as in the United Kingdom and Hong Kong, where government authorities encourage competitive entrants to provide both cable and telecommunication services in the same geographic area. Yet, even in those cases, the government disallows similar freedom of operation for incumbent firms. Only through symmetric application of the rules of competitive entry can the potential for full and fair competition be realized. The prevailing asymmetric entry policies make it impossible to determine whether competition and open entry are sustainable and, in any event, put any serious efforts for mass market deployment of an information superhighway on a relatively slow track by keeping major players out of the game. A government policy of free market entry for all players in all market segments and for all types of media may turn out to be essential to achieving the vision of media convergence on a mass scale.

It is high time for policymakers to bite the bullet on free market entry—there is no more powerful engine for investment than the freedom to pursue cash flow opportunities. The monopoly turf-allocation policies of the past have failed. Governments can try to stimulate private investment in public network infrastructure in essentially two ways: (1) by pursuing a policy of deregulation and open market competition, which means that government should get out of the way and allow the private pursuit of profits through productivity and innovation: and (2) government pro-investment initiatives such as flexible allocation of radio spectrum, tax breaks, universal service subsidies, and seed funding for achieving a critical mass for fledgling but promising technologies. In fact, some combination of those two basic policy options could provide an extremely powerful incentive for private investment in network infrastructure.[5]

Unfortunately, at least in the U. S., the government at both the state and federal level practices neither of the above, choosing instead to continue wallowing in the mire of inherently conflicting policies of competition and universal service. Why? The answer lies in the process of regulation. In other words, it

---

5. There are other novel competition policy alternatives which the government may try, like conducting auctions in which private firms bid for franchise rights to become a (temporary) monopoly service provider in a given market area.

is the stated government objective of increased investment that seems invariably to fall victim to the process for implementing it.[6]

At the highest levels of political discourse, everything sounds very pro-infrastructure and pro-consumer. Take the three basic pillars of the Clinton administration's telecommunications policy objectives for the NII: (1) America must remain the world leader in information infrastructure by speeding up investment through private sector incentives spurred by opening markets to competitive entry; (2) infrastructure networks must be compatible to maximize functionality and minimize the expense of interconnection; and (3) all Americans should have affordable access to an advanced network infrastructure.

Even to an economist, at this level anyway, the administration had it just about right. So what went wrong? Again, it's the process. The main theme of the new legislation, open market entry and competition, is barely mentioned when it comes to establishing the regulatory rules for implementing the new legislation. Instead, the focus of the rules is on maintaining cross-subsidies and so-called competitive safeguards.

It's not too late for politicians involved in telecommunications policymaking to do the right thing. They merely need to retreat on the detailed regulations and continue to pass rather broad and far-reaching legislative provisions mandating that market-based solutions be favored in lieu of regulated competition. They also need to specify that the FCC and the states handle the details of implementation by adopting new regulations in strict accordance with the principle of deregulation and competition.

In an effort to get going on the implementation of a progressive competition policy, many governments have commissioned blue ribbon panels of industry experts to solicit recommendations for speeding up infrastructure investment. These august panels are populated primarily with CEOs from the largest firms in the business, including incumbents, entrants, and would-be entrants. Their presentations in state and federal hearings are usually little more than a carefully choreographed spectacle pleading that competition is good but that their industry constituents, as competitors, for one reason or another, need special protections because someone else has a market advantage. That is exactly what is happening in the Congressional hearings currently being conducted to decide the fate of the RF spectrum now reserved for advanced television. The odd thing is that market advantage is not a bad thing. It is, after all, market advantage that drives the engine of economic investment in every other industry.[7] Let firms compete to build better mousetraps and you tend to

---

6. For an insightful discussion of the political economy of telecommunications regulation, see J. Wenders, "The Economic Theory of Regulation and the US Telecommunications Industry" (*Telecommunications Policy*, 1988).

7. Furthermore, a firm's market advantage may be very long lived depending on its ability to perpetually build better mousetraps. This paradigm of economic market disequilibrium provides

get the most investment. The beauty of this market process is that consumers benefit too, because quality is higher and, ultimately, the price is lower.

It is unfortunate that, in the rare cases when qualified academics are asked to present their relatively objective opinions, their testimony is largely ignored. Academics tend to favor market processes in practice, not just in theory, where competitive capabilities determine market advantage in lieu of gaming the regulatory processes.

The bottom line is that regulation, both the process and the practice, limits a firm's ability to take advantage of market opportunities, creates an environment of uncertainty because it is so fickle, hurts consumers by protecting competitors, and hurts investment because it limits the potential benefits from innovation.

The basic problem inherent in the process of policymaking via regulation, with all its political underpinnings, was put most succinctly by a retired academic who served a distinguished career as a politically appointed state and federal regulator. Professor Charles Stalon served on the *Illinois Commerce Commission* (ICC) and later on the *Federal Energy Regulatory Commission* (FERC). In a policy research paper presented just after he stepped down from FERC, Professor Stalon explained why he specifically and regulators generally often rendered uneconomic decisions, even though they certainly knew better. When two or more industry factions are directly involved in any pending regulatory decision, regulators who wanted to keep their jobs followed a sort of regulator's creed: "Do no direct harm" (to any direct party to the decision).[8] In other words, a regulator could render judgments that indirectly harmed many, even many millions, of ordinary consumers, so long as the decision did no direct financial harm to the parties to the debate. That perspective is particularly revealing of the fundamental flaws of the regulatory process.

---

for a realistic view of dynamic market processes and the long-term welfare gains that flow from the process. While the economics of disequilibrium is anathema to the traditional neoclassical view of the benefits of marginal cost prices, the latter is a rather naive and static view of markets. In other words, it is precisely the deviation of price from marginal cost, or at least the pursuit of high margins, that drives the engine of economic investment and long-term welfare gains. For applications of these notions to telecommunications markets, see Egan and Wenders, "The Costs of State Regulation."

8. The original quote came from Charles L. Schultz as he was describing the U.S. political system and the nature of political decisions made within it, in *The Public Use of Private Interest* (Washington, D.C.: Brookings Institution, 1977), p. 23. Professor Stalon was making an extension of the point that decisions of U.S. regulators, as part and parcel of the political system, would tend to be subject to the same unwritten rule. See C. Stalon, "Recent Developments in the Political Economy of Regulation: The Sometimes Conflicting Objectives of Efficiency and Fairness" (Lecture for the NARUC Advanced Regulatory Studies Program, Williamsburg, VA, February 21, 1992).

9.5 INFRASTRUCTURE COSTS AND CAPITAL RECOVERY PROSPECTS

Chapter 5 provided estimates of the average incremental costs on a per-household basis for network access line digital upgrades for telephone and cable TV networks using new digital fiber optic and wireless technologies. These costs, which are reproduced in Tables 9.1 and 9.2, are based on many industry sources and generally represent the consensus. For purposes of comparison, the tables also provide base-case estimates of current AICs of telephone and cable television company local networks using traditional analog technology (discussed in Chapter 3).

Table 9.1
Wireline Network AIC per Subscriber

Technology	Type of Wireline Service	Average Incremental Cost per Subscriber
Base case current cost	POTS: AIC of new telephone network access line	$1,000
	POCS: AIC of new cable network access line	$700
Narrowband ISDN N-ISDN	N-ISDN telephone company access line upgrade	$100–200
	N-ISDN upgrade including digital switch placement	$300–500
Mediumband digital service	ADSL	$500–700
Fiber optic network access line upgrades		
Fiber-to-the Home (FTTH)	Telephone company (FTTH) for POTS only	$3,000+
	Future (1998–2000)	$1,000+
	Telephone company FTTH (two-way broadband)	$5,000+
	Future (1998–2000)	$2,000+
	Cable network FTTH (N-ISDN + two-way broadband)	$1,500+
	Future (1998–2000)	$1,000+
Fiber-to-the-Curb (FTTC)	Telephone company (FTTC) for POTS only	$750
	Telephone company FTTC (POTS + POCS)	$1,350
Cable hybrid fiber/coaxial network for POCS only		$50–100
Cable hybrid fiber/coaxial network for POTS + POCS		$200–300

Table 9.2
Wireless System Capital Costs per Subscriber

Wireless System	Type	Capital Costs per Subscriber
Current AMPS		$700–1000
PCN/PCS	AMPS-D (TDMA)	$300–500
Macrocell environment	CDMA	
		$350 (for urban system with over 50,000 subscribers)
Microcell environment	TDMA	$500
	CDMA	
		$500
Wireless cable	MMDS (television only)	$350–$450 (50% CPE and 50% installation)
		$525 ($380 reusable if subscriber discontinues service)
	LMDS (television only)	$40 (cost per urban home passed)
		$110 (cost per suburban home passed)
		$700 (CPE cost per subscriber)
	Two-way MMDS, LMDS	Not available or experimental
Satellite	DBS (television only)	$300–$800 (includes CPE)

Telephone company access line upgrade costs for broadband FTTH or FTTC systems are much higher than those for narrowband systems (N-ISDN). It is useful and potentially instructive to consider the costs of a relatively inexpensive technological alternative that was originally (back in the mid-1980s) forecasted to be here by now but is not, when considering the costs of more expensive network systems forecasted to be forthcoming in the next decade but that may never materialize.[9]

---

9. Considering the hype about the mass deployment of very expensive broadband technologies, it is amusing to remember the hype from 10 years ago regarding the industry forecasts for widespread deployment of narrowband ISDN technology, which is still not deployed anywhere in the world on a mass scale. In the United States, residential ISDN service, while being widely tariffed and available for purchase, is virtually nonexistent in the marketplace. Thus, even when ISDN capability is deployed in the public network, subscribership is very low, if any exists at all. For an examination of the reasons for the lack of demand see B. Egan, "Benefits and Costs of Public Information Networks: The Case for Narrowband ISDN" (Columbia Institute for Tele-Information, Research Working Paper Series, Columbia Business School, February 1992).

The costs presented in Tables 9.1 and 9.2 do not include consumer terminal gear, fancy new digital TV set-top boxes, or the additional costs of interconnecting to the intercity networks of the telcos necessary for ubiquitous call completions two-way service. The costs for those excluded items are potentially substantial, making the costs of final service to households considerably more than the costs of upgrading only the supplier network. In the case of consumer terminals for integrated narrowband and broadband services, the incumbent supplier's costs to supply that equipment to its subscribers or the cost to subscribers who choose to purchase the equipment themselves will likely be similar for both telcos and cablecos. The cost to an individual subscriber will vary directly with the level of functional capability of the terminal itself (digital television receiver, set-top box, computer terminal, modem, videophones, etc.).

The second excluded item, interconnection costs, can be substantial given current tariff rates for access to the PSTN. For that reason, long distance companies, cablecos, wireless, and other private network providers are aggressively pursuing both wired and wireless bypass alternatives that provide direct connections to the nationwide network facilities of other carriers, like AT&T and MCI. If future cablecos cooperated to originate and terminate calls, this arrangement could eventually achieve ubiquitous intercity service capability. Coordinating such a level of cooperation is no small task, considering that there are over 11,000 local cable companies in the United States. Thus, at least in the near future, the costs of local telco interconnection will be an important financial consideration for would-be market entrants.

Relatively new modem-based mediumband technologies, like ADSL, which is capable of providing two-way narrowband digital service integrated with one-way video service, can already be used to provide VDT and VOD services over a normal phone line. This technology is continuing to progress to the point where it could potentially provide multiple high-quality video channels. Perhaps the biggest plus is that the cost performance of the required modems is continually improving.

Given the costs of all the multimedia infrastructure network alternatives, it is useful for purposes of illustration to point out what is implied for the demand side of the capital budgeting equation. As a rule of thumb, for every $1,000 ($1) of per-subscriber network access line upgrade costs, fully $14.00 ($0.014) per month of additional revenue per household served would be required to allow for full capital recovery of the original investment costs over a 10-year discounted payback period at a 12% ROR. This hypothetical case includes the rather heroic assumption that new revenues would begin flowing immediately on completion of the network construction (which is why the cost and implied capital recovery estimates represent a best case for cash flow analysis). The numbers are cause for alarm for any would-be infrastructure network providers planning to go it alone.

Even assuming a rapid mass market deployment, the additional monthly revenues required per household to pay for the original investment are staggering. The average household in the United States spends about $45 per month on telephone services and about $25 per month on cable television services. Advertisers pay another $30 per month per household to support over-the-air broadcasting ("free TV") and another $8 per month for broadcast radio.

Therefore, not including what an average household spends on electronic devices, just over $100 per household is available for communication services in a competitive marketplace. This amount has remained relatively stable, as have household disposable incomes. In fact, over the last decade, the percentage of household income spent on telecommunication services has been flat, at about 2%. The percentage of household income spent on cable TV service has also been flat in recent years. Per-household broadcast media revenues have been steady or slightly rising after slowly declining from the late 1980s to early 1990s. However, there are other potential revenue streams involving video media like movies, videotape, and video game rentals and sales, which could add another $20 billion in potential revenues. Revenues from information and transaction services like home shopping and banking and other advertising services also exist, but there are no solid data on their market potential. However, based on the rapid growth of direct mail advertising to over $20 billion annually (and growing), it is reasonable to assume that it could potentially be substantial. The phenomenal growth potential of the Internet, while still relatively low in terms of total revenue, speaks for itself.

Still, the demand and revenue data from the telecommunications sector is indicative of the uphill battle faced by a competitive service provider of two-way residential broadband network services. New revenue growth is always going to be subject to the ability of households to afford fancy new services and the terminal devices that support them. What is more, current revenue streams are supporting the payback for old and current capital investments and may not be immediately available to fund new construction budgets if alternative investments are more attractive. The bottom line is that, unless an integrated broadband telecommunications network operator is allowed to freely pursue all revenue opportunities, including partnering with other service providers to save on new construction costs, it will be difficult to justify mass deployment of the new broadband technology.

9.6 POLICY IMPLICATIONS

The upshot for public policy is that regulators must open markets so that all media companies can compete—or cooperate—to modernize the infrastructure. By effectively maintaining the regulatory status quo of limiting the market alternatives of major players to provide a wide range of services, the regulatory reform proposals currently under consideration make it increasingly difficult to justify the required investments for public network upgrades.

Two bureaucratic factions tend to dominate the regulatory debate. Despite the enormous investments required for advanced network infrastructures, these factions still harp about the potential for consumer abuse and price gouging if we let big media companies get together to do something creative for consumers. Even though these factions have recently prevailed in setting the competitive agenda outlined in the new law, on close scrutiny their reasoning is nonsensical.

First, there is the First Amendment faction that, in the name of diversity in media, wants at least two "pipes" into every home to ensure that no big, bad monopolist controls too much of the information. Of course, well-known regulatory principles of common carriage and other minor regulations ensuring nondiscriminatory interconnection to the public network infrastructure (discussed in Chapter 6) can handle the problem, but that solution is too easy for this faction. They insist on protecting incumbent member firms, which probably enjoy a local media monopoly like cable TV or being the only newspaper in town.

In bureaucratic circles, the First Amendment faction rules the policy roost. To the FCC, the *National Telecommunications and Information Administration* (NTIA), Congress, and even the courts, the risk of monopoly control is too great to let existing cablecos and telcos join together to provide integrated network services. In principle, this faction has no leg to stand on. Nowhere in the most basic of economic principles is it written that two (or more) local media monopolies are better than one. More important, based on capital budgeting analyses to date, the relevant choice for a widely available and affordable integrated broadband network infrastructure may be one provider or none, not one or more.[10]

Common sense would dictate that it may be preferable to have a single information superhighway that represents a cooperative effort over the alternative of several private dirt roads and other private roads paved with gold for those who can afford to drive on them. Why would policymakers want to prevent companies from investing to modernize parts of the infrastructure, like household connections to the public network, that virtually everyone wants to use but no one company can afford to build on its own (especially when the only other alternative is a taxpayer-funded and government-owned public infrastructure)?

---

10. One thing is certain: The presence of even an unregulated monopolist is welfare enhancing for consumers relative to a situation where regulation may cause there to be no service provider at all.

Market competition in the telecommunications sector has never been tested for its impact on consumer welfare. Should such a novel idea as open entry ever be tested in the real world, the worst that could happen is that consumer welfare losses would materialize to the extent that a natural monopolist would emerge that was able to sustain high markups of price over marginal costs or otherwise engage in unacceptable market discrimination against certain customer classes. Even if that were to happen, the last and biggest monopoly, the government itself, could always come to the rescue and reregulate, as it did in the case of cablecos, by capping rate levels for basic service. Based on the extensive literature to date, the source of huge consumer welfare losses is imperfect regulation and its attendant uneconomic price structures, not monopoly market power.[11] In the political economy of regulation, we as a society seem simply to have opted for a known bad alternative.

If anything, opening up markets to see if only one firm would in fact emerge would actually be a useful exercise to determine if having only one firm is socially efficient. In any event, it is certainly premature to assume such an outcome and opt instead for the status quo of many media monopolies and the politics of turf allocation. Even some of the most renowned experts in the industry cannot agree on which market segments, if any, have natural monopoly characteristics. For example, in his various writings in books and journals and in evidence submitted in regulatory proceedings, Peter Huber of the Manhattan Institute firmly believes that it is the intercity toll service market that is most likely to support natural monopoly and that basic local exchange service is most likely competitive. His belief stands in stark contrast to the views of some other experts. Professor Alfred Kahn, for example, has written and testified to roughly the opposite view. The point is that if such students of the industry and its underlying technology cannot agree on which markets are or are not naturally monopolistic or competitive, it is patently absurd to rely on the early judgments of bureaucrats and politicians. In any case, why would we assume that the predictions of bureaucrats, academics, or anybody else are superior to those of businesspersons in their attempt to meet consumer demand?

The other popular policy faction is the *not in my backyard* (NIMBY) faction. This faction says that competition is great as long as it occurs outside the faction's traditional monopoly service territory. The NIMBY faction would have us believe that it should not matter that, with no opportunity to expand one's market area by leveraging a strong financial presence in its traditional

---

11. See B. Egan and J. Wenders, "The Costs of State Regulation," and the references therein. See also A. Kahn and W. Shew, "Current Issues in Telecommunications Regulation" (*Yale Journal on Regulation*, 1987).

home market area, it may be difficult to enter another monopolist's traditional market area. The required investment to build a broadband network infrastructure is difficult enough to justify at home, let alone competing out of one's current service area with no customer base and no particular technology edge. In any given market area, the incumbent firm is not going to take entry by an incumbent from another part of the country lying down. There is no reason to expect that regulations prohibiting incumbent firms from expanding their multimedia business activities in their own traditional market areas will force them to compete with dominant firms in other areas to the ultimate benefit of consumers. It is more likely that companies will expand operations via horizontal mergers, as Pacific Bell, Southwestern Bell, NYNEX, and Bell Atlantic have recently chosen to do, or via investing in infrastructure network projects overseas, where they are not so constrained by regulation.

In fact, considering the potential costs and benefits of regional integrated multimedia network infrastructures, regulators should not only allow the incumbent firms to pursue this type of in-region scenario, they should encourage it. Abuse of monopoly power and consumer protection in this situation is best handled by a minimum set of price caps for basic service and nondiscriminatory interconnection rules for competitive service providers.

If strict regulation does create an out-of-region scenario, in which an incumbent from a distant market does decide to invade another incumbent's traditional market area without the benefit of being able to leverage the relatively strong presence in its home markets, the relative loser will be the small consumer of traditional services. Competition of this sort is only going to target benefits for big, high-margin customers with a national market presence. On the other hand, the mass market of small users will be left facing relatively lower quality and higher prices than would logically exist if regulators would allow all users, large and small, to benefit by allowing incumbents to upgrade the public network infrastructure facilities incrementally, with freedom to pursue any and all revenue opportunities from value-added services along the way. Surely this is preferable to a situation in which everyone is skimming everyone else's cream with relatively little incentive to cultivate and develop the mass market along the way.

If policymakers were really worried about stimulating infrastructure development, they would get out of the way and let risks and rewards accrue to those firms willing to accept the challenge in an unprotected market environment. Bureaucrats, continually preoccupied with the prospect that granting too much market flexibility to incumbent network operators might result in windfall profits at the expense of captive customers, need to look at the real-world evidence to date.

Pursuant to being granted the appropriate statutory authority to deregulate, many state regulators have implemented relaxed regulation or total regulatory forbearance policies only to observe after the fact that few market abuses occurred.[12] In fact, in several cases in which state commissions have granted incumbent telcos market flexibility, there was little noticeable market response in the form of creative new service offerings or radical price discrimination. That has led many industry observers, not the least of whom are state regulators themselves, to wonder what all the fuss was about in the first place.

Apparently, market abuses are more a figment of bureaucrats' imaginations and preconceived notions than actual experience. Yet to this day, the specter of market abuse continues to take center stage in arguments against relaxed regulation. Market abuse is a bogeyman that can be dispelled only by letting unfettered market entry have its day in court.

What about the failure of market competition to provide for the truly disadvantaged market segments, including high-cost rural consumers, often considered for public policy purposes as desirous and socially deserving of affordable access to the information superhighway? On this policy front there is relatively good news and some consensus emerging. The new law has called for federal and state regulators to propose and implement a competitively neutral funding mechanism specifically targeted at subsidizing network connections for very low income and high-cost consumers (see Chapters 6–8).

9.7 SUMMARY AND POLICY RECOMMENDATIONS

The prevailing regulatory paradigm of managed competition must be modified or eliminated to overcome the substantial financial hurdles that private sector firms face in committing funds to network infrastructure development.

The first item on the regulatory reform agenda must be to eliminate all the constraints that result from earnings regulations placed on incumbent public network operators, both cablecos and telcos. Incentive-compatible (i.e., price cap) regulation should replace earnings regulation.[13] That means that only the price levels of essential services provided on a monopoly basis (e.g., residential basic access lines and public network interconnection arrangements) should continue to be regulated. Everything else should be completely deregulated.

Second, regulators should eliminate all operating restrictions that prevent entrants and incumbent firms from entering into competitive or cooperative arrangements as market conditions dictate. Freedom of entry is essential. As in other industries, general business laws will continue to be available to protect

---

12. While there have been some documented abuses and questionable business practices by incumbent monopolists, they were not necessarily tied to any legislative approval for regulators to pursue deregulatory policies.

13. Recall that price cap regulation in this context refers to pure price regulation as it is described in Chapter 6.

against potential market abuses and anticompetitive behavior in the telecommunications market.

Third, incumbent monopoly network operators must be subjected to residual regulations regarding nondiscriminatory prices, terms, and conditions governing public network interconnection arrangements. These rules must be aggressively enforced. If the rules governing physical aspects of interconnection (e.g., standards and information disclosure rules) are properly designed and supplanted with an appropriate price imputation test imposed on the use of underlying essential facilities of incumbent monopoly network operators, this will still allow for the realization of economies of scope from vertical integration.[14] The purpose of a price imputation test is to provide an *ex post* indicator of the presence of an anticompetitive price squeeze on those competitive network suppliers that must rely on public network facilities. A regulatory and legal complaint process must also be established to prevent predatory pricing and undue discrimination in prices or terms and conditions of service for essential public network facilities of a monopoly carrier.

Fourth, to preserve affordable universal public access to an advanced information infrastructure, a competitively neutral and sustainable funding mechanism should be implemented.

Fifth, residual regulations governing network hardware and control software (as opposed to applications software) compatibility and basic service quality should be established. That will allow equipment manufacturers and service providers to efficiently engineer to public network standards, allowing for low-cost production of plug-in network peripheral devices and consumer terminals. Regulators should not set standards. Rather, they should specify a process and a timetable for industry standards development and be empowered and prepared to enforce the rules.

Full and fair competition in the consumer's best interest requires that all residual regulations be symmetrically applied to all competing carriers, incumbent and entrant alike.[15] In the case of firms that want to enter traditional mo-

---

14. The basic rule for a price imputation test is that an incumbent network operator supplying an underlying essential facility on a monopoly basis must make that facility available to its competitors on the same terms and conditions as it is available to the incumbent itself to provide like services, and that the incumbent's pricing floor for any competitive service offering be equal to the sum of its marginal cost of providing the service plus any profit or contribution it would have made from the provision of the service by a competitive (interconnecting) provider. For a comprehensive discussion of economically efficient imputation rules, see the articles by W. Baumol and G. Sidak; J. Meyer; and A. Kahn, and W. Taylor in the *Yale Journal on Regulation*, Vol. II, No. 1, Winter 1994.

15. The only exceptions would be the mandated interconnection requirements placed on incumbent monopoly carriers' unique essential network facilities and their mandated residual obligation to serve all consumers on a common carrier basis, including the network access subsidies paid to them for purposes of meeting such residual obligations. However, the overarching principle of competitive parity suggests that whenever a competitor's network represents a

nopoly public utility portions of the market (e.g., residential basic access and local network interconnection), all residual regulations and service and subsidy obligations (e.g., universal service and COLR obligations) should again be symmetrically applied. Of course, in any given geographic market, if and when there is a viable competitive alternative to the traditional public utility service provider, then it is no longer an essential facility provided on a monopoly basis and should simply be deregulated. In that case, there are not necessarily going to be any special service and subsidy obligations to worry about, except perhaps for issues of service red-lining to avoid serving certain classes of consumers within a geographic market area.

Beyond those regulatory reforms, which alone may effectively remove existing disincentives to private investment in broadband public network infrastructures, governments could implement a number of other policies to further stimulate investment, including tax breaks and investment credits, guaranteed loans, seed funding for R&D, and so on. Such a combined policy mix would provide powerful incentives and encourage a relatively rapid deployment of an advanced information infrastructure.

9.8 MARKET IMPACTS OF THE TELECOMMUNICATIONS ACT OF 1996

There has been considerable discussion and analysis of the Telecommunications Act of 1996 and the impact it will have on the various industry stakeholder groups. Following is a broad overview of major provisions of the law and predictions as to how those provisions will affect major industry players.

9.8.1 Major Provisions of the Act

Major provisions of the Telecommunications Act of 1996 include:

- Allowing telcos and cablecos to compete head to head.
- Eliminating existing U.S. Justice Department restrictions on RBOCs and allowing them to enter markets for long distance and manufacturing subject to a list of conditions and rules that must first be met. The rules require that telcos provide unbundled (i.e., loop/transport/switching) equal

---

market alternative to the access lines of the incumbent (e.g., two-way cable and cellular networks), then its network is no longer truly essential, and reciprocity of interconnection rules would be called for. To the extent that the competitor network is also subjected to residual regulatory obligations like common carriage and universal service, then it too should be able to receive a proportionate share of any subsidies associated with those obligations. However, whenever a unique essential facility is no longer involved, deregulation is the best policy option.

access and interconnection to PSTN along with (1) dialing parity; (2) number portability; (3) reciprocal compensation for traffic exchange; (4) wholesale discounts below tariff rates for unbundled PSTN network components purchased by interconnecting carriers; (5) colocation (physical and virtual); (6) nondiscriminatory access to databases and signaling network, rights of way, emergency services, operator services, telephone number assignment; and (7) provide white page listings for competitors.

- Vacating GTE and McCaw/AT&T consent decrees.
- Allowing telcos into information services and electronic publishing (but requiring a separate subsidiary for next three years among other rules and restrictions on operations).
- Allowing telcos into interLATA long distance (with a number of conditions for in-region service and via separate subsidiary for next three years).
- Allowing telcos to provide interLATA video/programming services and intelligent network services (no separate subsidiary required).
- Requiring long distance rate averaging.
- Allowing electric utilities to provide phone service.
- Eliminating equal access rules for mobile radio services (allows bundling of wireless and wireline services).
- Eliminating regulation of pay phones.
- That no state or local government may restrict market entry (except for some rural areas).
- That no state or local government may unduly restrict tower sites for wireless systems.
- Eliminating cross-ownership restrictions on telcos and cablecos (limit of 10% financial interest if in the same market area).
- Eliminating many cross-ownership restrictions on radio and television broadcasters, cablecos, and newspaper companies.
- Eliminating the FCC's proposed regulation of VDT systems (limits system owner programming to one-third of channel capacity and subject to local franchise fees and some operating rules governing cable systems).
- Eliminating cable rate regulation on March 31, 1999, or sooner if competing suppliers in same market area.
- Allowing rural telcos to be exempt or acquire waivers for meeting most requirements of the Act.

9.8.2 Winners and Losers Under the New Law

The following list provides an indication of relative winners and losers under the new law. All stakeholder groups had something to gain from the new law, but some fared better than others. If the net impact on a stakeholder group is positive, (+) appears next to the group name; otherwise, (–) appears. The reasons for the positive or negative designation is provided as well.

Large Incumbent Local Phone Companies (–)

The law places substantial and costly regulatory conditions on LEC network operations and services that are not imposed on anyone else (see specific conditions above). The RBOCs and other large LECs would have been freed from most of the historical restrictions by separate court order or other Congressional actions, which might well not have required such onerous asymmetric regulatory burdens. In a sense, the law forced the LECs to "open up the (network) kimono" and let all their competitors enjoy the benefits, including: (1) lower access and interconnection charges; (2) colocated interconnection facilities; (3) unbundled and convenient access and interconnections to the LEC's PSTN; (4) a competitive head start, since LECs have to meet a so-called competitive checklist to prove that they faced so much actual competition before they too could be deregulated; (5) most all existing and outdated asymmetric regulations at the state level (and many at the federal level) remaining in place until regulators see fit to change them; (6) continued and even expanded regulatory cost allocations imposed on LECs; (7) harsh asymmetric conditions imposed on RBOCs that wish to enter manufacturing business (expected to continue to hinder RBOC innovation in PSTN equipment and consumer devices); and (8) harm to LEC investment incentives.

Some other minor redeeming provisions favor the LECs, like elimination of the FCC's ridiculous proposed regulations of VDT service and the freedom to provide interLATA video and intelligent network services.

Incumbent Cablecos (+,–)

Some major provisions of the Act will serve to help cablecos and some will not. On the plus side, the law: (1) deregulates cable rates by 1999 or even sooner if cablecos prove that they face competition; (2) allows cablecos to deaverage rates to compete with niche providers in lucrative market segments (e.g., multiple dwelling units); (3) makes it easier to get into LEC's business without having to go through onerous state regulatory proceedings or certification processes; and (4) preempts state and local government regulations restricting interconnect to LEC PSTN.

On the minus side, the law: (1) imposes more competition on cablecos, probably resulting in less profit margin; and (2) exempts competitive services from satellite broadcasting systems from paying the same local franchise fees and taxes to which cablecos remain subject.

New Wired and Wireless Cable Networks (–)

The new law makes it more difficult for niche market cable service providers to compete with incumbents, who are now allowed to deaverage rates.

New Over-the-Air Broadcasters (–)

The new law makes it more difficult for niche market broadcasters to compete or otherwise coexist with incumbent service providers, who are now allowed to buy and own more stations in one geographic area.

Direct Satellite Broadcasters (+)

Direct satellite broadcasters do not have to pay any local franchise fees or taxes, which are required of other wired and wireless systems.

Rural Local Phone Companies (+)

Rural telcos are huge winners. The Act codifies the most important item on the rural telcos' regulatory agenda—price and service parity with urban areas.

Long Distance Companies (+)

Long distance companies gained a blanket provision forcing LECs to grant them the right to resell LEC PSTN services and facilities at discounted wholesale rates. The law also reduces the disproportionate financial burden on long distance carriers who have historically been saddled with paying for the LECs universal service subsidies. This will serve to reduce the access charges that long distance companies pay to LECs.

New Local Exchange Companies (+)

Cablecos, competitive access providers (CAPs), and other entrants benefit from better, quicker, and cheaper PSTN access and interconnection to LEC PSTN facilities.

Cellular Companies (+)

Cellular carriers were essentially deregulated regarding their network service operations, but not regarding state rate regulation, which is still allowed. Cellular carriers may bundle local and toll service for purposes of marketing and network operations and are not obligated to grant competing network providers equal access to their network system.

Manufacturers (+)

Manufacturers not affiliated with an RBOC gain by the retention of harsh restrictions placed on RBOCs wishing to enter the market.

Consumers (+, –)

Consumers will get more service choices and more price competition for many services, but they could have benefited much more if genuine deregulation had occurred. Fundamental unfairness still exists whereby certain subscriber groups are cross-subsidized by others without any requirement that the subsidies are really needed based on ability and willingness to pay. Costly and pervasive regulatory practices will still exist, and, in fact, a higher cost of regulation is expected for some time. Investment in public local network infrastructure for multimedia and broadband network is still discouraged because the incumbent network operators are not free to pursue revenue opportunities according to market demand conditions.

State Regulators (+, –)

The competition policies of some state regulators were preempted by a few provisions of the Act. But they got much more than they lost (or, more important, what they stood to lose), because the Act also specifically granted state regulators wide-ranging responsibilities for setting local policies and for creating their own regulatory structures and tools to implement them.

References

[1] "The Great Telephone Paradox," *The Economist*, March 23, 1996, p.70.

[2] "How Washington Corrupts Telecom," *Wall Street Journal*, October 11, 1995.

[3] Egan, B., and J. Wenders, "The Costs of State Regulation: In Theory and Practice," Columbia Institute for Tele-Information, Research Working Paper #443, Columbia Business School, 1995; Teske, P. (ed.), *Crossing Lines: American Federalism, State Regulatory Institutions, and the Telecommunications Infrastructure*, Mahwah, NJ: Lawrence Erlbaum, 1995.

[4] From a survey of television in "From Idiot Box to Information Appliance," *The Economist*, February 12, 1994, p. 6.

Appendix A
Glossary of Selected
Economic and Cost Terms

Average incremental cost (AIC)   The incremental cost of a product or service (network component, etc.) divided by the quantity of output of the product or service.

Avoidable cost   Costs that are saved (or otherwise not incurred) when a particular action is avoided. An example is costs that are saved when output levels are reduced (or eliminated entirely). These costs are direct costs specifically caused by a given action. Avoidable costs can encompass investment as well as noninvestment costs. For example, when long-run capacity additions are contemplated, avoidable costs would likely include future investment costs.

Barrier to entry   A barrier to entry exists when a firm is capable of operating with equal efficiency as an incumbent firm but is unable to profitably enter the market. Most barriers to entry are due to artificial constraints on entry or a preponderance of sunk costs.

Capacity cost   A method of determining the incremental costs of business activities. It is especially useful in the presence of lumpy, shared network facilities and measuring the advancement effect caused by a given increment in demand. The basic methodology is as follows:

Given the investment cost of a facility, I, and the total units of usable physical capacity over the lifetime of the facility, C, the cost per unit of the facility is I/C. If a new service requires N units of this capacity, an estimate of the service's incremental capacity cost (CC) is $CC = N \times (I/C)$.

For example, if the total lifetime capacity of a particular facility is 50,000 units and the cost of the facility is $100,000, the cost per unit of capacity would be $2. If a particular increment in demand would require the use of 10,000 units, the capacity costing method would estimate incremental *CC* of $20,000 for serving the increment in demand.

Note: The formula gives an average cost per unit of capacity under the implicit assumption that every unit (of the facility's capacity) costs the same to use. This "smoothed" cost per unit eliminates the steep increases in costs that occur with the periodic installation of lumpy facilities and is, therefore, convenient.

Capital cost (1) The initial one-time investment made in long-term assets (e.g., plant). (2) Recurring annual costs resulting from the fact that money has been invested in long-term assets. Capital costs include capital repayment (or depreciation), return on capital, and income taxes.

Common (overhead) cost Costs incurred for the benefit of the enterprise as a whole, such as the cost of corporate accounting, executive salaries, and general business licenses. Such costs may or may not be variable.

Compensatory pricing A situation in which pricing generates revenues sufficient to cover the incremental cost of the object to which the price applies; the absence of subsidy.

Contestable market A market is said to be contestable if there are no sunk costs or other barriers to entry that would prevent an equally efficient firm from competing with the incumbent firm(s).

Contribution The difference between revenues and costs associated with a given action or proposed action. Most often, the action considered is to offer a product at specified production levels compared to not offering the product at all. In such cases, the product's revenue in excess of its avoidable cost is the product's contribution.

Cost of capital (1) The cost of money that is a composite of the cost of debt and the cost of equity. (2) The cost of money associated with capital assets' book or market values plus depreciation expense, income, and ad valorem taxes.

Cross-subsidization (1) The circumstance that exists when one or more products produced by a multiproduct firm fail the burden test. (2) A situation in which the direct revenue of a cost object is exceeded by the direct cost, while other cost objects within the firm display direct revenues in excess of direct costs.

Economies of scale The situation in which production processes are more efficient at a larger scale of operation than at a smaller scale. Economies of scale implies a reduction in average unit cost as production levels or facility sizes increase. Economies of scale can be present in both the short and the long run. When it is present, the cost structure of the enterprise exhibits common costs.

Economies of scope The situation in which savings are acquired through simultaneous provisioning of many different services. Costs can often be saved by offering multiple related services that utilize shared or common cost objects. Often referred to as multiproduct economies of scale.

Essential facility A facility, capability, or business activity that, for technical, legal, or economic reasons, cannot be equally efficiently provided by other than a single source. An essential facility is sometimes called a bottleneck monopoly component. Ordinarily, the existence of an essential facility implies a natural or unnatural monopoly structure with respect to the supply of the component. The essential facility doctrine discussed in the economics of antitrust is most concerned with the situation in which a supplier has an essential facility on which its competitors are dependent. Under such circumstances, involuntary unbundling and pricing constraints may apply to the essential facility.

Excess (spare) capacity The situation in which the resources of an enterprise are designed for producing higher output quantities than current output levels. Excess capacity may increase the cost of production without increasing the value of the product.

Fixed cost (1) A cost that does not vary with changes in output or level of activity of an ongoing firm. In the short run, fixed costs are incurred even if the firm's volume of output is temporarily zero. Because of this, fixed costs are the minimum costs a firm can incur in the short run (at an output volume of zero). A long-run fixed cost is the cost of the minimum scale at which a production process can be operated (e.g., one set of rail tracks, one locomotive, and one passenger car). (2) In a multiproduct firm, there are two kinds of fixed costs. The first is a shared cost that does not vary when segments of the business (products or services) are added or dropped. The second form of fixed cost is caused by a specific segment of the business (product or service) but is invariable with respect to changes in the output level of that product or service so long as the output is positive.

Forward looking A descriptive term normally associated with incremental costing methodologies and techniques utilizing estimates of quantities, costs, and conditions that will prevail during a future period. It also implies costs yet to be incurred as opposed to costs previously committed.

Fully distributed cost An assignment of all costs of an enterprise to cost objects of the enterprise such that the total cost of the cost objects equals the total cost of the enterprise.

Imputation The establishment of a price floor for bundled services consisting of competitive components and essential facilities (or other bottleneck components) that takes one of two forms, the second being a special case of the first. The first form is to include in the price floor the sum of the incremental costs of all the components of the bundled service plus the contribution that would have been obtained had the essential facilities or bottleneck elements been sold to a competitor offering substitute bundled services. The second form, which derives from when there is a lack of economies or diseconomies of vertical integration, is to include in the price floor the incremental cost of all the elements of the bundled service that do not constitute essential facilities or bottleneck elements plus the price charged to competitors for the essential facilities or bottleneck elements.

Incremental cost Any change in cost due to a specific action. If the action involves producing one more or one less unit of output, the incremental cost is equal to marginal cost. If the specified action involves entering or exiting an entire product line, the incremental cost is total incremental cost.

Joint cost (1) A cost shared by two or more products that gives rise to economies of scope. (2) A cost that is a result of the production of multiple products or services produced in fixed proportions. For example, cows make both beef and hides.

Long run The length of time necessary for costs, revenues, or other measures to reflect the situation after having fully adapted to a change. For example, long-run cost presumes the ability to fully adapt the size of production facilities and levels of investment to optimally accommodate a change.

Long-run incremental cost (LRIC) The difference in costs over the long run between two alternate courses of action. Normally, LRICs are forward looking and reflect the costs incurred in the provision of an additional increment of service; including cross-elastic effects.

Net present value (NPV) A capital budgeting technique used to evaluate proposed projects to determine their financial impact on a firm. NPV analysis compares the initial investment cost to the present value of all future cash flows using a predetermined discount rate, usually the cost of money. According to that technique, investment projects with positive NPV should be accepted, while projects with negative NPV should be rejected. If NPV equals zero, accepting the proposed project is a judgment call, because the project will service the debt and equity holders and pay off the original principal, but it will not add value to the firm.

Opportunity cost The value of the benefit that would have been provided by the second best alternative action that was foregone by the action chosen. Since resources generally are limited, this concept is used to determine the cost of choosing one alternative over another. Opportunity costs can include items such as the value of one's time spent in one activity over another, the return one could obtain on equity if an asset were liquidated and the proceeds invested in the best alternative manner, or the new construction costs that could be avoided if existing plant were made available for use in lieu of new plant.

Predatory pricing The practice of driving prices down to unprofitable levels for a period to weaken or eliminate existing competitors. The legal standard usually requires that prices be set at or above short-run marginal cost (including product-specific fixed costs, if any).

Present value The current value of a past or future cash flow. The formula for determining a present value of cash flow is:

$$\text{Present Value} = \sum_{t=1}^{n} \frac{X_t}{(1 + k)^t}$$

where
    X_t = the future cash flow in time period t
    k = the discount rate (sometimes equal to a given interest rate, i)
    n = the number of periods in the future (past)
    For example: The present value of a project that produces $100 in Year 1, $300 in Year 2, and $200 in Year 3 using a 10% discount rate is calculated as follows:

$$PV = \frac{\$100}{(1.1)^1} + \frac{\$300}{(1.1)^2} + \frac{\$200}{(1.1)^3} = \$489.11$$

Public good (inputs) A good, service, or input characterized by both nonrival consumption and nonexcludability. A public good is said to exhibit nonrival consumption when one person's or firm's consumption of the good does not reduce its availability to anyone else. Furthermore, a public good is such that if it is provided, the supplier is unable to prevent anyone from consuming it, which is the characteristic of nonexcludability. Examples are air, water, national defense, police, and fire protection.

Residual shared costs Residual shared costs are present when one set of incremental costs is nested in a larger incremental cost. For example, the incremental cost of individual central office (CO) features are nested in the incremental cost of the group of CO features. The latter may contain right-to-use (RTU) fees not included in the former. In that situation, a residual cost is the difference between the incremental cost of the entire product group and the incremental cost of the individual members of the group. Residual shared costs are not affected by additions or deletions of services (cost objects), but they are affected by decisions about collections of services (cost objects).

Shared cost   The portion of incremental cost shared by two or more objects that is not a direct cost of the individual objects. For example, the shared cost of two services is obtained by subtracting the individual total incremental (direct) cost of each service from the total incremental cost of the group of (two) services. Shared costs may be joint or common costs. It is shared costs, and only shared costs, that are not allocated to cost objects in incremental cost studies.

Short run   An unspecified period of time for which it is assumed that the productive capacity of capital assets (items such as heavy equipment, buildings, and other major facilities) will not change as a result of an action. For example, short-run effects are appropriate to use in assessing the cost of a temporary interruption or expansion in production.

Short-run incremental costs (SRIC)   The difference in costs over the short run between two alternative courses of action. The time frame used in developing an SRIC is not sufficiently long enough to permit complete adaptation of facilities, plant, and expenses to the change in outputs being considered. Normally, SRICs are forward looking and reflect the costs incurred in the provision of an additional increment of service.

Stand-alone costs   (1) The total economic costs required in the production of a particular product or service that would be incurred by an efficiently operating firm that produces only that product or service. (2) The lowest cost of entry into a market or service offering. This may be the cost of stand-alone supply or the incremental cost of entry by a multiproduct firm plus the additional shared costs to be recovered to make entry profitable.

Subsidy   (1) The provision of financial assistance to activities failing the burden test for cross-subsidy. (2) Money, services, or assets provided to an unprofitable activity to offset losses.

Sunk costs   Historical costs, resulting from past decisions, that are not affected by current decision alternatives. A resource is considered sunk if the resource has no opportunity cost (e.g., no market value or economic value in an alternative use). Sunk costs are irrelevant to current decisions and should not be considered in the decision process, but they may be considered in current accounts for accounting purposes.

Total incremental cost   The cost of a product or service (or other cost object), including both variable (volume-sensitive) and fixed (volume-insensitive) costs. For an existing product, the total incremental cost is the cost savings that would result from reducing the volume of the product to zero, all else remaining constant. That is, it is the product's avoidable cost. For a product not currently produced, it is the total cost of increasing the volume of production from zero to some specific level, all else remaining constant. Total incremental cost is synonymous with economic *direct cost* and *total service long-run incremental cost.*

Traffic-sensitive cost   Costs that are dependent on the volume of usage of a service. These costs are determined for the components of the system that are used to carry the traffic.

Unbundling   Unbundling occurs when a product or a service is disaggregated into its component parts and one or more of the component parts are offered for sale as separate products or services. Normally, unbundling is voluntary and determined by competitive market opportunities. When a company has an essential facility or bottleneck element, the economics of antitrust may require that the component be mandatorily unbundled and offered separately.

Unit incremental cost (UIC)   The change in cost caused by a change in output expressed per-unit change in output. For example, if a production increase of 100 units causes a $200 increase in costs, the UIC is $2 (per unit). At one extreme, when the change in output becomes very small, UIC approaches marginal cost. At the other extreme, when the change in production involves discontinuing or adding a whole product offering, UIC equals average total incremental cost.

Appendix B
Subsidy, Cross-Subsidy, and
the Costs of Universal Service

In regulatory proceedings concerning issues of telephone service cross-subsidies, a *subsidy* is said to exist when the average revenue (i.e., unit price) of one or more firms' services (or family of services) is higher than its corresponding direct (i.e., incremental) cost such that it contributes revenue toward a portion of the total costs of another firm (i.e., taxing one firm to support another firm). A *cross-subsidy* is said to exist when the average revenue (i.e., unit price) from one or more of a firm's services (or family of related services) is higher than its corresponding unit incremental (i.e., direct) cost such that it contributes revenue toward covering a portion of the firm's cost of another service or family of services that is priced below its direct cost. This definition of cross-subsidy is used to evaluate the appropriate level of the subsidy "free" pricing floor (i.e., direct cost) for services and may properly be characterized solely in terms of unit price and incremental costs.

In some types of firms, even if all product and service prices are greater than direct unit costs (i.e., no cross-subsidy), they may still not break even because average cost may be greater than average revenue or price (which may be set equal to or greater than the economic pricing floor of direct cost). In a declining unit cost business, it is entirely consistent with economic and business theory that a business whose average costs are greater than its marginal or incremental costs cannot possibly price at incremental cost levels and make zero or greater economic profit. That arrangement is, therefore, unsustainable. In other words, the subsidy is properly evaluated in terms of average total costs or so-called stand-alone costs, in the case of multiproduct firms.

A business cost structure in which average total cost may be greater than incremental unit cost may occur as a result of natural market forces (e.g., a water company or other public utility) and is typical of so-called natural monopoly businesses. But such a relationship may also occur due to regulatory accounting conventions, which often compel a firm to measure its costs such that calculated average costs are, in fact, greater than calculated incremental costs. This is a typical result of the calculation of the unit costs of telephone companies. This leaves open the following question: If average total and incremental costs were properly (economically) measured for basic and nonbasic telecommunications services, would average costs exceed unit incremental costs for any given service or family of services? In practice, for a multiproduct enterprise, the proper estimation of such unit costs has proved to be exceedingly difficult. On that point, there is extensive applied economics literature.

Nevertheless, the resolution of this question lies at the heart of the current debate regarding the level of funding of the costs required to maintain and even promote universal service. Many industry participants believe that, for a given geographic market area, universal service may be a natural monopoly, especially in rural and remote areas, which are relatively expensive to serve.

One item not in dispute is that, for any local telephone utility or local telephone company operating division of a vertically integrated telecommunications company, unless regulators (or, in the case of a deregulated situation, the market itself) allow average prices to equal average unit costs overall (which is to say that unless total revenue covers total cost for basic local service), the business is not financially sustainable as an ongoing enterprise unless a subsidy is provided to allow it to break even. *Any business in which average revenue does not cover average total cost must be subsidized to be sustainable.* That subsidy can come from external sources (e.g., taxes) or internal sources (cross-subsidy from other lines of business that may be somehow "taxed" or marked up over direct cost to cover total company cost).

About the Author

Bruce L. Egan is executive vice president of INDETEC International, a business consulting firm that specializes in media and telecommunications. As an adjunct professor at Columbia Business School in New York, he teaches Business and Technology of Telecommunications and Information in the Executive MBA program, and he serves as special consultant and affiliated research fellow at the Columbia Institute for Tele-Information (CITI). Mr. Egan has 20 years of experience in economic and policy analysis of telecommunications in both industry and academia. Before joining CITI in 1988, he was an economist at Bellcore for five years and at Southwestern Bell Telephone Company from 1976 to 1983. Mr. Egan has published numerous articles in books and journals on telecommunications costs, pricing, and public policy. He has been a consultant to several *Fortune* 500 companies, the U.S. Congress, the European Community, the United Nations, and the Organization for Economic Cooperation and Development (OECD). His latest research has concentrated on the economics of technology adoption in telecommunications, and he published a book titled *Information Superhighways: The Economics of Advanced Communication Networks* (Norwood, MA: Artech House, 1990).

Index

The Artech House Telecommunications Library

Vinton G. Cerf, Series Editor

Understanding Networking Technology: Concepts, Terms and Trends,
  Mark Norris

UNIX Internetworking, Uday O. Pabrai

Videoconferencing and Videotelephony: Technology and Standards,
  Richard Schaphorst

Virtual Networks: A Buyer's Guide, Daniel D. Briere

Voice Processing, Second Edition, Walt Tetschner

Voice Teletraffic System Engineering, James R. Boucher

Wireless Access and the Local Telephone Network, George Calhoun

Wireless Data Networking, Nathan J. Muller

Wireless LAN Systems, A. Santamaría and F. J. López-Hernández

Wireless: The Revolution in Personal Telecommunications, Ira Brodsky

Writing Disaster Recovery Plans for Telecommunications Networks and LANs,
  Leo A. Wrobel

X Window System User's Guide, Uday O. Pabrai

For further information on these and other Artech House titles, contact:

Artech House
685 Canton Street
Norwood, MA 02062
617-769-9750
Fax: 617-769-6334
Telex: 951-659
email: artech@artech-house.com

Artech House
Portland House, Stag Place
London SW1E 5XA England
+44 (0) 171-973-8077
Fax: +44 (0) 171-630-0166
Telex: 951-659
email: artech-uk@artech-house.com